Human Flourishing
Across Cultures and Disciplines

Also Available from Bloomsbury:

How to Think about the Climate Crisis by Graham Parkes
Human Flourishing in the Age of Digital Capitalism edited by Andrius Bielskis
Deleuze and the Problem of Experience by Dror Yinon

Human Flourishing Across Cultures and Disciplines

Paradigms of Well-Being and Development

Edited by Andrej Zwitter

BLOOMSBURY ACADEMIC
LONDON • NEW YORK • OXFORD • NEW DELHI • SYDNEY

BLOOMSBURY ACADEMIC

Bloomsbury Publishing Plc, 50 Bedford Square, London, WC1B 3DP, UK
Bloomsbury Publishing Inc, 1359 Broadway, New York, NY 10018, USA
Bloomsbury Publishing Ireland, 29 Earlsfort Terrace, Dublin 2, D02 AY28, Ireland

BLOOMSBURY, BLOOMSBURY ACADEMIC and the Diana logo
are trademarks of Bloomsbury Publishing Plc

First published in Great Britain 2026

Copyright © Andrej Zwitter, 2026

Andrej Zwitter has asserted his right under the Copyright,
Designs and Patents Act, 1988, to be identified as Author of this work.

Cover photograph by Andrej Zwitter

This work is published open access subject to a Creative Commons Attribution-NonCommercial-NoDerivatives 4.0 International licence (CC BY-NC-ND 4.0, https://creativecommons.org/licenses/by-nc-nd/4.0/). You may re-use, distribute, and reproduce this work in any medium for non-commercial purposes, provided you give attribution to the copyright holder and the publisher and provide a link to the Creative Commons licence.

No part of this publication may be used or reproduced in any way for the training, development or operation of artificial intelligence (AI) technologies, including generative AI technologies. The rights holders expressly reserve this publication from the text and data mining exception as per Article 4(3) of the Digital Single Market Directive (EU) 2019/790.

Bloomsbury Publishing Plc does not have any control over, or responsibility for, any third-party websites referred to or in this book. All internet addresses given in this book were correct at the time of going to press. The author and publisher regret any inconvenience caused if addresses have changed or sites have ceased to exist, but can accept no responsibility for any such changes.

A catalogue record for this book is available from the British Library.

A catalog record for this book is available from the Library of Congress.

ISBN:　HB:　978-1-350-54780-3
　　　　PB:　978-1-350-54781-0
　　　ePDF:　978-1-350-54783-4
　　　eBook:　978-1-350-54782-7

Typeset by Integra Software Services Pvt. Ltd.
Printed and bound in Great Britain

For product safety related questions contact productsafety@bloomsbury.com.

To find out more about our authors and books visit www.bloomsbury.com
and sign up for our newsletters.

Contents

List of Figures — vii
List of Tables — viii
List of Contributors — ix
How to Read this Book — xv
Foreword *Ban Ki-moon and Patrick Verkooijen* — xvii
Acknowledgements — xx

Introduction *Andrej Zwitter* — 1

Part One Meta-Scientific Approaches to Understanding Flourishing

1 The Spectrum of Human Understanding: The Conceptual Basis of Human Flourishing *George Ellis* — 11
2 The Indispensability Thesis and the Expanded Notion of Nature *Markus Gabriel* — 19
3 Ontological Security and Worldviews for Human Flourishing *Dean Rickles* — 27

Part Two Non-Material Conceptions of Flourishing

4 Positive Psychology and Human Flourishing *Theo K. Bouman* — 43
5 Christian and Neo-Platonic Roots of Flourishing – Higher Virtues and the Pursuit of Happiness *Andrej Zwitter* — 61
6 Learning to Flourish with Spinoza *Harald Atmanspacher* — 77

Part Three Context and Culture – Flourishing as Being

7 The Arc of Hope *Richard D. Hecht* — 99
8 Folktales, Stories and Traditions – Conceptual Indigenous Wisdom and Knowledge for Human Flourishing *Wakanyi Hoffman* — 119
9 Confucian Language of Flourishing: Reading the Classic Daxue *Victoria Sukhomlinova* — 139
10 Towards a Feminist and Radical Democratic Theory of Human Flourishing *Maggie O'Neill* — 151

11 The King's Philosophy and a Happiness Train: A Conceptual
    Overview of GNH *Karma Ura*   173

Part Four New Imaginaries of Flourishing

12 Planetary Thinking for Planetary Flourishing *Frederic Hanusch
    and Claus Leggewie*   195
13 Reclaiming Care – *Homo Curans* as Vision for Human Flourishing
    and Sustainability Transformation *Ariel Macaspac Hernandez*   209
14 Reimagining a World Fit for our Humanity *Ian Hughes*   229
15 Flourishing in the International Sphere – From Sovereignty to
    Solidarity: Challenges to 'Modern' International Law *Hans-
    Joachim Heintze*   245
16 Human Flourishing: An Integrated Systems Approach to
    Development Post 2030 *Andrej Zwitter, Carole Bloch, George Ellis,
    Richard Hecht, Ariel Hernandez, Wakanyi Hoffman, Dean Rickles,
    Victoria Sukhomlinova, Karma Ura*   261

Conclusion *Andrej Zwitter*   277
Epilogue *Mary McAleese*   285

Bibliography   299
Index   347

Figures

1.1	An integrative view: relating narrative on the one hand, and science and technology on the other hand	16
5.1	Cardinal and spiritual virtues	71
10.1	Traveller rights are human rights. Image: Marcin Lewendowski. Feminist Walk of Cork 2, July 2024	158
10.2	Ask consent. Oliver Plunket Street. Image: Feminist Walk of Cork 1	165
11.1	His Majesty Jigme Singye Wangchuck (reign 1972-2006), founder of GNH	174
11.2	An expression of an early view of the web of life. Ink on parchment by Karma Ura	177
11.3	Idealized scenario of implementation of GNH, which varies from government to government	190
13.1	The *Homo Curans* pathway – transformation wheel	222
16.1	Multi-level governance for human flourishing combining global principles, meso-level strategies and local-level target and indicator definition and implementation	272

Tables

5.1 Justice, Charity and Mercy — 74
16.1 Regions and Countries — 264
16.2 Policy Recommendations Regarding Implementing Integrated Flourishing Measures — 269

Contributors

Harald Atmanspacher

Harald Atmanspacher is an emeritus member of the Turing Center at ETH Zurich and faculty member at the University of Essex. After his PhD from Munich University (1986), he worked as a research scientist at the Max-Planck-Institute for Extraterrestrial Physics until 1998, as head of the theory division of the Institute for Frontier Areas of Psychology at Freiburg until 2014, and on the executive board of the Collegium Helveticum until 2020. His fields of research are the theory of complex systems, conceptual and theoretical aspects of quantum theory and mind-matter relations from interdisciplinary perspectives. He is the president of the Society for Mind-Matter Research and editor of its journal *Mind and Matter*.[1]

Theo K. Bouman

Theo K. Bouman, PhD, is an emeritus adjunct professor of clinical psychology at the University of Groningen, the Netherlands. He has been the head of a post-master's training programme for health care psychologists for many years, as well as a cognitive behavioural therapist. Currently, he is an adjunct professor of clinical psychology at the Faculty of Psychology, Universitas Gadjah Mada, Yogyakarta, Indonesia. His main topics of teaching and research are the treatment of anxiety and related disorders, as well as cultural aspects of clinical psychology.

George Ellis

George Ellis is Professor Emeritus and Research Fellow at the University of Cape Town, South Africa, and a Fellow at The New Institute, Hamburg, Germany. He started his research career studying General Relativity Theory and Cosmology at the Department of Applied Mathematics and Theoretical Physics, Cambridge University, where he wrote *The Large Scale Structure of Spacetime* with Stephen Hawking. At the University of Cape Town in 1973, he started a research group in general relativity and cosmology, and started research in areas such as low-income housing policy, quality of life indicators and how complexity such as the brain emerges from the underlying physics. He has been visiting professor at many universities, including a period as Professor of Cosmic Physics at the International School of Advanced Studies (SISSA) in Trieste. He has written or co-authored fifteen books and some hundreds of papers. He was awarded the Star of South Africa medal by President Nelson Mandela. He is a Fellow of the Royal Society (FRS), London and of the Third World Academy of Sciences (TWAS).

Markus Gabriel

Markus Gabriel is an internationally acclaimed philosopher and holds the Chair in Epistemology, Modern and Contemporary Philosophy at the University of Bonn. He is the Director of the Center for Science and Thought as well as Chairman of the International Centre for Philosophy NRW. He has been awarded numerous prizes, fellowships and visiting professorships. His books have been translated into many languages. Markus served as Academic Director of THE NEW INSTITUTE from 2022 to 2024. Since 2024, he has been Senior Global Advisor at the Kyoto Institute of Philosophy and in 2025 founded the Academy for Deep Innovation – Business, Philosophy, Science.

Frederic Hanusch

Frederic Hanusch is Professor of Planetary Change and Politics at Justus Liebig University Giessen and co-founder of its Panel on Planetary Thinking. In 2024, he was awarded a Visiting Distinguished Fellowship at the Centre for Deliberative Democracy & Global Governance at the University of Canberra and appointed to the British Academy–Carnegie Endowment working group on Transnational and Planetary Challenges. His research explores how planetary conditions shape politics, including in his recent book *The Politics of Deep Time* (2023). Previous works include *Democracy and Climate Change* (2018) and the co-edited volume *Seeds for Democratic Futures* (2024). He earned his PhD from the Institute for Advanced Study in the Humanities (KWI) in Essen and worked for the German Advisory Council on Global Change (WBGU), contributing to major policy reports. He later led a research group at the Institute for Advanced Sustainability Studies (IASS) and was a fellow at THE NEW INSTITUTE.

Richard Hecht

Richard D. Hecht is Professor Emeritus in Religious Studies at the University of California, Santa Barbara. He has published four books, including a twenty-five-year ethnographic study of Jerusalem with his colleague Roger Friedland, also Professor Emeritus in Religious Studies and Sociology at the University of California, Santa Barbara, titled *To Rule Jerusalem* (1996, 2000). Hecht and Friedland collaborated on more than thirty chapters and articles on the politics of sacred places. Hecht is currently completing a lengthy study of religion and contemporary art and a series of projects on sacred space.

Hans-Joachim Heintze

Hans-Joachim Heintze is a distinguished scholar in the field of international law, with a career spanning over four decades. He obtained his first Doctor of Jurisprudence degree from the University of Leipzig in 1977. In 1991, Hans-Joachim Heintze assumed the role of Senior Researcher at Ruhr University Bochum. His dedication to education and capacity building is evident in his role as the President of the International Association for the Study of the World Refugee Problem from 2001 to 2005. Additionally, he has been instrumental in organizing and leading international law courses for young foreign diplomats, sponsored by the German Foreign Office, from 1995 to 2019. Throughout his career, Hans-Joachim Heintze has been actively engaged in collaborative research initiatives, including projects with institutions such as the Max Planck Institute of International Law, the EU Institute for Security Studies and the OSCE High Commissioner on National Minorities.

Ariel Hernandez

Ariel Macaspac Hernandez is currently a fellow at THE NEW INSTITUTE under the programme Conceptions of Human Flourishing and a Senior Researcher at the German Institute of Development and Sustainability (IDOS). Since 2021, he has also been a private lecturer at the University of Duisburg-Essen. His policy advice, research and teaching interests include international negotiations, mediation, stakeholder dialogue, knowledge diplomacy, transformation to sustainability, climate change, climate and environmental justice, sustainability standards and due diligence, international cooperation and development policies of rising powers, sustainable energy, sustainable mobility, game theory, systems analysis, and Global North–Global South cooperation.

Wakanyi Hoffman

Wakanyi Hoffman is the Lead Researcher for Sustainable African AI Systems at Inclusive AI Lab, Utrecht University's Centre for Global Challenges. Her work integrates Ubuntu philosophy into the design of responsible, sustainable AI systems and emphasizes community and relational knowledge as guiding principles for culturally inclusive AI innovation. Wakanyi is also the founder of the Humanity Link Foundation, home to the African Folktales Project – a curated archive preserving intergenerational African oral stories rooted in Ubuntu ethics. She is also the author of *Sala, Mountain Warrior*, the first children's book depicting the Samburu Indigenous peoples of Northern Kenya. Through storytelling and Indigenous Knowledge Systems research, she champions ethical technology grounded in collective, cultural intelligence. She holds an MA in Global Studies from University College London and is currently pursuing a PhD at the Institute for Cultural Inquiry at Utrecht University.

Ian Hughes

Ian Hughes is Senior Research Fellow at MaREI Centre, Environmental Research Institute, University College Cork (UCC), Ireland. He leads the Deep Institutional Innovation for Sustainability and Human Flourishing (DIIS) initiative at UCC which comprises a transdisciplinary community of over thirty academics and practitioners, and has developed collaborations with the OECD, Columbia University, the German Institute of Development and Sustainability, and the International Humanistic Management Network. His recent books include *Disordered Minds: How Dangerous Personalities Are Destroying Democracy* (2018) and *Metaphor, Sustainability, Transformation: Transdisciplinary Perspectives* (co-edited with Edmond Byrne, Gerard Mullally and Colin Sage).

Claus Leggewie

Claus Leggewie holds the Ludwig Börne Professorship at Justus Liebig University Giessen and co-founded its Panel on Planetary Thinking. He previously held visiting professorships in Paris-Nanterre and at NYU (Max Weber Chair), and fellowships at the Institute for Human Sciences (Vienna), the Remarque Institute (NYU), and the Wissenschaftskolleg zu Berlin. From 2007 to 2015, he directed the Institute for Advanced Study in the Humanities (KWI) in Essen, pioneering research on Climate and Cultures. He also founded the Center for Global Cooperation Research in Duisburg and served on the German Advisory Council on Global Change (WBGU). In 2021, he became an Honorary Fellow at the Thomas Mann House in Los Angeles. His work has earned major honours, including the Cross of Merit, First Class, of the Federal Republic of Germany. Leggewie contributes regularly to international media such as *Le Monde*, *The Guardian*, and *The New York Review of Books*.

Maggie O'Neill

Maggie O'Neill is Professor in Sociology and Criminology at University College Cork and Director of ISS21, Institute for Social Science in the 21st Century and UCC Futures: Collective Social Futures. An elected member of the Royal Irish Academy. Maggie has a long history of conducting innovative culture work at the intersections of Sociology, Criminology and Women and Gender Studies.

Dean Rickles

Dean is a Professor at the University of Sydney, specializing in the History and Philosophy of Modern Physics. He also serves as the Co-Director of the university's Centre of Time exploring the nature and psychology of time. His primary research focus is the history and philosophy of modern physics, particularly quantum gravity and spacetime physics. However, he also has strong interests in econophysics, population health and musicology.

He holds a PhD from the University of Leeds, focusing on conceptual issues of quantum gravity. Between 2005 and 2007, he took up a postdoctoral fellowship at the University of Calgary. In 2008, he received a five-year ARC Australian Research Fellowship and a four-year ARC Future Fellowship in 2014, researching quantum gravity. His recent books include: *Dual-Aspect Monism and the Deep Structure of Meaning*, co-authored with Harald Atmanspacher (2022) and *Life is Short: An Appropriately Brief Guide to Making it More Meaningful* (2022).

Victoria Sukhomlinova

Victoria earned a PhD in Philosophy of Culture in 2022. The title of her dissertation was 'Remastering the Notion of Culture in Contemporary Neo-Confucianism', and since then she has published in Russian philosophy journals as well as in a Springer journal 'International Communication of Chinese Culture'. Victoria's research focuses on pre-imperial and contemporary Confucianism, the history of Chinese cultural thought, and non-Eurocentric methodologies within comparative and cross-cultural philosophy. Her bachelor's and master's degrees were in Area Studies, received from Moscow State Institute of International Relations. She also conducted research at the traditional Confucian academy Wenli Shuyuan in 2017.

Karma Ura

Karma Ura is the president of the Center for Bhutan Studies & Gross National Happiness Studies in Thimphu, Bhutan. He completed his undergraduate studies at St. Stephen's College in New Delhi, as well as Magdalen College, Oxford, and has an MA from Edinburgh University in the United Kingdom. Furthermore, he received an honorary doctorate from Nagoya University in Japan. His research and policy advocacy have mainly focused on wellbeing indicators applied in the national socio-economic development plans. Although a development economist by background, he has published on history and social anthropology. Ura was awarded the red scarf and the title Dasho by the Fourth King of Bhutan, Jigme Singye Wangchuck in 2006. In addition, he is a Member of the Chief Economist's Advisory Panel, World Bank, representing the South Asia Region. He is also engaged in the visual arts, mainly in directing annual performances at the Dochula national monument where his paintings are also displayed. A painting of his is in the British Museum.

Andrej Zwitter

Andrej Zwitter is Professor of Political Theory and Governance at the University of Groningen, the Netherlands, and as of 1 February 2026, Professor of Humanities and Digitalisation at the Alpe Adria University of Klagenfurt, Austria. He has a PhD in International Law and Legal Philosophy. Zwitter was founding dean of the Faculty of Sustainability Transformation and Governance (Campus Fryslân) at the same university. His efforts led to the establishment of the interdisciplinary and multiple award-winning BSc programmes such as the University College Liber Arts programme on 'Global Responsibility and Leadership'. He furthermore founded the Data Research Centre, the Centre for Innovation, Technology and Ethics, and the Cyan Centre for Climate Change Adaptation. Zwitter's scholarship spans a wide array of subjects including Political Philosophy of Statehood, State of Emergency Politics and Law, Data Ethics and Regulation, Digital and Blockchain Governance, and Innovation in Humanitarian Action. His current work delves into Meta-Science as the Science of Meaning and Non-Materialist Conceptions of Human Flourishing.

Note

1   https://www.mindmatter.de/about/board.html (accessed 8 October 2025).

How to Read this Book

This book aims to explore the interaction between metaphysical, scientific, Indigenous, and political realms as foundational elements in solving complex problems concerning human flourishing and the non-material aspects of the human condition. It broadens the discourse of science to include Indigenous wisdom, happiness, care and storytelling, positioning these not only as modes of knowledge but also as sources of wisdom generation. This edited volume reflects these principles across its chapters. Contributions from an international community of scholars, practitioners, wisdom-keepers and policy-makers illuminate this progress towards a holistic understanding of human well-being and flourishing. This book is not just about rethinking policies but about reimagining our collective futures, where material and non-material dimensions of life are equally valued in the pursuit of a truly flourishing society. In a synchronistic way, this book converges with several initiatives in different disciplines and individual efforts that focus on the same goal surrounding a meta-scientific embedding of human flourishing as an act of meaning-making and discovery in the face of contemporary challenges.

The present edited volume is constructed upon two key cornerstones:

1. **Purpose-Driven Science**: The belief that all science is purpose-driven and ultimately serves an end. In the context of this book, it is the resolution of complex global challenges based on universal principles and collective as well as individual aspirations.
2. **Epistemic Diversity in Science**: The idea that metaphysical, spiritual, and religious conceptions can possess validity in their own right, consistent within their own ontological frameworks. Therefore, they deserve equal consideration alongside other ontological frameworks such as materialism or idealism for the purpose of distilling meta-scientific principles of universal human aspirations towards flourishing.

This book is organized into four parts, each of which addresses different dimensions of human flourishing. Readers may choose to engage with the content according to their specific interests and professional perspectives:

- Philosophers and scholars interested in the foundational theoretical frameworks and the epistemological dimensions of flourishing will benefit most by starting with Part I, which discusses meta-scientific approaches and foundational theories.
- Those whose primary interest is in policy implications and practical governance frameworks are advised to begin with Part IV, where new imaginaries of flourishing are explored in concrete policy contexts and institutional designs.
- Readers particularly interested in exploring non-material aspects of human flourishing, including spiritual, religious, and existential dimensions, will find Part II most relevant. This section delves into non-material conceptions of flourishing, including psychological well-being and spiritual traditions.
- Individuals who wish to explore how cultural contexts and identity shape our understanding of flourishing will find Part III particularly insightful, as it addresses flourishing through cultural lenses, feminist theory, Indigenous wisdom and community-centric narratives.

Of particular note is that the chapters have been crafted not only to provide overviews of their respective fields and to engage in reflective, critical, and innovative dialogues with existing discourses. Each chapter contributes unique reflections, recommendations and critiques, fostering a nuanced and diverse understanding of human flourishing.

Foreword

Ban Ki-moon and Patrick Verkooijen[2]

The Sustainable Development Goals (SDGs) have profoundly transformed global governance. They create a shared vision for addressing the world's most pressing challenges. SDG 13 on Climate Action, for example, has galvanized countries to commit to ambitious climate targets as well as climate adaptation, such as the Paris Agreement, to drive global efforts to combat climate change and promote renewable energy initiatives. These global goals have shifted the focus from narrow economic metrics to broader measures of human well-being. The SDGs thereby represent a first global framework that champions inclusivity, sustainability and equity alongside environmental protection and economic welfare. The hallmark of the SDGs is their ability to align global, national and local efforts. Thereby, the SDGs foster a collaborative spirit that aims to catalyse change and that inspires policies and initiatives for stewardship for social justice and economic prosperity.

While the SDGs represent a landmark achievement, as we approach 2030, it is, however, evident that the SDGs are not a final destination. They are a stepping stone towards a deeper understanding of human flourishing on a global policy level. The focus on material and measurable outcomes, though vital, potentially overlooks the non-material dimensions of well-being that might be essential for a fulfilled life. Heeding the rising calls from the developing world, for the lack of a better term, a paradigm shift is needed to address potential gaps such as the inclusion of Indigenous knowledge systems and regional and local socio-cultural specificities for human flourishing. Building upon the SDGs, it is crucial that we incorporate these insights to better reflect the richness of human experience.

This starts with the profound acknowledgement that true human flourishing extends beyond material wealth and physical health. It also encompasses psycho-social health, spiritual fulfilment, emotional resilience and cultural vitality. Consider that spiritual fulfilment can manifest through different practices like meditation or communal rituals that foster a sense of connection and purpose. And cultural vitality, important for the enrichment of community identity and cohesion, can be evident in initiatives that preserve endangered languages or celebrate traditional art forms. Reductionist approaches focusing solely on measurable outcomes, while important for global mainstreaming of policy efforts, risk marginalizing these intangible yet critical aspects of wellbeing. This can lead to a potential loss of depth in our understanding of the human condition that goes against the spirit of the SDGs. However, to integrate non-material dimensions into policy and practice is not an easy task. It requires the study of how different cultures, religions and disciplines explore the depths of non-material wellbeing. Only through such an inquiry can we foster a more holistic approach that reflects both the complexity and richness as well as the interconnectedness of the human condition, constituted of both material and non-material aspects.

Furthermore, cultural and Indigenous knowledge systems can offer invaluable insights into the history of how humankind attained flourishing in a symbiotic relationship with nature. Indigenous agricultural practices, such as agroforestry in the Amazon, for instance, demonstrate how traditional knowledge systems can enhance biodiversity and sustain livelihoods while adapting to the changes in our climate. This showcases the value of integrating these insights into global efforts aimed at human development that is not at odds with the limited resources and the fragile Earth system which we depend on. Harmony with nature, the interconnectedness of all life, and the importance of community cohesion and spiritual fulfilment, these are the hallmarks of a future in which the human condition is understood as deeply connected with or rather constituting a crucial component of the integrated system that is our planet. By integrating different epistemic models and worldviews, we can enrich global governance frameworks, such as the SDGs, and create pathways for more inclusive and sustainable development. The principles of reciprocity, respect and relationality are deeply embedded in these traditions, such as *Inochi* (Japanese: life) and *Ubuntu*. Using these principles in novel ways to challenge less humane, but dominant, paradigms and illuminate alternative ways of living in harmony with the Earth and each other remains our most important task.

At the same time, any such ambitious goals require effective governance. Human flourishing cannot be attained by sheer will; we need global, regional and local frameworks that accommodate both global principles and local specificities – frameworks that galvanize state and non-state actors as much as corporations and communities. A multilevel governance model, as presented towards the end of this book – combining top-down guidance with bottom-up innovation – can bridge the gap between universal aspirations and local realities. Such an approach ensures that policies are not only contextually relevant but also inclusive of non-Western epistemic realities that are vital for empowering communities to shape their own futures. This dynamic, participatory model reflects the diversity of human experience.

The concept of the sacred, as highlighted by former Irish President Mary McAleese, offers an important lens for understanding the complexity of the human condition. We all know something that is sacred to us, whether we are religious or not. Sacredness transcends religious boundaries; it evokes a sense of shared purpose and moral responsibility. This sense of sacredness highlights the profound moral dimensions that can guide future initiatives and can encourage policies rooted in care, mutual respect and the recognition of humanity's interconnectedness with nature and one another. By reflecting on this perspective, we can cultivate a deeper respect for life's interconnectedness and inspire policies that honour the intrinsic value of all beings.

This book is a call to action for policymakers, scholars and practitioners to envision a future where flourishing transcends boundaries – be they cultural, disciplinary or geographical. By embracing the lessons of the SDGs while improving on them, we can co-create a new framework for human development that harmonizes material and non-material dimensions and that acknowledges our deep connection with the planet. This vision of human flourishing, informed by diverse perspectives and grounded in shared values, holds an important promise: a world where all people and the planet can thrive together. Let this book serve as both a reflection and a roadmap for humankind's journey ahead in pursuit of a flourishing future.

Note

2   Former UN Secretary General Ban Ki-moon and Professor Patrick Verkooijen, CEO of the Global Centre on Adaptation, President of the University of Nairobi.

Acknowledgements

The present book and its open access nature would not have been possible without the generous institutional and financial support of THE NEW INSTITUTE (TNI) in Hamburg, where Andrej Zwitter had the privilege to serve as Programme Chair for the Programme on 'Conceptions of Human Flourishing' during his sabbatical in 2023/2024. This edited volume is one of the key outcomes of the collaborative efforts of that programme, reflecting the ideas, insights and friendships that emerged from our shared time together.

We are deeply grateful to THE NEW INSTITUTE, its academic directors Markus Gabriel and Anna Katsman, and its managing director Britta Padberg for their support and facilitation of an inspiring environment in the Warburg Ensemble in Hamburg, as well as for their personal engagement and intellectual openness. Particular thanks go to the institute's founder, Erck Rickmers, whose vision and unrelenting commitment to 'plant an apple tree' made it possible to convene such a diverse and inspiring group of scholars, activists, politicians and thinkers in Hamburg throughout 2023 and 2024. His belief in the importance of reflective spaces for transformative thinking deeply shaped this project. We also extend heartfelt thanks to the administrative and hospitality teams at TNI. Their warmth, professionalism and tireless care made us feel at home, creating a nurturing environment that allowed our work to flourish.

Profound thanks also go to our editors at Bloomsbury Academic, particularly Colleen Coalter and her team, who saw value and importance in our work and whose guiding hand brought this book into reality.

On a personal note, I wish to thank my partner, Anna, whose patience and love sustained me through many long months of absence and intense work. Her understanding and encouragement were a quiet, constant strength behind this endeavour. I also owe a deep debt of gratitude to my mother, whose unwavering support and belief in me have always guided my path. She has encouraged me to follow the work I truly love and continues to be one of my most important sources of inspiration.

Finally, I would like to thank the University of Groningen, and in particular Campus Fryslân, for facilitating my sabbatical after my term as dean of the faculty. Their support enabled the time and space needed to fully dedicate myself to this project and bring it to completion.

As with any acknowledgement, there are always people who deserve thanks and praise and who go unmentioned undeservedly. To all who contributed – through insight, kindness, or care – thank you!

Introduction

Andrej Zwitter

1 Overview of the philosophical ideas in this book

What is flourishing? In policy circles, the concept of flourishing is often equated with development. However, at the core of sustainable development lie primarily material values. When we trace the history of thought on flourishing, or *eudaimonia*, back to Aristotle, an entirely different image emerges. This perspective encompasses all aspects of human character and human needs; and it is not restricted to material aspects of flourishing, nor to the human as a solitary actor isolated from nature. It presents a holistic perspective that includes non-material, often non-measurable aspects that make the human condition. It is aspirational and it includes ethics and politics as *sine qua non* contextual and normative dimensions. Viewed from this Aristotelian perspective, contemporary (Western) scientific thought has undergone a tremendous shift from a science of 'ought' to a science of 'is', losing its ethical, normative imperatives and a view on the human condition in all its aspects in the process (Zwitter 2023b). While this perspective may offer solutions for addressing global challenges like climate change and poverty, it falls short in explaining why we should tackle these issues and what unmeasurable principles underlie the human condition and by extension, human purpose. Science can predict technological evolution, but this in turn can lead to techno-optimism and innovation-path dependency. However, in its current state, the sciences fail to provide a narrative for a world conducive to flourishing – a world we aspire to live in. This ostensibly neutral stance of the sciences leaves politics, religion and other non-scientific fields of social engagement to fill in the values. Consequently, the ethics embodied in current economic and legal systems remain fundamentally unchallenged. It is as if we have tried to answer only Kant's first of the three fundamental philosophical questions of 'what can we know' and relegated the remaining questions 'what shall we do' and 'what can we hope for' to the fields of ethics, metaphysics and religion. This creates a strange, and from an Aristotelian perspective untenable, dichotomy between scientific and non-scientific domains.

This dichotomy between scientific and non-scientific domains of existence is a peculiar feature, particularly of modern Western thought, and it is also evident in the Sustainable Development Goals (SDGs). The SDGs limit themselves to the material conditions of sustainable development and, by extension, to material human flourishing. This book, however, aims to delve deeper into the subject of flourishing and its complex linkage with nature, culture and religions. It is not restricted to human flourishing alone, nor does it emphasize a specific ontological or methodological framing, but it encompasses storytelling, the ethics of care, policy approaches to happiness, Indigenous wisdom and various other aspects increasingly prominent in global scholarly discourses. The ultimate aim of this book is to reflect on these findings and provide a meta-scientific framework that contributes to flourishing from the perspective of different *Lebenswelten* (life-worlds) and ontologies to aid future policy making.

The concept of flourishing, or *eudaimonia*, as explored by Aristotle, is a cornerstone in philosophical discussions about the good life. Aristotle's vision of *eudaimonia* was deeply holistic, emphasizing a life of virtue, reason and community engagement. In his Nicomachean Ethics, he argued that true flourishing is achieved not through material wealth or transient pleasures, but through the cultivation of virtue and the exercise of reason (Aristotle 2000). This conception of well-being extends beyond mere hedonic happiness; it involves realizing one's potential and living in accordance with one's nature and reason.

In contemporary times, however, the application of this concept faces several challenges. One primary issue is the dominance of Western values in defining what constitutes a flourishing life. Western perspectives often emphasize individuality and material success, marginalizing the importance of community, spirituality and a harmonious relationship with nature. This focus on individual achievement and material wealth as the benchmarks of a successful life has led to a skewed understanding of flourishing, one that often overlooks the intrinsic value of interpersonal relationships, community bonds and spiritual fulfilment (Zwitter and Dome 2023).

Moreover, the traditional concept of human flourishing is often criticized for its anthropocentric approach, which considers human well-being in isolation from the natural world. This view neglects the fact that human flourishing is deeply interconnected with, and often contingent upon, a healthy and functioning ecosystem (Galesic et al. 2023; Leaney 2012). The prevailing attitude tends to view nature and other species as mere externalities or resources for human use, rather than as integral components of a shared existence. This separation of humanity from nature fails to recognize that what is often considered an 'externality' is, in fact, a fundamental part of our own well-being.

Furthermore, contemporary discussions about flourishing tend to emphasize material values as the primary, if not exclusive, conditions for a good life. This materialistic approach equates success and well-being with the accumulation of wealth and possessions, overshadowing non-material aspects of life such as emotional well-being, intellectual growth and spiritual fulfilment. This focus on material prosperity often leads to a neglect of the deeper, more meaningful aspects of life that contribute to a sense of purpose and fulfilment (Dowson, Devenish, and Miner 2012; van Dierendonck 2012).

However, the *Lebenswelten* (life-worlds) of various societies, both across time and in different geographic regions (Husserl 1954), provide a broader perspective on what it means to flourish. Many non-Western cultures place significant emphasis on spiritual and non-material aspects of life. In these cultures, flourishing is often seen as a balance between material well-being and spiritual, emotional and communal health. These societies recognize that true well-being encompasses a harmonious relationship with others, with nature, and with the spiritual or metaphysical aspects of existence. Such perspectives challenge the Western-centric view of flourishing and highlight the importance of a more inclusive, holistic understanding of well-being that integrates material, social, environmental and spiritual dimensions.

2 Core tenets

Western scientific thought (that is specifically its popular reception in wider audiences), is often viewed to be predominantly rooted in materialist empiricism. This perspective has significantly shaped contemporary understandings of human flourishing. It prioritizes empirical evidence and measurable outcomes, often at the expense of more intangible or qualitative aspects of life. In the mainstream, this materialist viewpoint has been influential not only in scientific circles but also in the formulation of policies and governance structures. Governments and bureaucracies, tasked with creating conditions conducive to flourishing, frequently adopt this empiricist approach, focusing on quantifiable indicators and material outcomes. This trend is evident in the SDGs set by the United Nations. The SDGs represent a laudable, global effort to address various challenges, including poverty, inequality, climate change and environmental degradation. While these goals are comprehensive in scope, the emphasis often remains on measurable targets and indicators, reflecting the materialist empiricism of Western scientific thought (Zwitter 2023b).

For example, consider SDG 1: No Poverty. Its indicators focus primarily on the number of people living below the international poverty line, income levels and material resources. While these are crucial measures, they don't fully encompass the broader, more qualitative aspects of overcoming poverty, such as social inclusion, dignity and access to cultural or spiritual resources. Similarly, SDG 3: Good Health and Well-being. This goal is comprehensive in addressing various health issues, predominantly measuring health in terms of physical conditions, medical services and mortality rates. The indicators, however, overlook aspects like mental well-being, traditional healing practices or the role of community and environmental harmony in promoting health.

These examples highlight how materialist empiricism, as expressed in Western scientific thought, influences the framing and assessment of global objectives like the SDGs. While these goals are undeniably vital for global progress, their reliance on material and empirical indicators can sometimes narrow our understanding of flourishing. This approach may inadvertently sideline the non-material dimensions of well-being, such as spiritual fulfilment, cultural richness and environmental connectedness, which are equally essential for a holistic view of human flourishing. This has been acknowledged recently by the 'Agenda 2050 group', a community of scholars and practitioners promoting flourishing as a global policy agenda (Karthikeya, R. et al. 2024), as well as the Programme on Human Flourishing at The New Institute, Hamburg (2023-2024), bringing together more than twenty scholars, Indigenous leaders, policy makers, politicians and business leaders over the course of one year to work jointly on policies regarding the future of the SDGs post 2030 ('Conceptions of Human Flourishing – Programme' 2024).

This book aims to serve as a groundbreaking nexus where scholars, practitioners, wisdom keepers and policymakers converge to explore and construct alternative visions of what a flourishing life could encompass. By interweaving the threads of metaphysical, scientific, Indigenous and political thought, this volume aspires to lay the groundwork for new attainable and liveable utopias. These utopias, embracing the diversity of various *Lebenswelten*, while acknowledging their common goal of flourishing, will help in shaping future policy frameworks post-SDGs. It endeavours to broaden the scientific discourse by integrating Indigenous wisdom, happiness, care and storytelling, acknowledging these elements not merely as knowledge systems but as vital sources of wisdom generation.

At its core, the book is anchored in two fundamental principles. First, it advocates for purpose-driven science, that is, it acknowledges that all science is at least implicitly shaped by underlying values. This principle emphasizes that scientific inquiry should be cognizant of the often tacitly underlying values and thus also serve a greater goal – in this context, addressing complex global challenges through universal principles. Second, it upholds the validity of metaphysical and spiritual encounters with nature, recognizing that spiritual and metaphysical beliefs can coexist with scientific endeavours and contribute meaningfully to our understanding of human aspirations and flourishing.

Meta-science, as also developed by Zwitter and Dome (2023), posits that to comprehend the world as it is experienced, or can be experienced, one must accommodate multiple ontologies (and their contextual *Lebenswelten*) concurrently without subjecting one ontology to the epistemology and methodology of another. For instance, numerous religious and cultural ontologies, as presented in this book, posit the existence of a mental or spiritual world in addition to the material world. By applying the conditions of empirical evidence of materialism – namely the epistemology and methodologies of physicalist reductionism – to such ontologies, one is predisposed to reproduce only results that materialism is inherently capable of accepting as proof. In other words, each ontology can only sustain epistemologies and methods that are capable of validation within its own framework of thought. Consequently, they are only able to generate outcomes accepted as proof within that specific ontology. If the purpose is to understand how other cultures, religions, philosophies and world views conceptualize flourishing, then this requires us to take them seriously in their own right. The presence of ontologies that diverge from materialism also indicates that these ontologies must be regarded as potentially equally valid in describing their respective domains unless they are contradicted by more consistent proofs from other ontologies. Here, 'proof' refers to ontologically-internal logical consistency.[1] Specifically, regarding human flourishing, where material and non-material aspects of the human condition converge (including the need for a meaningful life), researchers need to take a meta-scientific stance that enables them to potentially see what a worldview dominated by Western values and popularized, doctrinal ideas of science (i.e. scientism), might have overlooked (Zwitter 2025).

This book focuses on a fundamental inquiry: What does it mean to flourish as a human being in the twenty-first century? Answering this question requires transcending the confines of disciplinary silos and challenging the narrow epistemic bounds of empiricist materialism. Our approach emphasizes the necessity of a meta-scientific perspective – one capable of critically examining various ontologies, methodologies and their interdependencies. Throughout this volume, we advocate embracing epistemic diversity, acknowledging the legitimacy of diverse cultural, spiritual and philosophical frameworks. We further suggest the necessity of reimagining global governance and policy frameworks, notably the SDGs, to include non-material, relational and culturally sensitive dimensions of flourishing. The volume highlights meaning, spirituality, intuition and care as indispensable, foundational elements for human well-being that must be recognized alongside rationality and material prosperity. Moreover, the diverse contributions gathered here reveal recurring, perhaps perennial themes across distinct cultures and intellectual traditions, like relationality, meaning, care and transformative processes. These themes suggest pathways towards a more integrative and holistic understanding of flourishing. Ultimately, this volume is not only a reflection on and a proposal to expand beyond contemporary reductionist and technocratic approaches. It is also an invitation to dialogue, collaboration and collective envisioning of a world in which all beings can truly flourish.

3 Synopsis

This volume begins with a fundamental, meta-scientific rethinking of science in the service of human flourishing and meaning (Part I), critiquing the prevailing scientism of our age – namely, the doctrinal, quasi-religious belief in key tenets of popularized science such as reductive physicalism. Physicists, philosophers and historians of science lay a new epistemic foundation for understanding flourishing that transcends materialist limitations. George Ellis (Chapter 1) argues for a non-reductive view of reality where meaning, narrative and emergence play a crucial role in human understanding. Markus Gabriel (Chapter 2) calls for an 'expanded notion of nature' that accounts for human consciousness and belonging within heterogeneous 'fields of sense', while Dean Rickles (Chapter 3) points to alternative scientific worldviews rooted in informational and immaterial paradigms as essential for restoring ontological security.

Building on this reorientation, Part II explores non-material conceptions of flourishing, drawing from positive psychology, classical and religious philosophy, and existential thought. Theo Bouman (Chapter 4) discusses flourishing through the lens of positive psychology laying the groundwork for a psychology of virtues, while Andrej Zwitter (Chapter 5) delves into the Early Christian and Neo-Platonic roots of eudaimonia, showing how specifically spiritual virtues inform both personal and political preconditions for well-being. Harald Atmanspacher (Chapter 6) interprets Spinoza's metaphysics as a guide to a broader, intuition-based mode of flourishing that integrates mind, body and nature.

Part III foregrounds the importance of culture, context and identity in shaping the lived experience of flourishing. It critiques overly individualistic or Western universalist models by offering pluralistic and embodied perspectives. Richard Hecht (Chapter 7) bridges Viktor Frankl's logotherapy and Ernst Bloch's utopian philosophy to articulate 'the arc of hope' as a vital, non-material driver of flourishing, illustrated through the eyes of Mexican muralist Diego Rivera and his ten-panel fresco mural *Pan-American Unity*. Wakanyi Hoffman (Chapter 8) draws from Indigenous wisdom narratives across Africa and the Americas to argue that planetary flourishing is inseparable from ecological reciprocity and the sacredness of the Earth. Victoria Sukhomlinova (Chapter 9) offers a Confucian perspective where learning, ritual and relationality are central to the formation of human agency and belonging. Maggie O'Neill (Chapter 10) develops a critical, transdisciplinary, feminist framework rooted in ethno-mimetic research, emphasizing storytelling, intersectionality and biographical methods as tools to dismantle violent hierarchies and foster conditions of flourishing. Karma Ura (Chapter 11) offers a conceptual overview of Bhutan's Gross National Happiness index, emphasizing its ethical and philosophical underpinnings beyond GDP.

With these foundations in place, Part IV envisions new imaginaries of flourishing that challenge prevailing paradigms in policy, law and planetary governance. Frederic Hanusch and Claus Leggewie (Chapter 12) propose 'planetary thinking' as a trans-scalar lens for understanding the entanglement of human and planetary systems; and Ariel Hernandez (Chapter 13) introduces the concept of *Homo Curans* – the caring human – as a new subject for sustainability policy, integrating Indigenous and feminist ethics into global governance. Ian Hughes (Chapter 14) calls for deep institutional innovation and cultural transformation in response to the global polycrisis, while Hans-Joachim Heintze (Chapter 15) investigates the emergence of the international legal principle of solidarity, drawing on precedents in Space Law and the Law of the Sea to suggest post-sovereign pathways for international cooperation.

The final chapter (Chapter 16), co-authored by the research team that worked together in the Programme on Human Flourishing (Andrej Zwitter, George Ellis, Ariel Hernandez, Richard Hecht, Wakanyi Hoffman, Dean Rickles, Victoria Sukhomlinova and Karma Ura), synthesizes insights from the book into a comprehensive systems approach to flourishing. It critiques the shortcomings of the current SDGs and outlines a post-2030 development agenda rooted in relational, non-material and culturally diverse understandings of well-being.

Note

1 For an excellent foundation to this, see the methodological explanations of Stace (1961) in his *Philosophy of Mysticism*.

Part One

Meta-Scientific Approaches to Understanding Flourishing

1

The Spectrum of Human Understanding: The Conceptual Basis of Human Flourishing

George Ellis

1 Introduction

Human beings are a symbolic species (Deacon 1997): this is what distinguishes them from all the other great apes. It has enabled us to develop language and so function as coherent social groups, develop technology and thereby conquer the world (Bronowski 2011). In doing so, there has been a progression of human understanding that has, broadly speaking, developed over time in the following way. Firstly, narratives of various kinds, centred in the possibility of agency, were used to understand our life situation and to guide our actions. While these were useful in many ways, they often lead to useless or even harmful practices, for example, trying to influence the weather in ways that could not in fact work. This is discussed in Section 2.

Then the power of scientific investigation based in mathematics was realized, with experimental investigation leading to the discovery of impersonal scientific laws, underlying all physical events including life. This underlaid the development of technologies that transformed societies (Arthur 2009). However, it also led to many scientists and philosophers denying essential aspects of humanity, for example, claiming we have no agency or free will. This is discussed in Section 3. But a middle view is possible, which is able to take both perspectives into account – it integrates the narrative view, acknowledging the depth of our humanity and agency, and the impersonal scientific one, acknowledging the underlying impersonal physical basis of our existence. This is discussed in Section 4. A key tension between these views is whether we have agency and free will. Despite strong statements to the contrary, one can claim that we do indeed have both.

I thank Dean Rickles and Andrej Zwitter for useful comments.

This is discussed in Section 5. Overall, the middle view provides a sound conceptual basis for human flourishing, because it acknowledges our integrative multidimensional nature, as discussed in Section 6.

2 Understanding via narratives of agency

Initially in our evolutionary path, humans understood causation only in terms of agency. Our experience of agency by ourselves and others explained much of what was going on around us – we were able to change things by deciding to do so, and then taking action. Awareness of individual agency led to a projection of this understanding into the world: '*We have evolved to see minds every damned place we look*' (Ball 2022: 5). We had no other plausible explanation to rely on.

What is the relation to narratives? The key development in evolutionary history separating us from the great apes was our development of the use of symbolism, and specifically language (Simon 1993). This allowed us to communicate our understandings of events with each other and to plan joint ventures and develop cooperative societies (Aberle et al. 1950). We share our understandings of what is going on and what is likely to happen via stories: narratives that have an explanatory nature because they are based in agency that we understood (Breithaupt 2025; Johnson, Bilovich and Tuckett 2023). Crucially, narratives convey and are based in meaning and purpose, and in sets of values that shape behaviour. These lie on a spectrum that ranges from totally coercive to caring and generous and kind (Noble and Ellis 2021). Basically, the stance is that the universe is based in meaning – purposeful direction of some kind. The brain evolved both to perceive meaning and to produce it.

Narratives make this explicit in specific cases. The brain is constructed by evolution to underlie this process. Armstrong (2019) explains this thus:

> How stories configure experience and organize events in time is an especially intriguing and important example of how literature plays with the brain. As I argued in my previous book on neuroscience and art (Armstrong 2013), reading a literary work typically sets in motion to-and-fro interactions between experiences of harmony and dissonance. A novel, poem, or play may reinforce and refine our sense of the world's patterns through the symmetry, balance, and unity of its forms, or it may disrupt and overturn our customary syntheses by transgressing established conventions and refusing to satisfy our expectations about how parts fit together into wholes. These kinds of interaction between harmony and dissonance in aesthetic experiences help to

negotiate a basic contradiction that is fundamental to our cognitive lives—the contradiction between our need for pattern and constancy and our equally crucial need for flexibility and openness to change. The ability of the brain to play with these competing imperatives is also evident in our capacity to tell and follow stories.

This understanding via stories led to the supposition of existence of a variety of gods who had the same kind of agency as humans and controlled things in the world. This has led, inter alia, to human sacrifice to placate these gods which has taken place on a horrifying scale ('Human Sacrifice' 2025), or for example hoping that praying for rain will alter the weather. An example is this statement by the Governor of Texas in 2011:

> Now, **therefore**, I, Rick Perry, Governor of Texas, under the authority vested in me by the Constitution and Statutes of the State of Texas, do hereby proclaim the three-day period from Friday, April 22, 2011, to Sunday, April 24, 2011, as Days of Prayer for Rain in the State of Texas. I urge Texans of all faiths and traditions to offer prayers on those days for the healing of our land, the rebuilding of our communities and the restoration of our normal way of life.

This is based on narratives promoting an understanding of weather as being due to the direct actions of supernatural deities. But this kind of activity is simply unable to influence the weather: it cannot override the laws of physics. Our present solidly based understanding (see the next section) is that the weather is controlled by atmospheric physics ('Atmospheric Physics' 2022) that cannot be altered by actions of this kind. This understanding is unlikely to change. Trying to mitigate global climate change is of course highly relevant – but that is a quite different course of action, based in what humans have done and can do within the context of these unchanging physical laws.

These stories early on assumed a political and economic dimension. Those with the best stories could seize control. In many cases, there came into existence a caste of priests who claimed to have preferred access to true stories, so they would guide the population at large in understanding and behaviour. Sometimes this was liberating, but often not. Many of them used this narrative as a power base from which they could control local populations, extract resources from them and even abuse them. Their claimed values are often diametrically opposite to those they preach. This has been the case over centuries with major world religions often killing and torturing people in the name of a loving God. Present day examples of such religions abound, for example in the United States today the stand of the MAGA churches on abortion and its political enforcement is literally leading to young women suffering greatly and even losing their lives. Social, economic, political and religious narratives may or may not be supportive of human flourishing. This depends on the values of their proponents.

3 Impersonal science and the power of technology

Our use of symbolism led to a further key feature: narratives developed into logical arguments, and then to the development of mathematics (Hodgkin 2005). First, this was used to enable record keeping and accounting in economic transactions, but then this was generalized to relations between abstract mathematical quantities. This led to a profound development: the understanding that what happens in nature, such as the fall of an apple, the swinging of a pendulum and the motions of planets, can be described by mathematical equations relating quantities that change with time. This led to the rise of the scientific method based in experimentation and careful measurement. The laws of physics thus discovered – Newton's laws of motion, his law of gravitation, the diffusion equations, Maxwell's equations for the electromagnetic field and so on – have two profound qualities. They enable quantitative as well as qualitative predictions of natural events, and they are completely unaffected by human feelings or stories. They represent a profound deep layer of existence that functions independently of the human mind, but which can be expressed by equations that the human mind can comprehend.

Thus, the rise of the scientific method led to a new vision: what happens is governed by impersonal laws of physics that can be represented in a precise mathematical way. This has enabled the development of technology (Arthur 2009) that has transformed society (Bronowski 2011). Mathematical symbolism underlies science and technology. Through this relation, seventeen equations – Newton's law of gravity, the wave equation, the Navier-Stokes equation, Maxwell's equations and so on – have changed the world (Stewart 2013).

This success has led to a new materialist philosophy: the universe is based in matter, space and time, and nothing more. A new class of priests arose with preferred access to true stories: reductionist physicists. Physics underlies all. The rest is epiphenomenal, it is physics that determines both what happens and does the work. On this basis, some of them claim there is no meaning in the universe (Weinberg 1993). However, another group of reductionist physicists are loudly proclaiming that, on the basis of their expertise in physics, they can explain to the public from this privileged position the meaning of the universe and everything in it. Thus, we have *The big picture: on the origins of life, meaning, and the universe itself* (Carroll 2017), *Until the end of time: Mind, matter, and our search for meaning in an evolving universe* (Greene 2021), and *Existential physics: a scientist's guide to life's biggest questions* (Hossenfelder 2022). Physicists will explain it all to you. But physics itself has nothing whatever to say about life's biggest questions, nothing as regards meaning and purpose. This includes issues such as 'What is good and what is evil?', 'What is ugly and what is beautiful?', 'How should I behave?', 'How should I spend my life?', and so on. This is classic physics imperialism. These authors are in effect claiming to be a new class of priests with special understanding of everything whatever due to their technical expertise, in particular explaining the meaning of life to ordinary mortals.

Some evolutionary theorists and neuroscientists proceed similarly, each claiming their specific topic explains everything about life – for example genes dictate everything you do (Dawkins 2016) or neurons do so (Crick 1994). Such views proclaimed by reductionist molecular biologists, geneticists and neuroscientists have led to many scientists and philosophers denying essential aspects of humanity: everything is determined by some underlying science or other, not by our emergent high-level cognitive capacities and choices. Thus, Crick (1994) said:

> The Astonishing Hypothesis is that "You", your joys and your sorrows, your memories and your ambitions, your sense of personal identity and free will, are in fact no more than the behaviour of a vast assembly of nerve cells and their associated molecules.

In particular, this leads to claiming that we have no agency or free will (e.g. Sapolsky 2023). This is a major threat to our understanding of ourselves that I return to in Section 5.

Andrew Steane (2018: 1–2) makes clear the problem of such viewpoints as follows:

> It is my opinion, and the central thesis of the book, that all such visionaries have a skewed vision of both science and the arts. I think that what poets do, and what literary critics do, and what musicians do, is every bit as truthful and insightful about the nature of reality as is what physicists and chemists and biologists do (and mathematicians, engineers, historians, philosophers, and so on). ... The overarching aim of this book is to get in view something that pays deep attention to the whole of what goes on in human life.

Zwitter makes similar points in his discussion of meta-science (Zwitter 2023).

4 The middle view

These are the two poles of the spectrum of understanding: on one end is nothing but matter controlled by physics as expressed in impersonal equations, and on the other is meaning as expressed in narratives and myths. Both are expressed in symbolic ways. However, neither viewpoint is a sound basis for understanding the real world: each leaves out an important aspect of our existence.

The truth is somewhere in the middle: a non-reductive view of the physical and social world that recognizes the nature and power of the underlying physics, the existence of complex systems with emergent powers in their own right, and the power of narratives in creating understanding of social life. Such a worldview, recognizing the existence of meaning in the world, is needed as a sound basis for human flourishing (Figure 1.1).

A key point is the recognition that reductive world views are not enough: they encompass only part of what is going on (Steane 2023). Emergent levels in the hierarchy of structure and causation are not epiphenomena: they have genuine causal powers of their own, expressed in verifiable effective laws at each emergent level (Ellis 2017; Noble 2012). In particular, this applies to the functioning of the human brain: thoughts have causal powers that result in changes in the world around us, such as existence of aircraft and iPhones and the Internet. This essentially is enabled by the four types of Aristotelian causation: Effective, Material, Formal and Final (Ellis 2023). The last three are forms of downward causation, with higher-level causes shaping lower-level outcomes but with a vast amount of freedom as to how this happens: there is multiple realisability of higher-level structures and functions at lower levels.

None of this overrides or alters the lower-level physics: it provides the context within which that physics operates, thus determining the specific outcomes that occur, such as creation of an artistically shaped porcelain teapot out of clay (Ellis 2005). Physics equations by themselves do not represent any specific outcomes

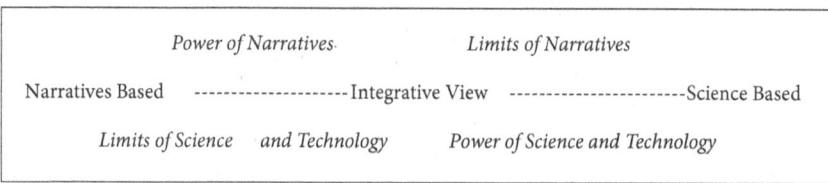

Figure 1.1 An integrative view: relating narrative on the one hand, and science and technology on the other hand. Each are powerful but limited. An integrative view recognizes the powers and limits of each.

whatsoever. You need to add to them statements of constraints, boundary conditions and what kind of matter is present for them to act on. This is where emergent dynamics comes in: it determines these features.

The power of narratives in this context is that they are a powerful integrative method of conveying effective ideas about the situation and causal effects in action in a particular situation. They are attuned to the predictive way the mind understands what is going on (Clark 2013). But they do not result in reliable quantitative predictions. The power of science is that it allows reliable quantitative predictions to underlie the technology that enables us to manipulate our environment as we desire, within limits such as those imposed by the First and Second Laws of Thermodynamics. Its prime limitation is that, of itself, it knows nothing of values and purpose: they are simply outside its domain. They have to be consciously imposed as a restraint on economic imperatives that only take technological possibilities – not costs – into account (Mayer 2024). A middle view is possible and has been developing, which is able to take both perspectives into account – it integrates the narrative view, acknowledging the depth of our humanity, and the impersonal scientific one, acknowledging the underlying impersonal physical basis of our existence.

5 The key tension between the views

A key tension at the interface between these views is whether we have agency and free will, with some scientists and philosophers claiming we do not. Examples are Wegner (2018) and Sapolsky (2021; 2023). A strong reply to the latter is given in Fischer (2023). Key references defending free will in a solid way are Ball (2022), Mitchell (2023), Murphy and Brown (2009), Roskies (2021), Seth (2021). I will not give a detailed defence here but rather make three key remarks.

Firstly, the groundwork for a defence of *free will* scientifically comes from the remarks in the previous section regarding the emergence of complexity based on downward causation in modular hierarchical structures. Causal powers thereby emerge at every emergent level, including that of the organism as a whole – which is where agency occurs. The immense stochasticity of biology at the molecular level (Ball 2023; Graham 2023) frees higher levels from slavish control by lower levels, enabling biology to extract order from chaos (Hoffmann 2012) and so opens up the road for agency to occur (Noble 2021; Noble and Noble 2021).

Secondly, the existence of human agency is obvious because of the consequences it has had, and continues to have, in the world around us. These consequences are apparent everywhere around us. They include the existence of aircraft, buildings, roads, digital computers, books about digital computers (Harris and Harris 2013), and simple things like teapots (Ellis 2005). None of these would exist if we didn't have agency: they did not come into being by magic or physicalist determinism of human behaviour. Furthermore, the existence of agency in others requires us to have a theory of mind in order to interact socially on a day by day basis (Frith and Frith 2005; Heyes and Frith 2014). This is solid evidence that agency exists.

Thirdly, if we did not have agency in a meaningful sense, then the practice of science itself would not be possible. Having it is a pre-requisite for carrying out a scientific investigation: planning experiments to test a hypothesis, carrying out the experiments and evaluating the results. If your theory of mind precludes agency, then you'd better work out a better one, for you can't develop any such theory without meaningful agency.

6 The conceptual basis for human flourishing

Overall, the middle view that takes seriously both present-day science and technology on the one hand, and the importance of purpose, meaning and values on the other, provides a sound conceptual basis for human flourishing, because it acknowledges our integrative multidimensional nature. Technology is about far more than just the development of physical artefacts (Brooks 1980). An integrative view relates narrative on the one hand, and science and technology on the other hand, recognizing the powers and limits of each. Both aspects are needed for a deep understanding of the world around us in general (Steane 2023) and in particular to underlie social actions we should undertake, such as those related to the SDGs.

2

The Indispensability Thesis and the Expanded Notion of Nature

Markus Gabriel

1 Introduction

I argue that we need an expanded notion of nature, one that includes aspects currently unknown to us. There is, therefore, a distinction between nature as known through science (which I refer to as 'the universe') and nature as including unknown and unknowable parts. This distinction between the universe and the expanded notion of nature runs through our very being as human animals and manifests in various forms (such as fundamental questions about our status as conscious or otherwise minded beings). Thus, an expanded notion of nature can integrate human mindedness into nature without reducing it to anonymous natural processes – that is, processes that can be known without considering the indispensability thesis.

Somewhat paradoxically, in the wake of the linguistic turn, the very existence of meaning has come under ontological suspicion. Since Wittgenstein's powerful arguments showing that meaning is neither tied to the psychological occurrence of intentions nor grounded in our capacity to follow rules, it has become increasingly difficult to find a proper home for meaning and understanding. Language has turned into 'a thin net over an abyss' (Cavell 1999: 178). Along with language, subjectivity, consciousness, the mind and everything associated with mental phenomena have also come under suspicion, as the world has been reconceived as a 'big physical object' (Lewis 1986: 1), thereby depriving the mental realm of ontological significance.

In short, a largely undertheorized naturalist or physicalist outlook began to conquer the world – without ever having demonstrated that mind and meaning are unreal, mere illusions generated by the brain or by local patterns of neural tissue.

In what follows, I will:

1. briefly outline the ontology of fields of sense, which I have developed elsewhere (in particular: Gabriel 2015; 2024c);
2. argue that both the mind belongs to nature and nature belongs to the mind, being entangled in relationships of mutual mereological containment;
3. use this to propose an account of human flourishing that integrates us fully into nature without reducing or eliminating the mind.

2 Fields of sense – not worlds

What exists always only appears in a context. There are no absolutely independent entities – entities completely isolated from all others. Even if there were such isolated entities, their splendid isolation would relate them to others through relationships of similarity. Instead of the broad notion of context, I speak of *fields of sense*. A field – much like in mathematics or physics – is a spatiotemporal structure in which every point corresponds to a specific value. I call such a value an *occupant* or *object*. What individuates a field of sense are the senses involved in mapping some objects onto it while excluding others.

For example, quantum theory involves no meaningful intentional behaviour because its objects are not of the kind that we describe using mentalistic vocabulary. Yet, this does not imply that meaningful intentional behaviour does not exist. Rather, such behaviour is the legitimate target of theories in other fields (like sociology, economics or common-sense explanations of action).

The universe, in the sense of defining physics as a science, is structured by the axioms of physical theories (e.g. quantum theory, relativity, cosmology). These axioms generate the governing sense of physics, allowing empirical inquiry to identify the specific senses that individuate physical objects (such as spin, mass, etc.). The fact that mind and meaning do not appear in the universe, so defined, does not mean they do not exist – it simply means physics is not a mind science and need not be.

Similarly, fermions and bosons do not appear in Renaissance art history, yet this does not mean they do not exist. They simply lie outside that field's target system. Fields of sense can also be objects in other fields of sense – or even in themselves.[1] To exist is to be an object within a field.

There is no overarching hierarchy or ordering of all fields of sense. In sum, there is no field of all fields – a conclusion I call the no-world-view.[2] This is not merely an epistemic limitation but the ontological fact par excellence. Fields of sense, unlike 'worlds', are not unlimited wholes. Neither the universe nor nature in the expanded sense is an all-encompassing domain. As a result, not all entities can be squeezed into the domain of physics or science more broadly.

Mind and meaning do not lose legitimacy because they do not appear in physics. On the contrary, on closer inspection, mind and meaning – including high-level phenomena such as solving complex equations or engaging in philosophical self-reflection – are indispensable to any account of reality. We develop theories from within a reality already inhabited by us as thinkers and agents. Any account that omits this fact is hopelessly incomplete.

We can only limn the structure of reality if we integrate our capacity for understanding into that very structure.[3] Once we accept the irreducible heterogeneity of an open multiplicity of fields of sense, we recover ontological space for ourselves – as thinkers, agents and human animals who are part of any explanation of our ontological vicinity.

3 Mind in nature – nature in mind

The universe is an *epistemic object* – a model of non-minded parts of nature (Gabriel 2024b; Voosholz 2024b). Like any model, it contains features not present in its target system (e.g. imaginary numbers used to describe wave-like phenomena). A gap remains between the model (mathematical physics) and its target system (nature without high-level mindedness). Physics progresses by narrowing this gap, though it can never close it entirely. This incompleteness is itself knowable through physics: for instance, the uncertainty principle sets a limit on what can be measured and is built into quantum theory's axioms.

Relativity too converts epistemic phenomena – such as receiving information in unsynchronized ways due to the speed of light – into ontological facts. In this sense, human mindedness indirectly participates in constituting the universe as an epistemic domain. This does not mean that all objects in the universe are epistemic. Electrons, bosons, galaxies and so forth exist independently of how we individuate them. One can be a scientific realist while recognizing the epistemic constitution of the universe as a field of sense created by human inquiry. Physics and science show us that scientific models do not exhaust nature. Moreover, we lack scientific means to precisely measure the difference between our models and nature itself.

At this point, we have generated three distinct objects:

1. The universe as an epistemic object.
2. Non-mental nature, which includes the universe (*natura naturata*) and exceeds it (*natura naturans*).
3. Mind, as an indispensable part of nature.

Mind is indispensable because we rely on it – our conceptual and epistemic capacities – to explain our belonging to nature. As Hilary Putnam succinctly said: 'the mind and the world jointly make up the mind and the world' – though I would substitute 'the world' with 'nature' (Putnam 1981: xi).

That mind belongs to nature means that we must expand our conception of nature to include the mind without reducing it to non-mental phenomena. I call this contribution to the philosophy of mind *conditionalism* (Gabriel 2018). If we analyse an event (such as an action), which essentially involves meaning seeking, intentional agents like us into its components, we will find both non-mental and mental conditions that turn out to be necessary for the occurrence of this individual event. Jointly, these conditions are evidently sufficient for the occurrence of the event. This version of the principle of sufficient reason is a principle of analysis of events.[4] Our account of non-mental events is epistemically grounded in action explanation without thereby being reducible to them, as this would amount to an untenably subjectivistic theory of explanation (and causation).[5]

A necessary mental condition (ontologically committing to mindedness) involves the capacity to live in light of a conception of oneself. At the highest level, agents act based on conceptions of themselves (their roles, duties, social identities, etc.). Bosons, by contrast, do not act in light of self-conceptions.

Human agency also involves non-mental conditions – like neural tissue within an organism embedded in ecological systems. Agency is a multi-agent system with both mental and non-mental parts. Mind and nature cohere in agency, though they are distinct. This is not metaphysical dualism: *conditionalism* separates mind and nature in its account without positing them as separate substances. Their relationship is *mereological*, not causal in some (antiquated) materialist sense of push-and-pull mechanisms. Mind, like the universe, is part of nature.

Yet nature is not an overarching category subsuming universe and mind. The universe, as an epistemic object, and the mind both belong to nature but are not identical with it. Otherwise, we risk falling back into Spinozist metaphysics.

To grasp this, consider: how do we know the universe and its place in nature? Answering this requires us to grasp the dazzling thought that both the universe and expanded nature, are part of the mind.[6] Both are, in part, *in the mind*. The mind encompasses the universe via mathematical models and encompasses expanded nature through philosophical thought – because nature, in turn, is part of the mind. This is a *mereological*, not ontic, relationship.

The point is not that nature is a mental illusion as argued by Hoffman (2019). Rather, the idea is that thought is not screened off from how things are by a layer of mental states (such as representations). Thought is out there, grasping how things are from within how things are. Denying this form of epistemic and ontological realism makes it impossible for us to explain the enormous epistemic success story of modern science (see also Deutsch 2012; Gabriel 2024c).

Mind and nature thus coexist in mutual mereological entanglement: each is part of the other in different mereologies. Each corresponds to a distinct field of sense. Neither field reduces to the other; they are logically orthogonal ways of understanding reality from a human standpoint.

4 Human flourishing as a spiritual category

Let us call the reality of the entanglement of nature and human mindedness *spirit*. Something is spiritual insofar as it concerns both our mental and non-mental nature, which are connected through our self-conception as human animals. Contemporary discussions of human flourishing benefit from this notion, as it legitimizes spiritual needs – such as the quest for meaning, including the meaning of life itself.

Human flourishing cannot be reduced to survival. While survival is essential, meaningful lives require more than mere existence (see the famous discussion in Giorgio Agamben's *Homo Sacer* [Agamben 1998]). The moral imperative to alleviate suffering for hundreds of millions speaks to this higher dimension – one that exceeds economic concerns.

Humans are the paradigmatic source of perceived moral and aesthetic value. Recognizing humans as value sources allows us to extend ethical concern to non-human animals and the environment (see famously: Jonas 1985; Singer 2011). To do this, we must overcome the dichotomy in modern science between valueless physical processes and meaning-producing agents. We must integrate alternative sources of human self-understanding – such as Indigenous knowledges – into our concept of humanity. The resurgence of religion and spirituality must not be overlooked in this context. Flourishing is a spiritual category encompassing the entanglement of mind and nature.

It is also necessary to reject simplistic conceptions of identity.[7] With the digital transformation of society, we witness a proliferation of simplified social categories – 'stereotypes' – which reduce human diversity to algorithmically filtered data. These data are socio-economically valuable because they simplify human self-representation for increased visibility online. This benefits tech oligopolies aiming to reduce us to predictable, controllable stereotypes.[8]

Spirit is the entanglement of mind and nature. Technology arises at this interface and can be part of human flourishing – *if* it aligns with the normative demands of historically diverse human beings. Cultural diversity, classified under the broad rubric of *cultures*, is central. Cultural difference requires its own framework for understanding meaning and value.

To flourish is not merely to survive. Human flourishing occurs on a spiritual level, deepening our sense of what it means to live a meaningful life. We return not just to nature, but to *culture* – understood as modes of integrating ourselves into nature. Cultures express our self-conception and are the forms nature takes through human articulation. Flourishing is cultural – it involves technology, art, religion, social practices and knowledge systems. Values generated by cultures reflect accounts of humanity, measurable in terms of norms set by those very cultures. Cultures are nature's internal ways of conceiving itself. For nature is not identical to the meaningless universe of physics. The universe is merely a *part* of nature – an articulation of it. Science, as part of culture, cannot be understood solely through its own practices.

Science and meaning are opposed only when we mistake nature for the skeletal structure of mathematical invariances. These symmetries indicate objectivity, but they must not obscure their spiritual foundation: only humans – part of nature – can discover such patterns. This insight has guided many cultures. Philosophically, it has taken the form of *objective idealism*, which holds that nature manifests in human mindedness (spirit) to become aware of itself.[9]

The idea of nature as a meaningless material system, alien to our capacities for sense and meaning, is a historically provincial self-image. It arose in modernity as a flawed response to epistemic, social and economic upheavals. From a broader historical perspective, scientism, physicalism, materialism and naturalism are all limited and ultimately inadequate accounts of the realm of meaning we inhabit as *spiritual animals*.

The indispensability thesis compels us to recognize that any account of reality is incomplete if it excludes the knower – us – from its ontological picture. This chapter has argued that such inclusion demands an expanded conception of nature, one that integrates mindedness without reduction. Nature, as we now understand it, is not exhausted by the universe described by physics; it includes the irreducibly heterogeneous fields of sense in which we, as human animals and spiritual beings, participate. Mind and nature are not opposites but mutually entangled: each is part of the other through a mereological rather than causal relationship. This insight not only restores ontological dignity to mind and meaning, long dismissed by a reductive scientism, but also grounds a richer conception of human flourishing. Flourishing is spiritual – it unfolds in the interplay of culture, meaning and self-understanding. In reuniting the sciences with the humanities through the indispensability of mind, we recover a vision of nature that is alive with sense, shaped by spirit and inseparable from the human endeavour to understand and inhabit it meaningfully.

This renewed philosophical orientation resonates with what I have elsewhere called a *New Enlightenment* – an effort to move beyond the limitations of both prescientific dogma and ill-conceived modern reductionism. Such an enlightenment must be more than a theoretical reorientation; it demands trans-sectoral action that bridges academia, politics, business and culture. We must cultivate institutions and practices that reflect the irreducible plurality of human sense-making. This includes rethinking education, policy-making and economic systems in light of our shared spiritual reality – where meaning, value and agency are not byproducts but foundational. Interdisciplinary cooperation, grounded in a recognition of the human being as both natural and spiritual, can offer a pathway towards a more humane, just, and sustainable world (Gabriel 2025).

Notes

1 For a discussion of the technical details, see Gabriel and Priest (2022).
2 For a recent presentation of the view with responses to objections and further discussion, see Gabriel (2024a); Voosholz (2024a).
3 On this argument in the spirit of what I previously called 'transcendental ontology' (Gabriel 2011), see Ellis and Gabriel (2021).
4 On this principle of sufficient reason restricted to events, see Gabriel (2017). Thanks to Jocelyn Benoist for pushing me to clarify the status of the principle. It is not analytic (or a priori in some other more demanding, synthetic sense), but an epistemological feature of our account of action explanation. On this basis, we

5   For a similar view of the relationship between action explanation and causation, see Price (2011).
6   For details of the two-pronged mereology proposed here (where mind belongs to nature and nature to mind), see Gabriel (2024d).
7   For a rejection of associated concepts of identity, see Gabriel (2022: 130–87).
8   This corresponds to the much-discussed surveillance capitalism diagnosed in Zuboff (2019).
9   For a much-discussed attempt of formulating an objective idealism in an age of science, see Nagel (2012).

[continuation from previous page] are entitled to extend it to natural events whose behaviour we can model without invoking human notions of agency.

3

Ontological Security and Worldviews for Human Flourishing

Dean Rickles

1 Quo vadis, humanity?

The recent explosive changes in the structure of society and technology point to a radical re-imagining of the world and our place in it. This would appear to be underway whether we like it or not. The Covid-19 pandemic left us with the potential for a radical rebound, and the new technologies provide the potential to do so. We see multiple signs in the terminology of the moment, especially 'resilience', which means nothing more than the act of rebounding or springing back (though with the implication of a leap *forward* into a new and better state). Thus, we find humanity poised in a region of high uncertainty, with many possible options for how to rebuild itself, which would lead to many possible timelines. We have to weigh up such futures very carefully, and doing so requires that we step back from the changes themselves and consider the landscape of possibilities from on high.

The first steps to consciously selecting some particular timeline must lie in answering certain core questions: what is humanity and what would we like it to be (i.e. what do we think it *should* be?). As Charles Taylor once expressed it, we are 'self-interpreting animals' (Taylor 1985)[1] and can tell all manner of stories about ourselves, where we came from and where we are going. Yet how we answer this fundamental question will determine what we become in the future, and so will also determine what projects, attitudes, policies and technologies make sense relative to this vision or goal. Indeed, without a firm grip on where all of these changes are heading, what they are for and their deeper purpose, they are meaningless and lose the kinds of amplifier effects they could otherwise exploit.

My thanks to George Ellis, Angé Weinrabe and Andrej Zwitter for their comments on earlier drafts.

At the core of our answers to the basic question is a specific worldview in which they are embedded. Our worldview circumscribes what is possible and advisable in terms of future development. It provides the background structure against which what we are and what we do, and should and must do, makes sense. We are, at least in the 'developed' world, currently told, from the practical point of view, that we ought to be 'more connected' to our planet. We ought to be reducing carbon, recycling and so on. At the same time, from a theoretical point of view, we are being progressively *detached* from the planet at a more fundamental level by theories and ideas that say we are not just insignificant from the perspective of the world (just one group of material entities amongst many others), but a mere accidental blot on its landscape which would do far better without us being here at all. There appears to be an anti-human element that makes it difficult to motivate human action. This worldview is seemingly entrenched in science and leaks far outside of it. Whether we consciously realize it or not, these ideas form the backbone of our engagement with the world and each other and form the picture we have of ourselves: our worldview forms the very ground on which our stories are built. But we can do better, and, despite its own claims to the contrary, 'the science' by no means necessitates this anti-humanitarian worldview. As I will seek to advance in this chapter, there are far more enlightened ways of proceeding in which we are not cast as the villains but, if handled well, the heroes, and there is every reason to proceed in the ways suggested.

2 Worldviews and their ontologies

Before turning to these ways, let us make clear that without a firm grasp of our worldview, providing the overarching story of our development, we have no way of assessing whether the *means* for change align with the end goals (which are themselves not necessarily reachable). And what good is undirected change? Such change threatens to cancel itself out by acting in opposing directions. We find this often in the realm of identity politics, in which a claimed desire for unity is matched by divisive action. We also find it in such grand endeavours as the United Nations' '2030 Agenda for Sustainable Development' ('Transforming Our World: The 2030 Agenda for Sustainable Development | Department of Economic and Social Affairs', n.d.) which lack such a deeper underpinning that incorporates a meaningful story about the very humans whose behaviour it aims to influence. In this way, it is verging on a light form of fascism, with its centralized control mechanisms, and at worst a deliberately contrived deception intended to get control of humanity. These goals speak of 'a world of universal respect for human rights and human dignity, the rule of law, justice, equality and non-discrimination' without dealing with the deeper issue of what those rights *are* at a fundamental level and, more importantly still, *why* they are (i.e. what are they *grounded in* and why should we take them seriously?). Moreover, the 'plan of action for people, planet and prosperity' does not provide an underlying vision for making it emerge naturally. Why should humanity be treated in this way rather than some other way? I take a worldview to provide just such a ground by providing an overarching ontology (i.e. a theory of what humans are and how they fit in the world). We can have ontologies and worldviews that make the universe a home for humans or one that renders them somehow alien within it. We can view a worldview in this way as a kind of narrative technology that can help or hinder our goals. Governance does not make sense without this foundation.

Yet now consider all seventeen Sustainable Development Goals [SDGs]. They are couched in mechanistic, economic, and, indeed, fear-based terms. They are not based on a common view of humanity against which we can evaluate them and consider whether they advance us towards where we would like humanity to be going. They do not consider, in fact, any part of the human *experience* at all, and amount to what this project considers to be a debilitating scientistic materialism that has every chance of doing more harm than good. All is focused on the exterior conditions with no thought given to what is surely paramount to humans: to affectively engage with other humans and their world. Of course, there is a reason for this that originates in the modern scientific method itself: to implement such goals requires that they be measurable. And to be measurable requires that they are objective, repeatable events, representable in abstract terms. And if they are objective in this way, we must let go of the more living, meaningful elements. Yet what is more human than these unmeasurable elements? Are these not the very hallmarks of humanity that we should be trying to preserve and enhance?[2] Purpose. Meaning. Empathy. Joy. Love?

Indeed, even taken on their own materialistic terms, we find SDG 8[3] even retains what most reflective people would see as chief amongst the culprits behind our problems: economic growth! To still be pushing this (rather than its exact opposite) as a virtue must surely point to a serious problem with whatever deeper principles[4] such goals are resting on. Indeed, it is in direct conflict with the very concept of sustainability itself and undermines all of the other environmental SDGs – 'sustainable growth' is, transparently, an oxymoron as

Bartlett (1994) puts it (in other contexts we would call such unstoppable growth a *cancer*).

Such a drive for objectivity and materiality has been the path science has been following for several hundred years. Yet this was never part of the Enlightenment, which embodied a far more balanced viewpoint in which the target for scorn was *dogma* rather than the human spirit or the human soul. The current scientific priesthood would have been as much on the chopping block as the religious dogmas. For example, Robert Boyle, one of the founders of the scientific method, still viewed the world as 'The Temple of God', and Isaac Newton viewed it as the 'sensorium of God'. There was no discrepancy to be seen (see, e.g. Harrison 2017: 111). Science was essentially the soul navigating God via embodiment, revealing this Great Work, the cosmos, that is our home.

We needn't (and, I believe, *shouldn't*) return to so directly religious a viewpoint (i.e. in the sense of organized religion), to realize that we have lost something of significance between then and now. We have eradicated from science what was for its founders the very reason for pursuing it: for figuring out our place and purpose in a miraculous cosmos and realizing that we ourselves are equally miraculous and an integral part of its workings. It was never a clockwork universe for the likes of Newton, with humans cast as any other part of the overall mechanism. This was a myth invented much later and has had dire consequences for humanity's self-esteem. Yet it forms the contours of the stories that we can then tell about ourselves, which can then be used to justify various kinds of (mis-)treatment. For example, if you are nothing but a cog in a mechanism, then all that matters is the functionality of that cog relative to the machine.[5] The SDGs are very much along the lines of such functionalism, in which the humans being reorganized, shuffled, tracked, surveilled and so on are at best secondary to the system being constructed. They are labelled and catalogued by simple terms, such as 'consumer', with all complexity and depth removed. I doubt such proposals would be devised, let alone enacted on an often unsuspecting population, were it not for the ontological terraforming (in this case, an all-encompassing scientific materialism) that prepared the way.

3 Ontological security studies versus popularizing disenchantment

Many of the more vocal, popular scientists and commentators on science seem to draw a degree of glee from stamping out any such notions of a free and purposeful humanity. Consider this heartwarming passage from Nobel laureate Steven Weinberg (1993: 1):

> It is almost irresistible for humans to believe that we have some special relation to the universe, that human life is not just a more-or-less farcical outcome of a chain of accidents reaching back to the first three minutes, but that we were somehow built in from the beginning … It is hard to realize that this all [i.e., life on Earth] is just a tiny part of an overwhelmingly hostile universe. It is even harder to realize that this present universe has evolved from an unspeakably unfamiliar early condition, and faces a future extinction of endless cold or intolerable heat. The more the universe seems comprehensible, the more it also seems pointless.

Yet it is often the same scientists responsible for such bleak viewpoints, seemingly denying agency to humans, that are also those calling for human action to combat various world problems, often after calling out humans for their poor behaviour. This is a conflict of the kind I referred to above, in which the ideas about what we *are* conflict with the vision about what we *should do*. It simply does not make sense to claim that humans do not have free will (are essentially biological robots)[6] while berating them for their *choices* with respect to their habits of consumption. If we wish humanity to make important decisions, then we need a worldview that frees up a space for this to make sense, so we can then hold individual humans responsible. In this responsibility comes a source of meaningful existence in the world, through a deep form of cosmic engagement. To internalize that one is a mere accident that is spoiling the Earth is just as bad for collective humanity as it is for a child to be raised in a household believing it should not have been born (cf. Benatar 2008, for an example of a philosopher advocating human extinction). Collective shame is just as debilitating as individual shame; perhaps more so, in virtue of complex amplifier effects. The effects are a victim mentality and learned helplessness.

To see the power of framing, consider the following two statements:

1. I AM a co-creator of the universe; a sovereign being, with free will and rights, responsible for my actions.
2. I AM an accident of biology, with no free will, no rights and no responsibility for my actions.

If one desires collective efforts then clearly the first is going to provide superior motivation, while the second will support nihilism. Simply put: the idea that the world is dead matter means we can treat it as such (including the humans that are, after all, made from nothing but the selfsame matter). If the whole cosmos is organic and we are part of it, helping it evolve as it helps us, then we will be much better custodians and will feel pride in playing our part.[7] Let me state, however, lest misunderstanding should occur: I am not against science. I am against a dogmatic science that has closed off its mind. Yet, and this is really the key point, the scientific foundations themselves, even presently, do not definitively support either of the above two statements: there are formulations consistent with either. And yet, on the whole, the scientific community, and those that take its claims as gospel, adopts the disempowering stance. I am against science like this, that damages humanity, whether intentional or not, through an inability to see that its claims might not be the final word – that is, through hubris. I am against a science that makes its members priests. I am for science that lifts the human condition. Either we must have science open the door to spiritual (non-material) matters (see, e.g. Steane 2018), or else we must be steadfast in our refusal to allow scientists to encroach on matters spiritual with its current materialistic underpinnings. Not to do so risks 'ontological security', which I understand here as a feeling of being a stable self at home in the universe (rather than alienated from it), central to which are the stories and myths we define ourselves by. Indeed, to paraphrase an old saying of Max Weber's, we might say that the current *mythos* is that there is no *mythos* and only *logos*. However, it is still being used to tell the story of humankind. That story needs to be a human story because humans are going to be required to change themselves to make it all happen. It needs to inspire humans, through affect, to *want* to collaborate for the betterment of themselves and the planet because it touches some inner striving that goes beyond the merely material. Sustainability comes from the human beings themselves wishing to keep something going and improve, rooted in an enlightened self-interest in the planet's survival.

This is what is missing, with all of the UN SDGs, couched as they are in unappealing, mechanistic, economic and fear-based terms. This very way of framing things is, in my view, the source of the greatest threat to humanity: we have lost our basic vision of what the concept of humanity even means. Worse, the current framework of thought constricts any suggestions to a limited span of decidedly uninspiring options. The modern world is essentially what happens when a civilization thoroughly detaches from anything other than materiality. Academics especially are scared, through the possibility of humiliation perhaps, of even speaking of anything that might transcend the level of inert matter, presenting the cosmos instead as a giant wasteland. This situation is relatively recent. The mathematical physicist Hermann Weyl wrote not so very long ago (Weyl 1989: 35):

To Plato, the mathematical lawfulness and harmony of nature appeared as a divine mind-soul. The following words are from the twelfth book of the Laws: There are two things which lead men to believe in the Gods: one is our knowledge about the soul, as being the most ancient and divine of all things; and the other is our knowledge concerning the regularity of the motion of the stars and all the other bodies.

The present opinion is just the opposite of what once prevailed among men, that the sun and the stars are without soul. Even in those days men wondered about them, and that which is now ascertained was then conjectured by some who had a more exact knowledge of them – that if they had been things without soul, and had no mind (*nous*) could never have moved with numerical exactness so wonderful; and even at that time some ventured to hazard the conjecture that mind was the orderer of the universe. But these same persons again mistaking the nature of the soul, which they conceived to be younger and not older than the body, once more overturned the world, or rather, I should say, themselves; for the bodies which they saw moving in heaven all appeared to be full of stones and earth and many other lifeless substances, and to these they assigned the causes of all things. Such studies gave rise to much atheism and perplexity, and the poets took occasion to be abusive.

It is as if we have reached fever pitch in the problem Weyl presents. Our 'poets' are now figures like Sam Harris and Richard Dawkins, who seem to be in competition for the 'occasion to be abusive'.

Weyl's way of speaking might seem hopelessly outmoded, but simply consider the implications. You don't have to look far: we witness it every day. A purposeless, anxious population. The reason is obvious, but I'll repeat it: if you view the planet as a hunk of matter, then you may do as you wish with it. If you view humankind as more of the same, then you can treat it too as you wish. There is no place for dignity, divinity, creativity and freedom here. But, by contrast, if we view these latter principles as important characteristics of humanity, and if we view the planet as a living organism with a kind of dignity of its very own, and to which we are deeply connected, then we become highly constrained in our dealings with it and each other.

On this theme, we seem indeed here to have returned to a social condition that Max Weber was speaking of in his 1918 Munich lecture, *Science as a Vocation* (Weber 1918). There too he was forced to state that scientists are not prophets, that they do not deal with matters of meaning and transcendence. Weber famously also spoke of the existential void left by the science of his age, through its 'transformation into a causal mechanism' of the world denying the 'claims of the ethical postulate, that the world is a divinely ordered cosmos with some kind of ethically meaningful direction'. If we base everything on such a science, then we lose much, if not all, that is distinctly human. We indeed leave a moral and spiritual vacuum. C. J. Jung famously expressed the resulting phenomenon as 'Modern Man in Search of a Soul' (Jung 1933). And so we find ourselves here again, in a meaning crisis, only with a more advanced technological backdrop capable of more damage.

Here, however, I somewhat disagree with Weber. It is not science itself that is the problem. It is a science that limits itself to the exterior, to the third person, to the passive object. Modern science involves what Edmund Husserl labelled 'the Galilean style', which, as Steven Weinberg (1976) puts it, in *approval*[8] of such a style (unlike Husserl himself who viewed it as a 'crisis in European science'), involves making 'abstract mathematical models of the universe to which at least the physicists give a higher degree of reality than they accord the ordinary world of sensation'. In other words, our science is, at its deep core, detached from what is human.

I believe we can do something that Weber was unable to see because of the development of ideas at the time, namely restore meaning and purpose to the scientific worldview. This leaves us in better shape in terms of a solution, which for Weber was to simply be stoical in the face of disenchantment. We turn to this next, showing what is waiting in the scientific wings to bring this shift about – given limitations of space, and the technical nature of much of the material, we provide more of a sketch here, with pointers to further reading.

4 The Copernican involution?

The philosopher Hans Blumenberg wrote in great detail on the decentering/demotion of the Earth, and so humanity, brought about by the Copernican revolution in his *The Genesis of the Copernican World* (Blumenberg 1988). Blumenberg speaks in terms of the 'concept of reality' rather than worldview, but the idea is ultimately the same: both ground what is possible to experience. He argues that Copernicus (and later Darwin) were 'people who delivered a blow to humanity' (Blumenberg 1988: 1). Blumenberg's idea was ultimately that worldviews (and our creations of mind) are ways of avoiding the terror of the absolute – we need to limit it in some way. A similar idea can be found in a marvellous little book by Lewis Mumford (1922: 10):

> This world of ideas serves many purposes. Two of them bear heavily upon our investigation of utopia. On one hand the pseudo-environment or *idolum* is a

substitute for the external world; it is a sort of house of refuge to which we flee when our contacts with "hard facts" become too complicated to carry through or too rough to face. On the other hand, it is by means of the idolum that the facts of the everyday world are brought together and assorted and sifted, and a new sort of reality is projected back again upon the external world.

In our words, our worldviews are supposed to provide a *refuge*, not a means to insult and demoralize. Fortunately, there is more choice than people realize when it comes to our basic scientific worldview. Even Copernicus (in the preface to the *De Revolutionibus*), on whose decentering of humanity we can trace much back to (whether he intended it or not), wrote his account because of the problem that 'the philosophers could by no means agree on any one certain theory of the mechanism of the universe, which was constructed o*n our behalf by the best and most orderly Maker of everything*' (my emphasis). That is, Copernicus' own presentation involved the idea that the world was created 'on our behalf'! Yet any notion of teleology generates a kind of hysteria amongst many in academia (biology included), since they have been raised precisely on the mother's milk of the Copernican revolution outlawing it.

We wish to place the concept of humanity under the spotlight, and in particular consider how much plasticity the world has relative to human action; how it can be carved to our ultimate vision of what we are and why we are here. My approach involves a viewpoint that I have been examining and developing for almost a decade (Atmanspacher and Rickles 2022; Rickles 2016; 2022). It is the idea that humans are not passive observers in the world, but active participators in what occurs; cosmic custodians in what Ilya Prigogine labelled a 'community with the universe' (Prigogine 2004: 1). The world is not a completed project, a block as the orthodoxy would have it, but something whose growth and development (and hopefully *evolution*) we are involved in. It's no good to simply state this: there must be some evidence. That evidence comes from at least two convergent sources: from quantum mechanics and from aspects of the information age.

Rather than attempting to describe these theories, let me simply use an analogy that the physicist John Wheeler employs (whose ideas I'm much indebted to) that makes my point. Wheeler considers the nature of reality in quantum mechanics as akin to a game of twenty questions in which there is no pre-agreed upon word – something that actually happened to Wheeler without his knowledge.[9] As each question was asked (e.g. am I a mineral?), the final answer was progressively constrained by Yes or No answers, until the final object was determined: it happened to be a cloud in Wheeler's game. Wheeler's questions alone weren't enough to generate the final word: it was a genuine case of co-creation. Likewise, when one considers the quantum world, there is no way it is until a question has been asked of it, and reality is built up from such questions and their answers by the world. There is co-creation at this level too. The world does not know what we will ask of it, and we do not know what answer it will give. This basically amounts to freedom and chance. There are both experimental and theoretical reasons for this which define the quantum world and distinguish it from classical physics:

> Each query of equipment plus reply of chance inescapably do build a new bit of what we call 'reality'. Then for the building of all of law, 'reality' and substance – if we are not to indulge in free invention, if we are to accept what lies before us – what choice do we have but to say that in some way, yet to be discovered, they all must be built upon the billions upon billions of such acts of observer-participancy? (Wheeler 1980: 359).

There is potentiality here that applies to humans and the world alike. It is realism of an unusual stripe, known as 'participatory realism'. There is a world which kicks back in chancy ways, but it requires our participation to do so. We have something starkly different to the clockwork universe here. And we find Wheeler writing in ways that exactly suggest the kind of viewpoint necessary for empowering humanity (Wheeler 1973):

> Today I think we are beginning to suspect that man is not a tiny cog that doesn't really make much difference to the running of the huge machine, but rather that there is a much more intimate tie between man and the universe than we heretofore suspected! The physical world is in some deep sense tied to the human being.

This is a physics that stops the machine (Wheeler and Zurek 1983: 183):

> Nature at the quantum level is not a machine that goes its inexorable way, [rather we] are inescapably involved in bringing about that which appears to be happening.

It is not the probabilistic features of quantum mechanics that do this, but the deep potentiality that lies at the root of our freedom and the world's: no participator, no world; no world, no participator. The universe does not sit outside of our heads, ready-made. We make the difference and breathe actuality into potentiality. Since the questions are of the Yes/No form, the approach is aligned with information. Wheeler summed it up as 'It from Bit' – that is, the world of things (Its) comes from the specific yes/no questions posed (bits, where yes/no corresponds to 1/0).

To sum this all up: our best foundation for an overarching theory of reality is based ultimately on information. This is, moreover, a perfect approach for the world entering the fourth industrial revolution [4IR]. Yet this whole scheme only makes sense relative to participators that imbue bit strings with meaning. Without Us (we the people), there is no way to distinguish one bit string from another; nor is there a way of seeing how one and the same bit string can represent an infinitude of possibilities. As with quantum, so with information: none of it makes sense without humans.

5 Concluding remarks

Recall Blumenberg's idea about the terror of the 'absolutism of reality' (Blumenberg 1988: 3). We push the absolute away through stories. For Blumenberg, we do this to tame our fears. This is true of the kind of participatory worldview I'm suggesting we might adopt. Rather than a cold, indifferent world of the kind described by Weinberg, we have a world that is sensitive to our actions. It is not a God of religion, but it can serve a similar function. That this can indeed be the case is presented nowhere better than in the following passage of C. G. Jung (1989):

> [M]an is indispensable for the completion of creation; that, in fact, he himself is the second creator of the world, who alone has given to the world its objective existence – without which, unheard, unseen, silently eating, giving birth, dying, heads nodding through hundreds of millions of years, it would have gone on in the profoundest night of non-being down to its unknown end. Human consciousness created objective existence and meaning, and man found his indispensable place in the great process of being.

To be indispensable in the great process of being. What could be more rewarding? What could be more motivational than this? There is something almost mythical in the idea, and myths have the power to move. I am suggesting that this new myth for humanity and its evolution is especially timely because of the nature of the worldview in which we have been for a long time mired like quicksand. This is a story of *cosmic* participation (and, therefore, potentially evolution too) in which humanity is involved that is well grounded in science as well as the cultural milieu couched in digital and information-based terms. This can be the linchpin securing a scientifically acceptable, non-materialistic, non-religious (though compatible with spiritual ideas) worldview that does not discount human beings (and their free will and agency) in the universe (far from it). This kind of active, optimistic viewpoint can be traced to William James and his pragmatism. We should certainly feel free to add whatever dose of pragmatism gets us to the best result when it comes to humanity and its future. I doubt pessimism will achieve anything remotely like what we would like. William James refers to pessimism as 'an essentially religious disease [in which] [p]essimists issue a religious plea to the world: for it to be good. And they are let down by it: no response is forthcoming' (James 1912: 31). We can provide a more compelling framework, one that has the potential to provide an answer to pessimism, though it requires action: if we want the world to be good, then that is on Us.

Notes

1   Though, I confess, I think viewing us as 'animals' is part of the problem here: we are *so* self-interpreting that we can choose *not* to be just some other kind of animal, even if it is one with a special gift of interpretation (or, to look back to Aristotle, an animal that happens to be rational – cf. Eldred (2024) for a tracing back of many of the current world problems precisely to the bundling of humans into the natural order of the world, following laws of nature, rather than the separate category of existence, outside of those laws, that Aristotle had in mind).
2   We find a similar phenomenon at work in the debate over 'Liveable Cities' in which the subjective states that are the core of the *feeling* of liveability are completely ignored in favour of objective, measurable features such as walkability, number of amenities and so on. I will give a personal example close to home. There was a beautiful piece of nature near my house which I often went for walks in, as did many others. It was self-tended by the walkers. There was no littering, and a general appreciation of the place. The local council had the bright idea of turning it into a more official nature reserve, and promptly chopped most of it down to install a playground, with toilets, and an ugly road. It now looks like a generic inner city park, since that is the template. Needless to say, I do not go there anymore. Yet the council views it as a success in making the area more liveable: it is another amenity.
3   'Promote sustained, inclusive and sustainable economic growth, full and productive employment and decent work for all': available at https://sdgs.un.org/goals/goal8 (accessed 21 August 2024).
4   I do not simply mean neoliberal capitalism here, which of course needs adjustment, but the deeper principles about humans that such systems also rest upon: that we

are consumers of the world rather than its custodians. The society that has been constructed almost corrals people into such behaviours, as the opening quote from Stafford Beer makes plain. This does not mean that they are in any way part of *human nature*.

5   Let me point out that I am not marginalizing other lifeforms here. The account is generalizable to animals and, if it eventually came to it, even artificial life that was reasonably seen to share in the specific features I am using humans to discuss here (namely, the ability to engage in a form of worldbuilding through free action and the creation of meaning). Hence, ultimately, my account might easily join Martha Nussbaum's (2023) call for the attainment of flourishing for animals (allowing them to express their full range of capabilities), and, as I mentioned, further, to all sentient life (i.e. life with the ability to *feel*). Even for lifeforms not so capable, a natural part of flourishing for humans is, in any case, surely tightly linked to a more general flourishing.

6   A recent notable example is Robert Sapolsky's book *Determined: A Science of Life Without Free Will* (2023). This book argues that there is simply no self/soul that can decide in a freely chosen way. This is a perfect example of a scenario that also undercuts responsibility and the means to make changes through will. Fortunately, his claims are simply false and based on a misunderstanding of the deeper workings of physics, and its gaps. His treatment of quantum mechanics, for example, targets the well-known mistake of treating the indeterminism as providing a habitat for free action. Yet this is not where the freedom lies. That freedom is in the deeper realm from which that very indeterminism also comes. We return to this later. For now, I want to point out how dangerous such books are in terms of how people view themselves. If true, I could accept that it is preferable to have this truth out there, although even then I would package it in an entirely different way that minimizes harm on minds. That it is not true and based on mistakes makes me puzzled as to the agenda of such writing. That he suggests his approach might lead to a 'more humane world' (by removing the very things that make us human) beggars belief!

7   Why communicate lack? As the novelist P. D. James once said, 'What a child doesn't receive he can seldom later give' (James 2000). In inputting lack, inertness, smallness, passivity and so on, that is what we can expect from our population, and perhaps that is the real goal of those that wish to retain control. We might view this along the lines of Robert Owen's remarks on Trevelyan: 'environment makes character and that environment is under human control' (as cited in Willmer 2024). Ultimately, this is the cybernetic vision of society and revolution. This is why cybernetics provides a useful framework for this project, but it can be interpreted in terms of the cultivation and elevation of humans rather than a mechanical approach to pulverize them into submission. Of course, this is in line with the basic theme presented here: once we have our vision of humanity, we can design our systems to best generate what we desire in future humans.

8   Contemporary academia appears to have turned such approval of the Galilean style into a *mandate* for participation in the community, with mechanisms in place for outlawing (through shaming, rejection of publication, and so on) attempts to push against it, and suggest that there might be something special to humans and the world they occupy.

9   Recall that twenty questions usually involves an answerer settling upon an object first, with a questioner then trying to identify it within twenty questions that can only be answered with a Yes or a No.

Part Two

Non-Material Conceptions of Flourishing

Positive Psychology and Human Flourishing

Theo K. Bouman

1 Introduction

People with all kinds of backgrounds have been pondering on the question 'What makes life worth living?'. Interestingly, this question receives remarkably little attention in the discipline of psychology, and is mostly addressed by psychologists from the humanistic and existential traditions. Despite this, when the discipline of psychology was still in its infancy, the American psychologist and philosopher William James wrote an essay entitled 'Is life worth living?' in which he concludes (James 1895: 22): 'This life *is* worth living, we can say, *since it is what we make it* (...)', thereby giving a sense of agency to the individual. In the 1950s, Abraham Maslow formulated the hierarchy of needs, at the top of which he placed self-actualization (i.e. the realization of a person's full potential), emphasizing the importance of human aspirations and virtues (Maslow 1943). At that time, Maslow coined the term 'positive psychology'.

It was Martin Seligman, an American professor of clinical psychology, who put the concept of positive psychology on the scientific agenda during his presidency of the American Psychological Association in 1998. He observed that traditional (and in particular clinical) psychology had disproportionately focused on alleviating suffering and addressing mental illness, which left the study of positive aspects of human experience largely neglected. While acknowledging the importance of addressing mental health challenges, Seligman advocated for a more balanced approach – one that investigates the conditions and practices that enable individuals, communities and societies to thrive. As Seligman and Csikszentmihalyi wrote in their seminal paper (Seligman and Csikszentmihalyi 2000: 5):

> The field of positive psychology at the subjective level is about valued subjective experiences: well-being, contentment, and satisfaction (in the past); hope and optimism (for the future); and flow and happiness (in the present).

The framework of positive psychology represents a transformative shift in the field of psychology, emphasizing the scientific study of human strengths, virtues and factors that contribute to human flourishing. Positive psychology moves away from a pathology-centric model that focuses on diagnosing and treating mental illness, instead promoting the exploration of what makes life meaningful, fulfilling and worth living. This approach seeks to empower individuals to lead flourishing lives, and it does so by combining empirical research with practical applications.

In this chapter, we will mainly refer to the work of Seligman and his co-workers as an illustration of the theoretical and applied aspects of positive psychology. It should be noted, however, that over the years many other scholars have also extensively contributed to this field (see e.g. the edited volume by Joseph 2015). Furthermore, the field has grown considerably, as is shown by the sharp increase in publications on positive psychology over a 20-year period, with a prominent role of several research institutions in the United States (Wang, Guo and Yang 2023).

2 Features of positive psychology

Concepts that are often related to positive psychology are *eudaimonia*, happiness and well-being. *Eudaimonia*, an ancient Greek concept often translated as 'flourishing' or 'living in accordance with one's true self' is rooted in Aristotle's philosophy. Eudaimonia emphasizes the pursuit of a virtuous and meaningful life rather than mere hedonic pleasure (temporary gratification). It should be noted that flourishing represents just one facet of the broader construct of well-being, together with positive emotions and life satisfaction. Seligman's positive psychology incorporates and expands upon this concept of eudaimonia by empirically exploring the components that contribute to human flourishing (Seligman 2011b). This vision of human flourishing goes beyond transient pleasures, emphasizing virtue, meaning and growth. Positive psychology acknowledges the value of hedonic happiness but prioritizes eudaimonic well-being, focusing on long-term fulfilment through growth, relationships and meaning. In a survey in the Netherlands (Schotanus-Dijkstra et al. 2016), flourishers were defined as having high levels of both hedonic well-being and eudaimonic well-being, revealing that 36.5 per cent of the Dutch were flourishers. They also found that social support and positive life-events contributed significantly to flourishing.

Happiness is a subjective positive emotional state and as such one of the key components of positive psychology. Positive psychology emphasizes the cultivation of positive emotions such as joy, contentment and gratitude, which are essential elements of happiness. It is considered an important part of a broader concept of *well-being*, which encompasses both affective well-being (the balance of positive and negative emotions) and cognitive well-being (life satisfaction). Happiness is a multifaceted concept influenced by both internal factors (such as emotional and cognitive processes) and external factors (such as relationships and social contexts). The social psychologist Jonathan Haidt (2006) explored the 'happiness formula', suggesting that happiness is influenced by three primary factors, i.e. *'genetic set point'* (a largely genetically predisposed baseline level of happiness), *'circumstances'* (external factors such as wealth, social status, relationships and health), and *'intentional activities'* (conscious actions individuals take to enhance their well-being). Intentional activities connect directly to the core interventions of positive psychology such as practicing gratitude, cultivating positive relationships, and engaging in meaningful activities (Haidt 2006).

2.1 The PERMA framework

Well-being (and not life satisfaction or happiness) is the focal topic of Seligman's conceptualization of positive psychology. For that reason, the PERMA framework (Seligman 2011) was developed, which identifies five building blocks considered to be essential for well-being, namely *Positive emotion, Engagement, Relationship, Meaning* and *Accomplishment*.

Experiencing *positive emotions* such as joy, happiness, optimism and contentment enhances life satisfaction and resilience. While positive psychology acknowledges that transient emotions like joy and pleasure are important (*hedonia*), it is emphasized that cultivating enduring feelings like hope and gratitude sustain well-being over time (*eudaimonia*). This topic is theoretically and empirically addressed in the Broaden-and-Build Theory (Fredrickson 2001) which posits that positive emotions broaden cognitive and behavioural repertoires, fostering creativity, problem-solving and resilience. Empirical studies (see Fredrickson 2013) indeed demonstrated that positive emotions help people to broaden their thinking and build psychological resources over time. These resources contribute to enhanced resilience, improved social relationships and better coping strategies during times of stress or adversity. Positive emotions can undo the harmful effects of negative emotions and promote resilience (Tugade and Fredrickson 2004). People who experience frequent positive emotions tend to have better mental health and longer lives.

Engagement refers to being deeply involved in meaningful activities and is often described as 'flow'. The concept of flow – coined by Mihaly Csikszentmihalyi in 1975 – is a subjective state that people report when they are completely involved in something to the point of forgetting time, fatigue and everything else but the activity itself (Csikszentmihalyi 2014: 230). Flow and its correlates have been extensively studied over the years (see Csikszentmihalyi 2014), and were found to be associated with well-being, personal growth, creativity, satisfaction and the fulfilment derived from challenging yet rewarding experiences.

Relationships, i.e. having supportive and meaningful social connections characterized by trust, empathy and mutual respect, are crucial for emotional health, resilience and overall well-being, as well as a buffer against adversity. Research has consistently shown that having strong, supportive social connections is one of the best predictors of happiness and life satisfaction (e.g. Diener and Seligman 2002). A longitudinal study by Rohrer, Richter, Brümmer, Wagner and Schmukle (2018) showed that actively engaging in social relationships predicts higher life satisfaction one year later.

The pursuit of *meaning* is central to positive psychology, whereby meaning refers to having a sense of purpose and direction in life, which helps contribute to a feeling that life is valuable and worth living. Living with purpose and aligning one's actions with something greater than oneself – such as family, community, faith, religion or social causes – provides a profound sense of fulfilment. In positive psychology, finding meaning often involves contributing to the well-being of others or engaging in activities that resonate with personal values. In general, empirical studies support the notion that having a sense of purpose is associated with better physical health, longer life expectancy, reduced risk of mental illness, higher levels of satisfaction, positive emotions and feeling more connected. A meta-analysis including sixty-six studies found small to moderate associations between meaning in life and health indicators (Czekierda et al. 2017), although no specific conclusions can be drawn about causality. Another meta-analysis (Li, Dou and Liang 2021) showed a robust positive link between the presence of life meaning and subjective well-being.

Accomplishment refers to having feelings of mastery and achievement. Achieving goals, whether large or small, instils a sense of pride, competence and self-efficacy. Positive psychology encourages individuals to set and pursue meaningful objectives that align with their values and passions. Research has shown that setting and achieving goals significantly boosts self-esteem, confidence and happiness. A study by Sheldon and Elliot (1999) found that people who set goals that are intrinsically motivated (aligned with their core values) experience greater satisfaction and well-being. The Self-Determination Theory (Deci and Ryan 1985) posits that when individuals are able to achieve goals that satisfy basic psychological needs for autonomy, competence and relatedness, they experience higher well-being and motivation.

The five pillars or building blocks laid down in the PERMA framework are interrelated, and collectively, they provide a comprehensive framework for understanding and enhancing well-being. Later, specifically for organizational contexts, four additional building blocks were added (referred to as PERMA+4), i.e. physical health, mindset, environment and economic security (Donaldson, Van Zyl and Donaldson 2022). A systematic review of empirical studies indeed found the PERMA building blocks to be significant predictors of subjective well-being (Cabrera and Donaldson 2024).

2.2 Strengths and virtues

A key component of Seligman's theory is the identification and cultivation of character strengths and virtues. For that purpose, Seligman and Peterson (mentioned in Seligman, 2011: 243-65) developed the *Values in Action (VIA) Classification of Strengths,* identifying twenty-four strengths grouped under six virtues, namely:

Virtues:	*Strengths:*
Wisdom and Knowledge	Cognitive strengths, such as creativity, curiosity and love of learning.
Courage	Emotional strengths, including bravery, perseverance and honesty.
Humanity	Interpersonal strengths, such as kindness, love and social intelligence.
Justice	Civic strengths, including fairness, leadership and teamwork.
Temperance	Strengths that protect against excess, such as humility, prudence and self-regulation.

Transcendence	Strengths that provide meaning, such as gratitude, hope and spirituality.

As can be seen, these virtues are closely related to the Aristotelian cardinal virtues: prudence, justice, temperance and courage. Psychological strengths have been broadly defined as ways of behaving, thinking or feeling that an individual has a natural capacity for, enjoys doing, and which allow the individual to achieve optimal functioning while they pursue valued outcomes (Quinlan, Swain and Vella-Brodrick 2012). Strengths are thus viewed as pathways to achieving well-being. By identifying and using one's top five (or 'signature') strengths, individuals can build resilience, foster positive relationships and lead more meaningful lives. Working on one's strengths rather than one's weaknesses produces greater benefits for the individual. Empirical studies suggest that people who engage in activities that align with their strengths report higher levels of happiness and well-being (Schutte and Malouff 2019).

3 Positive psychology and psychopathology

3.1 Complementarity

One of the central ideas of positive psychology is that mental health should not merely be defined as the absence of mental problems, but as the presence of positive qualities like well-being, resilience and personal strengths, thereby emphasizing what people can do to thrive, regardless of whether they have mental disorders. Positive psychology and psychopathology, while often seen as opposite ends of the psychological spectrum, are closely related and can complement one another in understanding and promoting mental health. Psychopathology is the study of mental disorders, focusing on the diagnosis, treatment and prevention of psychological conditions such as depression, anxiety and schizophrenia. Positive psychology, on the other hand, is concerned with studying and fostering the positive aspects of human experience, such as happiness, resilience and strengths. While psychopathology traditionally focuses on pathology and dysfunction, and on *vulnerability factors* leading to these conditions, positive psychology seeks to enhance well-being and personal growth. Positive psychology might also play a crucial role in preventing mental illness by fostering *protective factors* that reduce the risk of developing psychopathological conditions. By promoting positive emotions, strong social connections and personal strengths, positive psychology helps individuals build psychological resources that can buffer against the onset of mental disorders, reduce the severity of existing conditions or enhance positive psychological states.

3.2 Post-traumatic growth and resilience

An interesting case in which a pathology-oriented approach contrasts with a positive psychological approach pertains to post-traumatic stress disorder. Most research on traumatic experiences, such as being confronted with death and dying, focuses on the ensuing mental health problems, such as hypervigilance, physical sensations, traumatic re-experiencing, nightmares, emotional numbness and avoidance. Positive psychology, on the other hand, investigates how individuals may recover and thrive following adversity. Tedeschi et al. (2018) defined *post-traumatic growth* as a 'positive psychological changes experienced as a result of the struggle with trauma or highly challenging situations'. Research suggests that people who experience significant adversity or trauma often emerge from these experiences with greater resilience, stronger relationships and a deeper sense of meaning in life (e.g. Brooks et al. 2020).

A related concept is *resilience*, commonly described as the ability to bounce back or overcome some form of adversity and thus experience positive outcomes despite an aversive event or situation. Positive outcome refers to individuals maintaining, regaining or surpassing their prior level of functioning (Vella and Pai 2019). Tugade and Fredrickson (2004) found that resilient individuals tend to experience more positive emotions and better psychological adjustment to stress. Studies in various samples (e.g. older adults, adolescents, military, school children, corporate organizations) have demonstrated that resilience can be cultivated and improved through positive psychology interventions such as strength-based approaches, gratitude practices, and mindfulness training (Liu et al. 2022).

3.3 Integrating positive psychology with treatment approaches

The integration of positive psychology in clinical practice is becoming increasingly recognized as beneficial for promoting overall mental health. By incorporating positive psychology interventions into mental health care, professionals can provide a more holistic approach to mental health that not only alleviates suffering but also promotes flourishing and personal growth. Bohlmeijer and Westerhof (2021), for instance, make a plea for integrating, and thereby balancing, complaint-oriented and strength-oriented treatment approaches. It is interesting to notice that well-established and evidence-based types of psychotherapy actually do incorporate positive psychology elements.

Cognitive Behavioural Therapy (CBT) is based on the premise that psychological distress is largely influenced by distorted or negative thinking patterns and ineffective coping behaviours. The goal of CBT is to help individuals recognize these harmful patterns, reframe their thoughts and adopt healthier behaviours. CBT has been extensively studied and appears to be effective for the treatment of many mental disorders (e.g. depression: Cuijpers et al. 2023). CBT and positive psychology have distinct but complementary approaches to mental health. Both recognize the power of thought processes and emphasize the role of behaviour change. While CBT focuses on identifying and correcting maladaptive thoughts and behaviours to alleviate distress, positive psychology emphasizes strengths, positive emotions and well-being. Integrating both approaches can create a more comprehensive therapeutic strategy that not only treats mental illness but also fosters long-term well-being and resilience. The integration becomes explicitly apparent in the emergence of Positive CBT, in which the focus is not only on reducing symptoms and complaints, but also on increasing strengths and well-being (Bannink and Geschwind 2021).

Mindfulness-Based Interventions. Mindfulness, rooted in ancient contemplative traditions, involves cultivating present-moment awareness with a non-judgemental attitude, as well as acceptance and self-compassion. For instance, mindfulness enables individuals to savour positive experiences by helping them focus on the present, rather than being distracted by past regrets or future projections. Self-compassion helps individuals develop a kinder relationship with themselves, reducing self-criticism and promoting emotional healing. Mindfulness aligns with the eudaimonic perspective in positive psychology, which prioritizes meaning, purpose and self-actualization over transient happiness. Research (Zhang et al. 2021) has shown that mindfulness-based interventions can reduce stress, improve emotional regulation and enhance overall well-being, resilience and life satisfaction, all of which are essential for flourishing.

Acceptance and Commitment Therapy (ACT), which is grounded in mindfulness and acceptance, aligns with many principles of positive psychology, such as living in accordance with one's values and building psychological flexibility. It has been shown to be effective in treating various mental disorders (Gloster et al. 2020), while promoting well-being and flourishing. One significant overlap between positive psychology and ACT lies in their shared emphasis on the individual's values and meaning in life. Positive psychology highlights the importance of living a meaningful and purpose-driven life as a central component of well-being. This concept aligns with ACT's focus on identifying and committing to personal values as a guide for action. Both approaches encourage individuals to reflect on what truly matters to them and to align their behaviours with their values to enhance life satisfaction and fulfilment.

Although these forms of psychotherapy contain positive psychology elements, other therapies have been specifically developed to promote well-being and positivity. *Solution-Focused Brief Therapy* (SFBT) is a short-term goal-focused evidence-based therapeutic approach, which incorporates positive psychology principles and practices, and which helps clients change by constructing solutions rather than focusing on problems. The main idea is that problems are best solved by focusing on what is already working, and how a client would like her life to be. According to a meta-analysis (Vermeulen-Oskam et al. 2024), SFBT shows substantial effects in reducing psychosocial problems, in particular marital dysfunctioning.

In Italy, Fava (1999) developed *Well-Being Therapy* (WBT) characterized by its reference to the concept of euthymia, where lack of mood disturbances is associated with positive affect and psychological well-being. WBT is designed to be delivered in addition to another treatment (such as CBT) rather than as a stand-alone therapy. In structured treatment sessions and using homework assignments, patients are monitoring the course of well-being episodes, identifying and tackling interruptions therein, and pursuing optimal experiences (Guidi and Fava 2021).

4 From concepts to practice: interventions and applications

4.1 Positive psychology interventions

The concept of flourishing is central to many positive psychology interventions that are designed to enhance well-being. These interventions are structured activities or exercises designed to increase well-being, promote positive emotions, enhance strengths and foster personal growth. They do so by focusing on the development of positive traits and the cultivation of aspects like gratitude, mindfulness, strengths and positive relationships. While the primary goal of these interventions is not to cure mental illness, they have been shown to be effective in complementing traditional therapies and improving overall well-being. A meta-analysis of their effectiveness shows that specific positive psychology interventions can successfully improve subjective and psychological well-being while also assisting in the reduction of depressive symptoms (Bolier et al. 2013; Carr et al. 2024).

Some of the most widely studied and implemented interventions are presented below.

4.1.1 Gratitude interventions

These are one of the most commonly used and researched interventions in positive psychology. One variant, called gratitude journaling, consists of writing down three things one is grateful for each day, which can boost happiness and decrease stress. Another variant involves writing a letter to someone who has had a positive impact on your life, expressing sincere appreciation for their actions. Delivering this letter can improve the writer's well-being and foster stronger, more positive relationships. Studies have shown that people who engage in gratitude exercises report higher levels of well-being, improved mood and greater life satisfaction. For instance, a study by Emmons and McCullough (2003) found that participants who kept gratitude journals showed significant improvements in their mood and well-being compared to those who wrote about daily hassles. Gratitude journaling may need modification in collectivistic cultures where gratitude is often expressed publicly rather than privately. Encouraging individuals to share their gratitude with family or community members may align better with cultural values.

4.1.2 Acts of kindness

Performing acts of kindness (as a form of prosocial behaviour) is another widely used intervention in positive psychology. A common intervention involves asking participants to perform five acts of kindness in a week, with the goal of increasing social connection and boosting mood. These acts can be anything from helping a neighbour to giving compliments to a colleague. Random acts of kindness, whether big or small, can enhance social bonds, promote feelings of happiness and boost overall well-being. The overall effect of acts of kindness on the well-being of the actor was found to be small-to-medium in a meta-analysis (Curry et al. 2018).

4.1.3 Savouring positive experiences

Savouring refers to the process of fully experiencing and appreciating positive moments. Engaging in savouring exercises helps individuals amplify positive emotions and enhances their ability to experience pleasure. Individuals may, for instance, be asked to reflect on a positive event, such as a family gathering or personal achievement, and focus on the emotions, sensations and details of the experience. This enhances the ability to appreciate life's pleasures and promotes long-term well-being. Taking a walk with the intention of savouring the present moment by paying attention to sensory details (e.g. sights, sounds, smells) is another intervention to foster mindfulness and amplify positive emotions.

4.1.4 Strengths-based interventions

Positive psychology emphasizes identifying and using personal strengths, which helps individuals experience greater fulfilment, engagement and accomplishment. By recognizing and developing their strengths, individuals are more likely to achieve goals that are meaningful and aligned with their values, increase their self-esteem and foster positive relationships. Over the years, various strengths-based interventions have been developed. An early systematic review showed them to have small to moderate effects on increasing an individual's well-being (Quinlan, Swain and Vella-Brodrick 2012). Strengths-based interventions, which encourage individuals to identify and use their core strengths, must account for cultural preferences. In collectivistic societies, for instance, emphasizing relational strengths like kindness or teamwork may be more impactful than focusing on individualistic traits like leadership or creativity.

4.2 Application contexts

Positive psychology findings and practices have been applied in various domains, including education, healthcare, workplace and public policy (see Seligman 2011b for elaborate examples).

Positive *education* integrates well-being principles into traditional school curricula. It aims to equip students with skills such as optimism, resilience and emotional regulation alongside academic knowledge. Programmes inspired by positive psychology have shown improvements in students' mental health, academic performance, and social behaviour. For example, teaching gratitude practices or encouraging students to identify and use their strengths has been linked to increased motivation and happiness.

In *health care settings*, positive psychology complements traditional therapies by focusing on strengths rather than solely addressing deficits. Interventions such as practicing gratitude, mindfulness or savouring positive experiences have been shown to reduce symptoms of depression, anxiety and stress. The emphasis on resilience and optimism helps individuals better manage chronic illnesses and improves overall quality of life.

Organizations have increasingly adopted principles of positive psychology to enhance employee well-being and productivity. Strategies such as recognizing employee strengths, fostering a positive work culture and encouraging work-life balance contribute to higher job satisfaction, engagement and retention. Leaders who adopt a strengths-based approach are more likely to inspire and motivate their teams.

Public policy initiatives aimed at measuring and improving citizens' happiness, such as Bhutan's Gross National Happiness index, reflect the growing recognition of well-being as a critical aspect of national governance, emphasizing collective flourishing over economic growth.

5 Cultural aspects of positive psychology

Well-being is not a one-size-fits-all concept. Cultural context profoundly shapes how individuals perceive, pursue and experience well-being. By incorporating cultural diversity, positive psychology evolves from a universalist perspective into a more inclusive and nuanced field, capable of addressing the unique needs and values of different societies.

Culture serves as a lens through which people view and interpret the world. It influences values, social norms and behaviours, all of which affect how well-being is understood and pursued. What constitutes a 'good life' in one culture may differ significantly from another.

5.1 Individualism vs collectivism

One of the most significant cultural dimensions influencing positive psychology is the distinction between individualistic and collectivistic societies. Individualistic cultures, such as those in North America and Western Europe, prioritize autonomy, self-expression and personal achievement. Well-being in these contexts is often associated with hedonic happiness – pleasure, material possessions, positive emotions and self-fulfilment. In contrast, collectivistic cultures, found in many regions in the world, prioritize interdependence, harmony and group cohesion, as well as a strong relationship with the natural ecosystem. In these societies, well-being is tied to fulfilling social roles, maintaining relationships and contributing to the collective good. For example, happiness may stem from meeting familial or community duties and responsibilities rather than individual desires (Kiknadze and Fowers 2023). Interestingly, Ng and Ong (2022) argue that prosocial behaviours, defined as actions of goodwill that are performed to benefit other individuals, might be universally beneficial, i.e. irrespective of their cultural context.

5.2 Diverse conceptions of happiness

Cultural variations also shape how happiness is understood. In Western cultures, happiness often aligns with high-arousal positive emotions like excitement, joy and pride, reflecting an emphasis on personal achievement and self-actualization. However, non-Western cultures frequently value low-arousal positive states like contentment, serenity and harmony.

Similarly, while Western frameworks often distinguish sharply between positive and negative emotions, many Eastern traditions embrace the coexistence of both. These traditions advocate the dialectical nature of well-being (e.g. happiness and unhappiness are complementary) and recognize that flourishing encapsulates both positivity and negativity (Kiknadze and Fowers 2023; Ng and Ong 2022). Such a holistic approach contrasts with Western tendencies to prioritize positivity and minimize adversity, thereby underscoring the importance of cultural sensitivity in, for instance, designing positive psychology interventions.

5.3 The role of meaning and purpose

Meaning and purpose, central to eudaimonic well-being, are culturally constructed. In Western contexts, finding purpose often involves personal passion, career goals or individual achievements. In contrast, many non-Western cultures derive meaning from fulfilling societal or familial obligations. Religious and spiritual beliefs also play a significant role in shaping meaning. In many cultures, faith traditions provide frameworks for understanding life's purpose, offering rituals and practices that promote well-being. Moreover, religion and spirituality are inherent aspects of flourishing for many people in the world (Kiknadze and Fowers 2023). As an example: an Islamic understanding of health and well-being maintains that a good life is primarily achieved through living in accordance with Islamic teachings (Saritoprak and Abu-Raiya 2023). Positive psychology's focus on meaning must therefore accommodate these diverse sources of fulfilment, recognizing that purpose may emerge from both individual pursuits and collective traditions.

5.4 Universal virtues and cultural expressions

Seligman and Peterson identify six universal virtues: wisdom, courage, humanity, justice, temperance and transcendence. While they assume these virtues to be universal, their expressions vary across cultures. For example, humility – a strength associated with temperance – is highly valued in many Asian societies, reflecting Confucian principles of modesty and deference. In Western cultures, however, self-confidence and assertiveness may be more celebrated, aligning with individualistic values. Similarly, the concept of courage may emphasize different behaviours: physical bravery in some cultures versus moral fortitude in others.

6 Some critical reflections

Positive psychology gained significant popularity both inside and outside academia, but it is not without criticism. A systematic review of critiques and criticism (Van Zyl et al. 2024) yielded six overarching themes found in critical literature. Three of these pertain to (1) the lack of proper theorizing and conceptual thinking, (2) problems in measurement and methodologies and (3) the lack of novelty and self-isolation from mainstream psychology. We will briefly discuss and integrate these with some other issues.

6.1 Theoretical and conceptual concerns

Conceptually, positive psychology is still a challenging field. According to many authors, flourishing represents just one facet of the broader construct of well-being, together with positive emotions and life satisfaction. However, definitions of these concepts, as well as the ways they are assumed to be related differ substantially between researchers. For example, the way in which virtues are described, operationalized and used is not in line with the Aristotelian premises from which they are derived (see Van Zyl et al. 2024). Furthermore, an overarching metatheory is lacking, making it difficult to converge and synchronize understanding.

6.2 Methodological concerns

Many studies in positive psychology depend heavily on self-reported data, which may be subject to all sorts of biases (e.g. cultural, recall, or factor bias). Hardly any objective measures have been developed in addition to the many subjective questionnaires. Both measurements and research designs have been seen as the product of 'quick and dirty' procedures, leading to failure to replicate findings, thereby raising questions about the robustness of the research. Furthermore, the general empiricist and quantitative approach in positive psychology research runs the risk of reducing broad and subjective experiences (e.g. virtues) to small measurable atoms, which seems paradoxical, considering the holistic claims of the field.

This can for instance be observed in the multitude of definitions and operationalizations of the concepts of 'flourishing' and 'well-being'. Some of these definitions are not very sophisticated and refer to for instance rather general factors such as high levels of emotional, social and psychological well-being. In a recent extensive study covering 145 countries, with almost 400,000 participants over a 3-year period, flourishing emerged as a complex, multidimensional phenomenon that varies across individuals and nations. The researchers evaluated thirty-eight well-being metrics, including life evaluation, daily emotions, balance and harmony, and other well-being factors (Lomas et al. 2025). Interestingly, many nations ranked highly on certain metrics but fared poorly on others.

6.3 Lack of novelty

Critics point to the fact that the main premises of positive psychology are not entire new, and can be easily traced back in the history of psychology as well as in contemporary and adjacent disciplines. For one, positive psychology is neither a subdiscipline nor a speciality within psychology. Rather, elements, concepts and procedures from positive psychology can be found in many areas of psychology, both in theory and in practical applications. See for example the various forms of psychotherapy mentioned earlier in this chapter. Therefore, contrasting 'positive psychology' with 'negative psychology' might be too polarizing and at variance with theory and clinical practice. In a different area, authors emphasize the positive and even flourishing aspects in religious contexts, such as in Christianity (Charry and Kosits 2017) and Islam (Saritoprak and Abu-Raiya 2023).

6.4 Overemphasis on positivity

Some argue that the emphasis on positivity risks minimizing the complexities of human experience, including the importance of addressing negative emotions and systemic challenges. Others caution against 'toxic positivity', where individuals feel pressured to suppress negative emotions in pursuit of constant happiness. Critics argue that positive psychology sometimes downplays the importance of negative emotions, which are essential for learning, growth and resilience. Encouraging relentless positivity may lead to the suppression of natural, valid feelings like sadness or anger, which can be counterproductive and harmful. In particular, the commercialization of positive psychological methods and interventions run the risk of presenting a caricature of the original and more balanced idea.

6.5 Cultural bias

Positive psychology has been critiqued for being rooted in Western concepts, such as individualism and self-actualization, which may not align with the values or experiences of collectivist or non-Western cultures. In addition, most research has been carried out in Western, educated, industrialized, rich and democratic samples. Furthermore, quite some findings are generalized as universal truths without sufficient cross-cultural validation (Kiknadze and Fowers 2023). Recent efforts aim to address these critiques by incorporating indigenous knowledge systems, examining the impact of inequality and social justice, and exploring the interplay between culture and psychology. These efforts also underscore the importance of respecting local values and avoiding the imposition of Western ideals.

6.6 Neglect of systemic factors

Positive psychology often emphasizes individual responsibility for well-being while paying far less attention to structural or systemic factors like poverty, inequality, discrimination or environmental challenges that significantly impact mental health. Recently, Lomas, Waters, Williams, Oades and Kern (2021) indicated that positive psychology had entered a third wave of research aimed at broadening beyond the individual to include systems, contexts and cultures, using a wider range of methodologies and a more interdisciplinary approach.

7 Concluding remarks

Positive psychology is considered by many an important transformative approach potentially promoting a good life, and a life worth living. In that sense, it is thought to be able to significantly contribute to human flourishing. The relatively recent focus on positive psychology has certainly influenced the way psychologists, educators and policymakers approach mental health and well-being. By shifting the focus from pathology to potential, Seligman and others have inspired a wealth of research and practical applications that empower individuals to lead more meaningful, engaged and fulfilling lives.

According to its advocates, the empirical foundation of positive psychology is robust, rooted in validated constructs, reliable measurement tools and evidence-based interventions. Despite this enthusiasm, as we have seen, many critics point to conceptual and methodological issues and imperfections in positive psychology research. Here lies a task for the field to be self-critical and to further develop the theoretical, conceptual and methodological underpinnings. While challenges and critiques persist, positive psychology's emphasis on understanding and nurturing what makes life worth living ensures its relevance in addressing the complexities of human experience.

Studies on the components of positive psychology have provided empirical evidence supporting the idea that cultivating positive emotions, engagement, relationships, meaning and achievement leads to greater well-being and life satisfaction. These studies have not only highlighted the importance of these elements in promoting flourishing but have also shown that individuals can actively cultivate them through intentional practices. Positive psychology interventions targeting gratitude, mindfulness, resilience and strength-building are effective tools for improving psychological well-being and preventing mental illness. Together, the findings from these studies demonstrate the power of fostering positive psychological states to enhance both individual and collective flourishing.

From this, the impression arises that the general atmosphere in positive psychology feels like future-oriented, optimistic and doable. For some, however, this might sound too positive, because of the many obstacles, adversities and challenges that are part and parcel of many people's lives. If taking the wrong turn, one might put the blame on people for not fully realizing their potential, or their failure to flourish. This would be a serious accusation to say the least, because it presupposes that individuals can be fully responsible for each aspect of their lives. Macro-economic circumstances, climate change, societal and political unrest have been shown to have detrimental effects on individual psychological well-being. In recent times, researchers in this field seem to be more aware of the complexities of the real world, with its serious challenges for many people. The focus seems to be broadening from individual approaches to flourishing to flourishing on the level of communities and the collective.

Positive psychology's exploration of well-being must account for cultural diversity to remain relevant and effective. Culture shapes how individuals perceive happiness, meaning and strengths, influencing their pathways to flourishing. By embracing cultural nuances, positive psychology can evolve into an inclusive discipline, capable of addressing the diverse ways humans thrive across the globe. Recent large-scale studies show that the profile of cultures and nations differ widely when it comes to well-being, happiness and flourishing. These phenomena come in very different forms and shapes, which makes research interesting and challenging at the same time. Data show that in some economically poor regions, people can feel happy and content for reasons of connectedness and interdependence, that are totally different from a Western experience.

Reaching the end of this brief overview of positive psychology and its contribution to human flourishing, the conclusion may be that there is a great potential for individual and collective factors to a life worth living. Positive psychology inspires us to think of flourishing not as a final destination but as a teleological process. Rather than focusing on a pathological starting point towards normalcy, it posits an individual and collective human potential towards which one can aspire and in which character strengths and virtues can be instrumental.

5

Christian and Neo-Platonic Roots of Flourishing – Higher Virtues and the Pursuit of Happiness

Andrej Zwitter

1 Introduction

The roots of the concept of *eudaimonia* (Greek: flourishing or well-being) in the context of Virtue Ethics go back to the Platonic and Neo-Platonic tradition. Deriving from Neo-Platonists like Plotinus and Porphyry, in Augustine and Aquinas the distinction between cardinal and spiritual virtues emerges. This is also when a fundamental shift of understanding of the higher virtues takes place: the practice of self-perfection becomes part of a larger ethics and politics recognizing the necessity to providing societal preconditions for *eudaimonia*. Aquinas identifies the cardinal virtues as prudence, justice, fortitude and temperance, drawing from the virtues emphasized by Plato and Aristotle. The spiritual or theological virtues are faith, hope and love. Beyond the Christian tradition the concept of flourishing in the Neo-Platonic discourse, specifically in Plotinus, strongly focuses on the transcendent, making spiritual virtues a prerequisite to attain a flourishing that transcends the immanent. Unlike cardinal virtues, which can be developed through human effort, spiritual virtues by their very nature transcend the material and temporal boundaries of the physical world. *Eudaimonia* or flourishing is then not an end in itself but a means to partake in the Augustinian pilgrimage from the earthly city to the *Civitas Dei* (the City of God).

Thomas Jefferson included the phrase 'the pursuit of happiness' in the US Declaration of Independence. This phrase is often interpreted as the pursuit of wealth in the sense of John Locke's 'life, liberty and estate (property)'. However, some scholars argue that the 'pursuit of happiness' and specifically the term 'happiness' is wrongly interpreted and should rather be seen in the attainment of human flourishing (Conklin 2014; Schlesinger 1964). The contemporary understanding of happiness as a material condition of well-being or flourishing comes as no surprise. It is the result of a several-century-old shift in philosophy and political science that had emancipated itself from religious doctrines towards the strict application of logic, reason and philosophical materialism (Zwitter 2023). *Eudaimonia*, if stripped from all non-material conceptions, does indeed become equivalent to physical and psychological well-being. This Enlightenment turn towards rationalism and physicalism can be traced up to today's conception of human development and human rights, and it is the philosophical worldview behind many socio-political and economic doctrines and practices. In this worldview, any non-immanent goals are strictly an individual and thus a private affair. The practical implications can be directly observed in how human development is conceptualized, measured and pursued through international and national policies as is evident in many contemporary conceptions of human development, such as the SDGs and the Human Development Index.

However, if we trace *eudaimonia* back towards its original conceptions as part of a larger ethical philosophy of Virtue Ethics, the mere immanent aspect of *eudaimonia* becomes questionable. As this chapter will show, *eudaimonia*, in its original conception, is not restricted to material conditions nor is it an immanent goal in itself.[1] Rather, it must be understood as intermediary conditions enabling a process of human growth that goes beyond the immanent and aims towards the transcendent. Such a metaphysical rethinking of flourishing, which at the outset might seem indeed metaphysical and thus impractical, in fact, has immense practical implications. In part, it requires us to rethink the 'goal' of human flourishing (Section 2). Furthermore, this change in perspective prompts justified questions of whether the tools we use to attain flourishing are indeed suitable. Spiritual virtues as theorized in this chapter regain importance in the context of the immanent when the goals of human flourishing transcend the immanent and are pointed towards the fundamental Kantian question of 'What may I hope for'.[2] The spiritual virtues of *faith*, *hope* and *love* then become quite relevant and practical in an applied ethics that aims not only towards harm avoidance but also self-perfection (Section 3). To attain the non-material conditions of human flourishing as proposed by Aristotle in the human *ergon* as the contemplation of the divine, such spiritual (also sometimes referred to as transformative or infused) virtues become indispensable (Section 4).

2 Virtue Ethics

2.1 Agency and character

The physiological, psychological and spiritual conditions that translate into morals and ethics can be linked to the necessities that exist within communities to regulate their survival as groups and of each individual through limiting and requiring certain practices and rules of conduct (Zwitter 2013). At the same time, virtues seem not to derive from the same needs, although they may be linked. This becomes apparent when we look at the role of criminal law. Criminal law is law that limits behaviour so that everyone can live without fear of being harmed – so, criminal law norms are ethical norms that prohibit certain actions, which if permissible would put the individual and the whole social group at risk (jeopardizing the ability to survive). These norms are clearly linked to a general principle of do not harm. Virtues are more than the observance of negative duties. They are an expression of excellence of a human character.

Virtue Ethics in the contemporary discourse was rediscovered and popularized by scholars such as Slote, Hursthouse, Foot, and prominently MacIntyre, who takes a Thomistic and less modern approach in discourse on Virtue Ethics. Fundamentally, Virtue Ethics is agent centred rather than act centred. That means, it focuses on the features of what a person should be rather than of how they should act. However, Virtue Ethics is by no means limited to general recommendations about character traits – it can be concrete enough to give action guidance. A definition would state: 'P.I. An action is right iff it is what a virtuous agent would characteristically (i.e. acting in character) do in the circumstances. (...) P.Ia. A virtuous agent is one who has, and exercises, certain character traits, namely, the virtues. (...) P.II. A virtue is a character trait that ...' (Hursthouse 2002: 28–29). Virtues, then, are distinguished not merely as traits but as markers of excellence within character. They transcend simple acts of honesty or beneficence, embodying the essence of being an honest or benevolent person, respectively. This distinction is elucidated through an illustrative example: a person who refrains from lying does not necessarily possess the inherent traits of honesty. Such an individual might simply adhere to a utilitarian framework, wherein lying is deemed acceptable if the outcomes of deceit surpass those of truthfulness – essentially, if the lie serves a greater good for a larger number of people. Conversely, a deontologist, who upholds the principle of 'do not lie' as a universal norm, perceives intrinsic wrongness in lying, arguing that it is never permissible except when overridden by higher-order norms and duties.

Virtue Ethics occupies a nuanced position between these perspectives. It acknowledges that virtues, such as honesty and charity, may at times conflict. For instance, blunt honesty that results in harm may be less virtuous than a tactful omission of the truth. Thus, Virtue Ethics adopts a stance that, while seemingly flexible, is far from a mere eclectic selection of ethical principles. This is because (1) virtues represent intrinsic features of an individual's character, (2) they are oriented towards the ultimate goal of *eudaimonia*, and (3) they necessitate a form of practical wisdom (Greek: *phronesis*) capable of navigating between conflicting virtues. Consequently, embodying virtues such as charity and honesty may require navigating compromises, demonstrating that excellence in character often involves the judicious balancing of virtues. For this compromise to be reached, the agent requires *phronesis*. *Phronesis* enables the person to manoeuvre between the virtues in order to achieve *eudaimonia*:

> 'Each of these things can be done in many ways, and are not done in the same way in all circumstances. That is why one needs practical wisdom in one's judgement to determine the correct way. That is how correctness and full goodness in all the other virtues depend upon practical wisdom.'[3]

The emphasis on the centrality of the individual in Virtue Ethics highlights a crucial aspect: the moral quality of an individual is not defined by isolated actions but by the character that underlies these actions. Morality is assessed not solely based on individual deeds but also on the character of the person performing them. Consequently, a single act of wrongdoing (maleficence) does not render a person morally reprehensible (malevolent), unless that act emanates from the individual's will and reflects a trait of their character.

The principle that an excess of any virtue can lead to detrimental outcomes is vividly illustrated through the example of honesty devoid of temperance and mercy, which can compel an individual to speak the truth even when it is harmful. Similarly, justice, when pursued without mercy, ceases to be a force for flourishing and well-being; it becomes inflexible and indifferent to compassion. The biblical account of King Solomon's judgement regarding the two mothers further elucidates this concept: Two mothers demand from King Solomon to decide whom a newborn child belongs to. Both had given birth, but one child had died and in the night one mother had exchanged her dead child for the living one. On the next morning the mother of the dead child denies having exchanged the babies. They come before the king for settling the issue:

The king said, 'Get me a sword'. So they brought a sword before the king. The king said, 'Divide the living child in two, and give half to the one and half to the other'. Then the woman whose child was the living one spoke to the king, for she was deeply stirred over her son and said, 'Oh, my lord, give her the living child, and by no means kill him'. But the other said, 'He shall be neither mine nor yours; divide him!'[4]

Since Solomon has no other means to decide the issue, he decides to use a deliberately misplaced application of equitable justice as a means to discern who the true mother is by triggering the virtues of love and compassion for the child. The real mother immediately forfeits her claim and accepts her own loss just to save the child's life. Solomon's judgement, which initially appears to lack mercy, is actually a profound application of *phronesis*, aiming to uncover the truth in a situation where conventional justice seemed insufficient.

2.2 Eudaimonia and the *Ergon* of human life

In exploring the concept of *eudaimonia*, a term often translated as 'happiness' or 'flourishing', Aristotle presents a nuanced perspective in his ethical treatises, the 'Nicomachean Ethics' (Aristotle 2000) and the 'Eudemian Ethics' (Inwood and Woolf 2012). He defines the 'human good' as 'the activity of the soul in accordance with virtue'.[5] If there are multiple virtues, then this activity aligns with the most superior and complete among them. In the Eudemian Ethics, he further refines this idea, suggesting that the best way of life facilitates the contemplation of God. Anything that interferes with this goal, whether by deficiency or excess, is deemed detrimental.[6] This requires a contemplative life, which to Aristotle represents the pinnacle of human existence; and it requires minimizing awareness of the irrational parts of the soul. Nagel highlights the complexity inherent in defining *eudaimonia*. Aristotle's challenge, according to Nagel, lies in identifying the specific function or *ergon* of humans. He posits that to understand *eudaimonia*, one must understand the essential work or role (*ergon*) of humans, which in turn defines human excellence (Nagel 1972: 253).

Contrasting perspectives within Aristotle's own works are noted by Kirchner, who points out that while the Eudemian Ethics associates *eudaimonia* with the activity of a complete life aligned with complete virtue, encompassing all aspects of the rational soul, the Nicomachean Ethics seems to narrow this down to a singular, dominant end, that is, the activity of wisdom (Greek: *sophia*) (Kirchner 1986: 42). *Eudaimonia* is not simply about living well, which would be a common aspiration of all living things, but it is deeply tied to fulfilling the unique capabilities that define us as humans. The interplay between the rational and the irrational, the individual and the divine, and the singular versus the pluralistic conceptions of virtue, all contribute to the multifaceted account of Aristotle's ethical philosophy.

Plotinus, in his work 'Enneads' (Plotinus, n.d.), further deepens the discussion of the nature of *eudaimonia*, which he interprets ultimately as attaining likeness to God. This attainment is described as becoming just and holy, living by wisdom, with one's entire nature grounded in virtue.[7] For Plotinus, likeness to the divine is achievable only through virtues that are beyond the civic ones, indicating that while civic virtues have their place in human life, they are insufficient for the higher goal of divine assimilation. The civic virtues are seen as principles of order and beauty within the human life as long as we exist in this world. They provide bounds to our desires and sensibility, thereby elevating us by their very nature of limiting the boundless and preventing false judgement.[8] However, Plotinus distinguishes between these virtues and the higher virtues. Higher virtues are not concerned with the regulation of passions but with wisdom and dialectic, relating to the higher, intelligible part of the soul. McGroarty, commenting on Aristotle's and Plotinus's views, notes that Aristotle sought to redefine *eudaimonia* as an active function, a noble living in the active sense, challenging the idea that living well is sufficient for *eudaimonia*. Plotinus extends this argument by questioning why this living well would be exclusive to humans if other living beings also accomplish it effectively (McGroarty 1994: 104). Instead, Plotinus further emphasizes the contrast between the lower virtues related to passions and the sense-perceptive part of the soul, and the higher virtues connected to wisdom and dialectic, which refer to the soul's higher intelligible part, its inner self. This distinction is crucial as Plotinus aims to stress the superiority of the soul's higher part over the sufferings and passions associated with the lower sense-perceptive part, linked to the outer, physical self ('Internet Encyclopedia of Philosophy', n.d.). McGroarty concludes that *eudaimonia* is not merely happiness, but the outcome of a life characterized by both practical and theoretical wisdom. It involves an ascent to intellect, leading to a union with the totality and fullness of life, beyond the physical realm into that of 'real being'. This ascent transforms the wise person, making him a part of the greater intellect, a 'one-many' that transcends the simplicity of happiness as a concept (McGroarty 1994: 107).

Augustine, in his work 'Retractiones' (Saint Augustine 1968) addresses the pursuit of happiness in relation to human nature and reason. He asserts that while the mind and reason are the apex of human nature, they alone are not the path to happiness. Rather, to live happily, one should not live solely according to human nature but should instead live in a manner akin to the divine. This divine way of life involves not being self-contented with one's own mind and reason but subjecting oneself to God (Saint Augustine 1968, vol. 60, sec. 1.1.2). This perspective is reflected in the broader framework of Neoplatonism, as noted by Dodaro (Dodaro 2004). Augustine's thoughts parallel those of Neoplatonist philosophers, who emphasized a hierarchy of virtues. Political virtues, which are concerned with the governance of the city or state, are considered to be at a lower level compared to the purificatory or higher virtues, which aim at the purification of the soul and the philosopher's divinization. This divinization is a process that aligns the individual's soul with the divine, transcending the mundane virtues associated with political life. Augustine's viewpoint, Dodaro suggests, aligns with the Neoplatonist tradition, which includes philosophers such as Plotinus, Porphyry, Proclus, Macrobius and Marinus. These philosophers collectively maintain that the political virtues must be viewed in relation to the higher virtues, which are essential for the soul's elevation to divine likeness (Dodaro 2004: 433).

Thomas Aquinas also offers a nuanced view of the pursuit of happiness in his theological and philosophical works, particularly in the 'Summa Theologica' (Saint Thomas of Aquinas, 1947). He defines happiness as the attainment of the 'perfect good', which is possible because humans have the intellect to apprehend this universal perfect good and the will to desire it (Saint Thomas of Aquinas, n.d., sec. I-II.5:1c). Aquinas thereby synthesizes Aristotelian philosophy with Christian theology, maintaining that human actions are directed towards the *telos* (Greek: purpose or end) of *eudaimonia*, or happiness, understood as completion, perfection or well-being. This *telos* is achieved through the exercise of intellectual and moral virtues, which guide us to understand and consistently seek happiness. However, Aquinas also posits that complete happiness cannot be achieved in this life. True happiness, for Aquinas, is synonymous with *beatitudo* (Latin: happiness, blessedness) or supernatural union with God, an end that exceeds natural human capacities. He argues that humans require divine grace to transform and 'deify' human nature, thus making a person capable of participating in divine beatitude, which is beyond our natural abilities to attain. Contemporary scholar of Thomist philosophy, Wang, explains that humans are characterized by their capacity for infinite good. The 'perfect good' is not only a theoretical construct for Aquinas but also a real possibility. This is supported by the notion that the mere existence of a desire or capacity indicates that its fulfilment is possible, even if it is not actualized in the current circumstances (Wang 2007: 329). Nevertheless, as Wang clarifies, while perfect happiness in

God is unattainable in this life and by natural means, the philosophical dilemma of human happiness persists even with the presence of God. Aquinas asserts that seeing the divine essence is beyond not only human nature but also that of any creature, and such perfect happiness (*beatitudo perfecta*) can only be bestowed by God alone (Wang 2007: 331). In this life, Aquinas holds that we can attain an imperfect form of happiness (*beatitudo imperfecta*) through the intellectual and moral use of our intellect. This imperfect happiness involves the speculative and practical applications of intellect, which in turn are dependent on sense experience and, therefore, on the body. The body plays an integral role in this imperfect happiness, as the intellect's operations in life are inexorably linked to physical sensations and perceptions (Trabbic 2011: 557–8). Aquinas' view of happiness, therefore, encompasses both an imperfect, attainable form of happiness (*beatitudo imperfecta*) based on intellectual and moral virtues in this life, and a perfect, unattainable happiness (*beatitudo perfecta*) that lies in the beatific vision of God, achievable only through divine intervention. His theory of the pursuit of happiness, however, points to the necessity of both lower and higher virtues to be present.

The history of thought regarding *eudaimonia* from Aristotle to Aquinas illustrates both a continuity and a development in the concept's understanding. Aristotle lays the foundational groundwork by identifying *eudaimonia* with the activity of the soul in accordance with virtue, emphasizing a life of rational activity and contemplation. This definition persists through the ages but undergoes significant refinements and reinterpretations in the light of later philosophical and theological insights. Continuity in the concept of *eudaimonia* is found in the consistent association of the pursuit of happiness with virtue and the activity of the soul. Aristotle's original conception of *eudaimonia* as an active state of being, a fulfilment of the soul's potential in accordance with virtue, is a thread that runs through subsequent interpretations. The idea that happiness is not merely a passive experience, but an active realization of a life lived in accordance with virtue is echoed by later thinkers. However, the development of the concept is evident in how later philosophers like Plotinus and Augustine, influenced by Neoplatonism, and Aquinas elevate the goal of *eudaimonia*. For Plotinus, *eudaimonia* is not just the realization of human potential but the assimilation to the divine, achieved through higher virtues that transcend the civic virtues concerned with the governance of the city-state. Augustine builds on this by asserting that true happiness comes from living a life akin to the divine, not just by rational thought but by subjugating the self to God. Aquinas further develops the concept by integrating Aristotelian and Augustinian thought, proposing that while happiness is the result of the 'perfect good' and is accessible through intellectual and moral virtues, complete happiness or *beatitudo perfecta*, i.e. the supernatural union with God, is beyond human capacity in this life. Throughout this evolution, the core Aristotelian principle that *eudaimonia* involves fulfilling the *ergon*, or essential function of humans, remains central. However, the understanding of what constitutes this function expands from a focus on rational activity to include the purificatory virtues and ultimately, the necessity of divine assistance. While the association of *eudaimonia* with virtue and the activity of the soul remains constant from Aristotle to Aquinas, the scope of what constitutes this activity broadens significantly. The ultimate goal shifts from a human-centric view of rational contemplation to a theologically informed vision of union with the divine. This development reflects a deepening and widening of the understanding of human purpose and the nature of happiness, infusing the classical conception with theological significance.

3 Virtues

3.1 Lower and higher virtues

Starting with Plato, the canon of the four cardinal virtues has relatively firmly been established in the discourse on Virtue Ethics. The development of a hierarchy of virtues was however progressive, as was the idea of the nature of such a hierarchical structure. Plato gives an account of the four cardinal virtues in the Republic where he identifies: wisdom (*sophia*), courage (*andreia*), temperance (*sophrosyne*) and justice (*dikaiosyne*). Aristotle speaks of lower and higher virtues with *phronesis* (practical wisdom) taking a central role amongst the virtues – the highest of all virtues that moderates the use of all other virtues and between the general (theory) and the practical (judgement) (Massingham 2019). As Aristotle considers as true human fulfilment the contemplation of the divine, it comes as little surprise that he complements the set of four moral virtues with a set of four intellectual virtues besides *phronesis* in book IV of the Nicomachean Ethics: wisdom (*sophia*), knowledge (*episteme*), rational intuition (*nous*), practical skill (*techne*). The emergence of hierarchy of virtues, however, must be understood in the progressive development of the concept of *eudaimonia* from Aristotle to Augustine. This development reflects both continuity in the pursuit of happiness and the good life as well as a shift in the ultimate understanding of what constitutes this, namely a contemplation of the divine in this life (Aristotle) or an experience of the divine in this life (Plotinus).

For Plotinus, the political virtues serve as a preliminary stage in the soul's journey towards assimilation to God, marking the beginning of moral and spiritual perfection. Yet, he argues that true perfection is only achievable when the soul transcends the body and its passions. He introduces in this regard the higher virtues, or purifications. These are necessary to cleanse the soul from evils associated with the body (Dodaro 2004: 446). This marks a shift from Aristotle's conception of *eudaimonia* as an activity of the rational soul to a more transcendent ideal, where the soul must undergo a process of purification to align with the divine. Porphyry, a pupil of Plotinus, extends this notion of purification by identifying four guiding principles or virtues: faith, truth, love and hope. He argues that it is these virtues that enable the soul to draw near to God. They are seen as a pathway to theurgic union, derived from the spiritual practices outlined in the Chaldean Oracles (Dodaro 2004: 451).[9] These spiritual virtues are also referenced by Paul the Apostle in 1 Corinthians 13: 'And now faith, hope, and love abide, these three; and the greatest of these is love (*agape*)'. This represents a further development in the understanding of *eudaimonia*, incorporating elements of faith and divine grace that go beyond Aristotle's virtues.

Augustine further diverges from Aristotle's view by emphasizing the role of divine grace and the transformation of traditional virtues through faith, hope and love. He contends that while God's attributes are supreme, human virtues are imperfect, and public officials, in their governance, should not consider their virtues as equal to God's. Instead, they should recognize their limitations and seek divine guidance and forgiveness. Furthermore, Augustine stresses that the virtue of hope is crucial in sustaining oneself through the trials of temporal life by focusing one's desires on the eternal benefits of the heavenly city, rather than on earthly happiness alone. He warns that statesmen who focus solely on temporal peace and prosperity, without aiming for the eternal peace offered by the love of God, corrupt their political virtues (Dodaro 2004: 435–40). In this transformation from Aristotle to Augustine, the core elements of *eudaimonia* as a life of virtue and rational activity remain. However, the ultimate aim of these virtues is reinterpreted. While Aristotle's virtues are directed towards living well as a human being (*ergon*) in the here and now, for Plotinus, Porphyry and Augustine, these virtues are stepping stones towards a higher, transcendental goal. Aristotle's conception of *eudaimonia* as fulfilment of the soul's potential through virtues is not discarded but rather is transmuted into a more profound journey of self-perfection that culminates in a union with the divine. The political virtues of Aristotle's ethics are seen as necessary but insufficient for achieving the ultimate good; they must be transcended by spiritual virtues that orient the soul towards eternal truths and the divine essence.

It is with Aquinas that we arrive at the traditional heptalogue of virtues and a discernment of these between natural, cardinal virtues (practical wisdom, temperance, courage and justice) and supernatural or spiritual virtues (faith, hope and love) as a fixed canon that persists until today (see Figure 5.1) (Porter 1990: 67). The cardinal virtues serve the natural order and contribute to natural happiness, which is a proximate goal of human life. In contrast, the supernatural virtues pertain to spiritual matters and orient the soul towards eternal life and supernatural happiness. These supernatural virtues, according to Aquinas, are not developed through human effort but are infused by divine grace (Aquinas 2005: 66). They endow the soul with a new spiritual character, allowing the intellect to comprehend supernatural truths and the will to aspire to the supernatural good.

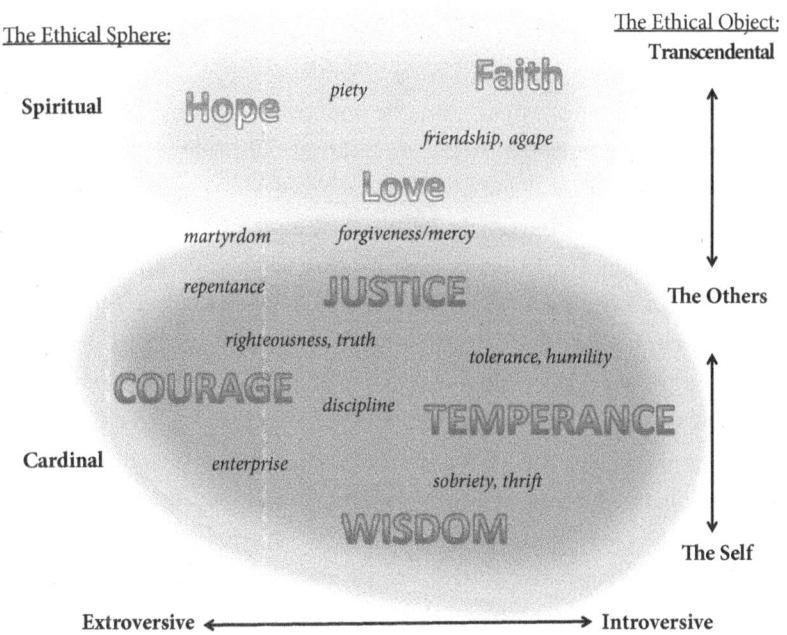

Figure 5.1 Cardinal and spiritual virtues.[10]

3.2 Extemporal nature of higher virtues

Eventually we can identify a development of Aristotle's higher virtues to transformative virtues (Neo-Platonists), to the spiritual virtues in Augustine and Aquinas, with the capability to infuse also cardinal virtues with a supernatural quality. Augustine indicates in his letter to Macedonius that the virtues of public officials must be transformed by spiritual virtues such as mercy, a principle derived from charity or love (*agape*), to transcend and solve immanent disputes which justice alone cannot solve. Similarly, Aquinas argues that infused virtues like charity can transform the actions of cardinal virtues, directing them towards the ultimate end of divine love. In this framework, virtues like temperance and courage remain essentially the same in their natural functions, but when governed by charity, their form changes; they are infused with an extemporal orientation that aligns with aims beyond the immanent life. This transformative effect of higher virtues on cardinal virtues is a key element in understanding the moral landscape that lasted until Adam Smith, where natural virtues were seen as incomplete without their elevation by metaphysical principles aimed towards more transcendent goals. The processes of Enlightenment and Humanism that secularized philosophy and the sciences, necessitated new ethical frameworks that were capable of action guidance on a purely rational basis, irrespective of theological ideas or references to the transcendent (Zwitter 2023b).

In the exploration of ethical life, the concept of spiritual virtues emerges as a transformative force. As Augustine illustrates in his letter to Macedonius, these virtues, rooted in divine love, are not constrained by the temporal and spatial limitations that bind cardinal virtues like justice. This transformative – or purificatory – nature of spiritual virtues, also elucidated by Plotinus and Aquinas, elevates human judgement beyond the temporal (i.e. reason). Spiritual virtues, so the argument, allow for the resolution of conflicts that exceed the grasp of human reason alone. The cardinal virtue of justice operates within the framework of 'each their due', a concept encapsulated in the *lex talionis*, 'an eye for an eye', found in Exodus 21:24. This principle of retributive justice, though foundational for the maintenance of social order, often finds itself inadequate in the face of conflicts that demand a higher form of resolution. Here, the spiritual virtues of mercy and charity, deriving from the virtue of love (*agape*), come into play. This insight also illustrates why many forms of justice beyond retributive justice emerged particularly after the enlightenment period, e.g. distributive justice, procedural justice, social justice, environmental and economic justice etc., to compensate for a lack of higher-level virtues in the enlightenment discourse.

Aquinas further clarifies this distinction by introducing the concept of infused virtues – those bestowed by divine grace, enabling humans to act in ways that align with the supernatural good. While the cardinal virtues aim to perfect the human within the natural order, infused virtues like charity reorient the soul's trajectory towards the supernatural order (Aquinas 2005: 66). The actions of temperance or courage, when governed by charity, are elevated from mere self-discipline or bravery to acts of divine love. The transformative effect of these spiritual virtues on cardinal virtues can thus be understood as a process of divinization. Mercy and charity do not simply complement justice; they elevate it, infusing it with a spiritual dimension that points towards an aspirational governance in the sense of a collective pilgrimage of the early city (read: imperfect society – *societas imperfecta*) towards the city of God as the template for a perfected society (*societas perfecta*) in an Augustinian reading. This reorientation is not a denial of the cardinal virtues but a fulfilment of their highest potential.

From the perspective of political theory, the examination of interpersonal betrayal through the lens of retributive justice reveals a dichotomy between the pursuit of retribution and the aspiration for reconciliation. When analysed solely within the framework of justice, the aggrieved party may seek retribution as a means to redress the harm suffered. This approach, while seemingly rectifying the imbalance, fails to mend the relational fissure birthed by betrayal, serving only to recalibrate the scales of justice temporally. In contrast, the virtue of charity, rooted in the Christian edict to 'love your neighbour as yourself' (Matthew 22:39), proposes an alternative paradigm. Charity endeavours to heal the relational breach, not by demanding recompense but through the unconditional extension of love and forgiveness. This perspective eschews the necessity for balanced exchange, favouring a gratuitous outpouring of goodwill devoid of anticipatory reciprocation.

Within the realm of governance, the application of mercy, as articulated by Augustine, suggests that a ruler should infuse justice with compassion. This integration transforms governance from a mere execution of laws into a manifestation of divine emulation, mirroring the ethos espoused by Christ in the Sermon on the Mount, particularly in the injunction, 'If anyone slaps you on the right cheek, turn to them the other cheek also' (Matthew 5:39). This precept of non-retaliation transcends the limitations of temporal justice. It encapsulates the eternal essence of mercy, which is predicated on a divine rationale of love and forgiveness. Such an approach not only redefines the concept of justice within the political sphere but also elevates it to reflect a higher form of justice that incorporates mercy and charity as foundational pillars (see Table 5.1).

Table 5.1 Justice, Charity and Mercy

Concept	Definition	Guiding Principle	Example
Retributive Justice	Seeking retribution for harm incurred	Equitable exchange	Retribution for interpersonal betrayal
Justice under Charity	Restoring relationship by extending love and forgiveness	Love your neighbour as yourself (Matthew 22:39)	Neutralizing conflict caused by interpersonal betrayal
Justice under Mercy	Tempering justice with compassion	Non-retaliation, Sermon on the Mount (Matthew 5:39)	Elevating governance to an act of divine imitation

Thereby, the spiritual virtues can be understood as carrying an extemporal effect that extends the reach of human action beyond the immediate confines of temporal justice, offering solutions that are not only restorative of human relations, extemporal in scope, but also directed towards perfection of the self and towards the divine. As such, they provide the moral and spiritual depth necessary for the comprehensive resolution of human conflicts. The overarching argument, as also indicated by Augustine's letter to Macedonius, is that the transformative nature of spiritual virtues lifts otherwise immanent virtues out of the spatio-temporal confines of reason to a higher-order logic that helps dissolve conflicts, which cannot be solved with justice or other cardinal virtues alone.

4 Conclusion: towards eudaimonic societies

Philosophical discourse on *eudaimonia* as the pursuit of happiness underwent some profound changes from Plato to Aquinas. Drawing upon the historical and theological nuances that have shaped the understanding of human flourishing beyond mere material conditions, we come to understand *eudaimonia*, as a concept that transcends the physical and temporal realms. It emphasizes a life that flourishes on the grounds of the right application of virtues and that is attuned not just to human rationality but also inspired by higher virtues that escape the spatio-temporal entrapment of immanent reason. It is in this blending of the cardinal virtues – prudence, justice, fortitude and temperance – with the spiritual virtues of faith, hope and love, that pre-enlightenment philosophers find a roadmap for a society striving for a kind of happiness that is as much about individual well-being as it is about societal harmony and self-perfection.

The virtuous individual, from this perspective, seeks eudaimonia – a state of flourishing achieved through a harmonious balance of virtues, with practical wisdom, or *phronesis*, crucially guiding their actions towards the attainment of a good and fulfilling life. This good life is deeply rooted in the human experience and confined by the human condition, and it is intertwined with others within the community. It reaches its zenith in a state of *Glückseligkeit* imperfect in this life (*beatitudo imperfecta*) but constantly progressing with and working on the harmony within oneself and the deeper structures of culture and society.

Practical wisdom, or *phronesis*, then is not merely an academic exercise; it is cultivated through individual and collective memory and the experiences of humanity. It is enshrined in heroic sacrifices in literature, codified in laws and embodied in institutions that strive for a peaceful and harmonious coexistence. It is in these repositories of collective wisdom that the potential for societal *eudaimonia* resides, a potential that is continually unfolding as each generation navigates the interplay of virtues and vices within the ever-evolving nature of human society and politics. To foster the development of *phronesis* and *eudaimonia*, society must engage in a reflective and interpretive process that values the insights of humanities and philosophy, understanding that the true measure of progress is not captured solely by material achievements but by the intangible quality of human experience and the human need for self-perfection.

It is evident that our exploration of the pursuit of happiness calls for a reimagining of what it means to live well. This approach challenges the global community to expand its vision of human development to encompass not just material conditions but also the deeper, metaphysical processes that contribute to human flourishing. Integrating these insights into global development agendas presents fundamental challenges of interfaith dialogue, conflict transformation and international solidarity that are hitherto lacking. But it also holds the promise of a more meaningful approach to building a world that values the journey towards individual and collective self-improvement and societal flourishing as much as the destination itself.

Notes

1. One can argue that modern conceptions of human development and human rights are not goals in themselves but conditions to attain the ultimate goal of modern Western political thinking that is *freedom*.
2. A805/B833.
3. (Aquinas 2005: 35) (Article 6).
4. 1 Kings 3:16-28.
5. Nicomachean Ethics, 1098a17-18.
6. Eudemian Ethics, 1249b17-24.
7. Enneads I.2, 1.
8. Enneads I.2, 2.
9. The Chaldean Oracles are a set of mystical-philosophical texts, traditionally attributed to Julian the Chaldean and his son Julian the Theurgist, said to have been received through inspired trance and later influential among Neoplatonists.
10. Modified from McCloskey (2008).

6

Learning to Flourish with Spinoza

Harald Atmanspacher

1 Introduction

Flourishing belongs to the central concepts of a humanistic psychology that, briefly speaking, aims at a truly fulfilling life. One of its pioneers is Abraham Maslow, who published his first ideas about it under the notions of 'human motivation' and the 'human potential' (Maslow 1943).[1] From the 1990s onwards the term 'positive psychology' took over under the leadership of Martin Seligman and colleagues. Positive psychology is about the conditions that contribute to the optimal functioning of individuals, groups and institutions to their overall benefit. According to Gable and Haidt (2005), it studies 'positive subjective experience, positive individual traits, and positive institutions ... it aims to improve quality of life'[2] (see also Dakwar 2024).

In parallel with humanistic and positive psychology, a related line of thinking originated with the work of Antonovsky (1987) and his concept of *salutogenesis* as a complex theory of health. A landmark handbook of salutogenesis has been compiled by Mittelmark et al. (2017). The key theoretical construct of salutogenesis is a 'sense of coherence' that makes individuals confident that their lives are comprehensible, meaningful and manageable. Accordingly, various notions of meaning are of crucial significance in salutogenesis and related approaches, such as existential analysis (Frankl 2004).

However, all this is not a completely new idea in the history of humankind. It has been a concern of all cultures East and West, at all times, to find out how life

---

I am indebted to Michael Hampe and the participants of our joint seminar at ETH Zurich in the spring semester 2023 for in-depth discussions of Spinoza's philosophy and its impact. In addition, helpful comments on an earlier version of this essay by Alexander Borbely, Elias Dakwar, Bob Moody and Jan Zwicky are much appreciated.

for oneself and with others can be as beneficial and as little harmful as possible, not only among humans but also in their relationship with nature as a whole. Admittedly, there have been periods in history during which this is hardly recognizable. Plus, there are human traits that seem to stand exactly opposed to peaceful coexistence and mutual support: the seven cardinal sins of greed, pride, wrath, envy, lust, gluttony, sloth, are only examples.

Many thinkers devoted much work on how to minimize affects and passions that are counterproductive to the human potential of living a good and flourishing life. In this brief chapter I highlight one important and influential example – dating back 350 years – that seems oddly underrepresented in human-potential movements: the *Ethics* by the Dutch philosopher (of Jewish origin) Baruch de Spinoza (1632-1677). In occidental philosophy, Spinoza is known as the pioneer of dual-aspect monism, a metaphysical worldview in which the mental and the physical are aspects of an underlying domain of reality that is psychophysically neutral (see Atmanspacher and Rickles 2022 for details).

Section 2 of this chapter discusses Spinoza's views on three different kinds of knowledge that he outlines in the later parts of the *Ethics*. They play different roles for living a life in recognition of and respect for others, a life that may liberate the human potential from needlessly impeding bounds. Section 3 focuses on Spinoza's third kind of knowledge, the most desirable one, through the lens of creative insight in the sciences, based on some impressive biographical and introspective material and its evaluation. Section 4 considers creative insight more broadly as an example of how knowledge of the third kind may arise: as the emergence of meaning, not in linguistic respects but in its metaphysical shape (Frege 1892; 1918). On this basis, Section 4 also speculates about the experiential origin of Spinoza's dual-aspect monism and the ethical guidelines that it suggests.

Section 5 presents an attempt to transform and apply what has been said so far to two different modes of education. The corresponding deliberations are the most publicly relevant part of this essay, and they are intended to demonstrate how Spinoza's deep philosophical insights and their ramifications may have a conducive impact on the status of contemporary societies and cultures. More precisely: impact that encourages and facilitates the unfolding of a flourishing way of life in resonance and coherence with all of nature (not only with other humans). The epilogue in Section 6 presents an excursion into the work of Jan Zwicky (2019) that offers some astonishing parallels with Spinoza's thinking.

2 Spinoza's kinds of knowledge

The major opus of Spinoza is without doubt his *Ethics*, posthumously published in 1677 (Spinoza 1994). The ultimate idea of the *Ethics*, in a nutshell, is to present a guide for humanity towards living a life that liberates all creatures from harm and thus minimizes suffering for all creatures. As the full title *Ethica Ordine Geometrico Demonstrata* indicates, it is designed *more geometrico*, i.e. in a geometric style, inspired by the literary form of Euclid's *Elements*. Much has been written about possible reasons for Spinoza to use this form of presentation (see, e.g. Wolfson 1934), and I will add another one as this chapter unfolds.

According to its mathematical format, the *Ethics* is constructed on the basis of definitions, axioms, theorems, propositions, corollaries and explanatory remarks (*scholia*). It consists of five parts. Part 1 is on Spinoza's impersonal version of God (which earned the *Ethics* an entry in the index of banned books of the Catholic church) and God's relation to nature. Since due to Spinoza's pantheism (or panentheism?) God is everywhere in nature, including in humans, it is in principle possible for humans to grasp the idea of God, yet in an imperfect and fragmentary fashion. Part 2 criticizes the Cartesian dualism of mind and matter and replaces it with a divine monism of which the two Cartesian substances of the mental and the physical are deflated into nothing more than aspects (*modi* in Spinoza's terms).

Parts 3 to 5 discuss the ways in which emotions, affects and passions lead humans to lesser or greater perfection, so to suppress or unlock the human potential towards growth and prosperity. In these parts of the *Ethics*, an essential distinction is made between inadequate knowledge and adequate knowledge. The highest form of adequate knowledge, for Spinoza, is knowledge that elucidates how things really are, irrespective of our inadequate opinions about them. And since God is in all things, this knowledge at the same time contributes to our – partial and imperfect – knowledge of God.

Inadequate knowledge is knowledge based on mutilated and deficient information, often creating confusion and noise, which Spinoza addresses as *imagination*, the first and lowest kind of knowledge. One way to express Spinoza's notion of inadequacy is to appeal to the difference between how things appear and how they in fact are. In this spirit, the inadequate idea that the sun appears smaller than the moon is correct insofar as it is based on how the sun and the moon *appear* to us. Since all knowledge about the appearance of objects and their relations is contextual, such knowledge is by definition partial (at best) and cannot be adequate.

Obvious examples of inadequate knowledge in everyday life are *opinions* such as rumours, hearsay and gossip. Preferred means of their distribution today are so-called 'social' media, apparently with no limits to indecency. But more generally all information that depends on appearances, and thus includes contextuality and perspectivity, would count as inadequate knowledge. For instance, this is reflected in the fact that hearsay is not considered as binding in legal court hearings. Ultimately Spinoza would classify every kind of empirical knowledge in this category insofar as it is based on information that, in one or another way, relates to how objects and relations appear to us and, thus, is contextual, partial and incomplete.

By contrast, adequate knowledge is assumed beyond the restriction of contextuality – Spinoza's formula *sub specie aeternitatis*, which features in almost every theorem in books four and five of the *Ethics*, expresses this independence. Adequate knowledge is knowledge from the viewpoint of eternity and does not depend on the inevitable perspectivity in the world of finite objects and relationships. Adequate knowledge is about the absolute essence of objects and comes in two kinds: reason and intuition.

Spinoza's second kind of knowledge is *reason*, yielding adequate ideas. Reason is the instrument with which affects leading to suffering can be depleted and affects entailing liberation can be cultivated. Reason also provides arguments to conclusions based on premises. An example is the well-known logical syllogism of Socrates being mortal because Socrates is human and humans are mortal.[3] Another example is the manipulation of mathematical objects in quantitative calculations or in proofs of theorems. Yet another example is causal explanations in science: everything that exists has a cause or explanation (principle of sufficient reason).

But where do premises come from? If they come from inadequate ideas, i.e. from opinions, they often lead to false conclusions, even if reason is applied properly to them: *ex falso quodlibet*. Since premises are the starting point for reason to act upon, they, or at least some of them, cannot themselves belong to the second kind of knowledge that follows from premises. So, it is a key question how appropriate premises as preconditions for chains of arguments can be found, which are not themselves already results of reason. This is where Spinoza's third kind of knowledge kicks in.

Spinoza's third kind of knowledge is *intuition*, subject to a *scientia intuitiva*, which may unveil truths that are unavailable through reason alone. Such intuition is the most powerful and most desirable kind of knowledge. For instance, the premises, from which all conclusions originate that reason may derive, belong to this kind of knowledge. It is an immediate insight into the truthful nature of all things. Since all things are expressions, or manifestations, of the infinite and eternal divine, this means – to some imperfect degree, because only the divine itself has full perfection – knowledge of the divine nature itself. That's why Spinoza also calls it *amor dei intellectualis*, the intellectual love of God.

What Spinoza does not tell us is a suitable way of getting there. How is knowledge of the third kind achievable? Is there a set of recipes, like a masterplan or user manual, that sets intuitive insight in motion if applied properly? The answer is negative, or at least near enough to negative, if recipes are understood as sufficient conditions that deterministically and predictably yield the desired outcome. Yet there may be necessary but insufficient conditions which favour and facilitate intuition without guaranteeing it. The next section addresses some ideas along those lines in more detail.

3 Creative insight in science

Spinoza's highest kind of knowledge aims at understanding in its broadest sense, and through understanding a slew of practical goods can be achieved, such as inner freedom, virtue, blessedness. Rather than addressing understanding in its full scope, this section focuses on specific examples of understanding that pertain to the areas of science and mathematics. It begins with some anecdotal phenomenological material of creative insight and classifies it as a four-stage process. Based on this, some explanatory ideas are outlined that embed the phenomenology in the metaphysical framework of dual-aspect monism, for which Spinoza is known as one of the pioneers.

3.1 Phenomenology of insight

Based on a wealth of biographical material on creative insight in science, Hadamard (1954) suggests four stages, each of which is inevitable for genuinely creative work. He calls these stages preparation, incubation, illumination and verification. The first and the last of them mainly function at the level of conscious, analytical thinking. The second and third stages, however, strongly involve unconscious processes as the core of actual insight. Let me cite two quotations, by Henri Poincaré and Albert Einstein, as examples who are clearly beyond any suspicion of mediocrity. In his essay on 'Mathematical Creation', Poincaré writes (1910):

> One evening, contrary to my custom, I drank black coffee and could not sleep. Ideas rose in crowds; I felt them collide until pairs interlocked, so to speak, making a stable combination. ... It seems, in such cases, that one is present at one's unconscious work, made partially perceptible to the over-excited consciousness, yet without having changed its nature. Then we vaguely comprehend what distinguishes the two mechanisms or, if you wish, the working methods of the two egos.

> Most striking at first is this appearance of sudden illumination, a manifest sign of long, unconscious prior work. The role of this unconscious work in mathematical invention appears to me incontestable. ... Sudden inspirations ... never happen except after some days of voluntary effort which has appeared absolutely fruitless and whence nothing good seems to have come, where the way taken seems totally astray. These efforts then have not been as sterile as one thinks. They have set going the unconscious machine and without them it would not have moved and would have produced nothing.

Einstein responded to a questionnaire by the French psychologists Claparède, Flournoy and Fehr, published in *L'Enseignement Mathématique*, with the following words (Einstein 1954):

> The words or the language, as they are written or spoken, do not seem to play any role in my mechanism of thought. The psychical entities which seem to serve as elements of thought are certain signs and more or less clear images which can be "voluntarily" reproduced and combined.

> There is, of course, a certain connection between those elements and relevant logical concepts. It is also clear that the desire to arrive finally at logically connected concepts is the emotional basis of this rather vague play with the above mentioned elements. But taken from a psychological viewpoint, this combinatory play seems to be the essential feature in productive thought – before there is any connection with logical construction in words or other kinds of signs which can be communicated to others.

> The above mentioned elements are, in my case, of visual and some of muscular type. Conventional words or other signs have to be sought for laboriously only in a secondary stage, when the mentioned associative play is sufficiently established and can be reproduced at will.

These two selected quotations could be supplemented by many others by leading lights in science and elsewhere, including the arts (see, e.g. Zwicky 2019).

Here is a compact characterization of Hadamard's stages with some additional commentary.

1. *Preparation*: As Poincaré (1910) emphasizes, no creative insight can 'happen except after some days of voluntary effort which has appeared absolutely fruitless'. Intense conscious work on a problem, sometimes even for years, precedes the final solution. Frustrating efforts without success characterize this stage.
2. *Incubation*: At some point the problem is removed from conscious focus, intentionally or by distraction, but the preceding conscious work 'has set going the unconscious machine'. Unconscious elements 'rose in crowds; I felt them collide' (Poincaré 1910), and 'this combinatory play seems to be the essential feature in creative thought' (Einstein 1954).
3. *Illumination*: When 'the mentioned associative play is sufficiently established' (Einstein 1954), 'pairs [of unconscious elements] interlocked, so to speak, making a stable combination' (Poincaré 1910). A particular configuration of unconscious elements stabilizes and, thereby, becomes conscious. This is the crucial moment in which an insight reveals itself as an experience of meaning. Often this happens holistically, not successively unfolded in time.
4. *Verification*: Finally, this insight has to be reconstructed in a logical way, i.e. by a succession of rational arguments which can be communicated. 'Conventional words or signs have to be sought for laboriously only in a secondary stage' (Einstein 1954). In this phase the understood meaning is transferred into language.

The first three stages cover what the philosopher of science Reichenbach (1938) called the context of discovery in scientific work, a first-person perspective as it were, which is usually not reported in the eventual publication of the discovery. Stage four is the context of justification, which reconstructs the result of the illumination in a rationally understandable way, by logical inferences and mathematical deductions, and ignores its origin. A quote from a dinner speech by the physicist and physician Helmholtz in 1891 expresses this nicely (Koenigsberger 1906: 180–1):

> Without knowing the way, slowly and tediously proceeding, often forced back by obstacles, discovering new paths leading forward – sometimes due to careful thinking, sometimes simply due to chance – and finally, when the goal is reached, much to my humiliation discovering the royal road which I could have

used if only I had been smart enough to see the proper starting point. Then, in my publications, I did of course not entertain the reader with my aberrations, but only described the paved way to him, along which he can now effortlessly reach the summit.

Likewise, the mathematical physicist Dirac (1977) in the same vein:

> The practicing scientist who made a discovery ... went a troublesome way, followed wrong paths, and wants to forget about all this. Perhaps he even feels a little ashamed and dissatisfied that it took so long. He asks himself, how much time did I waste on this path, as I should have seen immediately that it doesn't lead to anything. As soon as a discovery is made, it usually seems so evident that one is surprised that no one thought about it earlier. From this point of view nobody wants to recall all the work preceding the discovery.

The quotes by Helmholtz and Dirac insinuate that the long-winded road to a discovery might be due to a regrettable and humbling inability of the scientist to find the proper solution right away. But Hadamard's phenomenological taxonomy suggests otherwise: stages one and two might be necessary – but not sufficient! – conditions to eventually make the discovery. Where the quotes agree with Hadamard is that the communication of an insight (stage four) is crucially different from its original experience, from the illumination itself (stage three). This difference will concern us again in Section 4, as an attempt to understand the structure of Spinoza's *Ethics*.

3.2 Metaphysics of insight

While the phenomenological examples in the preceding section may seem somewhat profane in view of Spinoza's allusions to the divine, the present section (and the next) tries to relate them to a metaphysical framework that offers space for notions such as revelation or even epiphany. For such a framework the term dual-aspect monism has been coined, outlined in a recent monograph by Atmanspacher and Rickles (2022) and briefly sketched by Atmanspacher (2024). However, let me first introduce a quasi-cognitive account of creative processes by the psychologist Dean Simonton.

Simonton (1988) presented a 'chance-configuration model' for the second and third stages of creative processes, which recognizes the concept of stability as a central issue. The permutations of unconscious elements during incubation are not (asymptotically) stable, but float freely, coming and going randomly. Only a particular one among these configurations has stability properties that lead to its transition into a conscious idea. In this analysis of creative processes, stability provides a selection criterion among many chance possibilities.

The question of why and how particular configurations are distinguished by their stability remains unresolved though. In this respect, some speculative ideas addressed by Pauli and inspired by Jungian depth psychology are of interest. The Pauli-Jung conjecture, a contemporary version of dual-aspect monism, proposes that meaningful correspondences (synchronicities) between mental and physical domains arise from an underlying psychophysically neutral background reality (Pauli 1952: 111–12, translation by the author):

> The process of understanding nature, as well as the blissful experience in this process, when a new insight becomes conscious, seems to be based on a correspondence, a kind of congruence, of inner images pre-existing in the human psyche with external objects and their behaviour.
>
> At this point it seems most satisfactory to me to introduce the postulate of a cosmic order, eluding our direct access, which is distinct from the world of appearances. ... The relation between sensual perceptions and ideas would then follow from the fact that both the soul of the observer and the observed object are governed by the same objective order.

The origin of the stability properties addressed above could then be conceived at the level of this objective, psychophysically neutral order and stable configurations would manifest themselves in the selection of particular correspondences out of many possible ones. Examples of serendipity as described by Simonton (1988), resembling features of Jungian synchronicity, are interesting candidates fitting into this picture. Importantly, it is generically an experience of meaning that accompanies the resulting moment of insight.

Scientific work as a human activity is much subtler than a purely empirical-rational enterprise. Achieving understanding is a laborious process, guided by unconscious elements long before their result can be consciously formulated in rational terms. Current research on creative work focuses on psychological investigations of the solution of insight problems (Öllinger and Knoblich 2009), and Simonton (1988) studies biographical sources along similar lines. This does, however, not address the full scope of Pauli's views and concerns (Pauli 1952: 38, translation by the author)[4]:

> I hope that no one still maintains that theories are deduced by strict logical conclusions from laboratory-books, a view which was still quite fashionable in my student days. Theories are established through an *understanding* inspired by empirical material, an understanding which is best construed, following Plato,

as an emerging correspondence of internal images and external objects and their behaviour. The possibility of understanding demonstrates again the presence of typical dispositions regulating both inner and outer conditions of human beings.

Although referring to Plato, 'external' does not relate to the external world of Platonic (mathematical) truths but to external physical reality in this quote. 'Following Plato' means for Pauli to understand Jung's archetypes as ordering principles akin to Plato's ideas. And when Pauli speaks of the 'possibility of understanding', he obviously points to the meaning that arises from the depths of the psychophysically neutral. It refers to moments of insight which Frege (1892) means when he talks about sense as the mode in which psychophysically neutral structures present themselves mentally or physically. See Atmanspacher (2024) for a compact outline of how all this can be construed in context.

This may remind us of Gabriel's account of understanding meaning as a relational sense beyond the ordinary sense modalities (Gabriel 2020b). This sense of understanding meaning can be less or more developed and refined. And it has much more stringent boundary conditions in science than in the arts or in everyday life. After all, Gabriel (2020a) sees the faculty of understanding, rather than other forms of cognitive activity, as the key indicator for general intelligence. Zwicky (2019) argues forcefully and with numerous examples from literature and music that meaning is prior to language, not the other way around. And Hadamard (1954) presents lots of examples of scientists and mathematicians expressing the same.

Pauli's quote refers to the common ground of internal images, external objects and their correspondence, in other words, the origin from which the mental, the physical and their mutual correlations arise. In the Pauli-Jung conjecture this psychophysically neutral domain hosts archetypal structures (akin to Platonic forms) and symbols that express themselves mentally and/or physically. The most fundamental one among these archetypal structures is the so-called *unus mundus*, the one world that has no distinctions at all: Spinoza's domain of the divine.

And here the link to his second and third kinds of knowledge becomes evident again. The blissful experience of understanding is what accompanies Spinoza's intuition, the illumination at stage three of creative work. There is no recipe that guarantees it, although stages one and two may increase the chances for it to happen. Once the intuition is there, however, it may serve as the starting point, the premise as it were, for the rational reconstruction that communicates the insight step by step through reason, the second kind of knowledge.

When gifted mathematicians get to 'see' connections between structures in the mathematical world, they arguably 'see' with other senses than their eyes. And their insight comes from a domain beyond their mental system with its capacities and faculties. The moment of illumination yields knowledge of Spinoza's third kind, which then gets mentally processed as knowledge of the second kind during the verification phase. The following quote by Alain Connes, grandmaster of non-commutative algebra and geometry, is a perfect example (Connes 2001: 26):

> I maintain that mathematics has an object that is just as real as that of the sciences, but this object is not material, and it is located in neither space nor time. Nevertheless this object has an existence that is every bit as solid as external reality, and mathematicians bump up against it in somewhat the same way as one bumps into a material object in external reality. Because this reality cannot be located in space and time, it affords – when one is fortunate enough to uncover the minutest portion of it – a sensation of extraordinary pleasure through the feeling of timelessness that it produces.

4 Understanding the *Ethics* backward

Spinoza's third kind of knowledge is most closely related to the deep insights that arise out of a connection with the one psychophysically neutral domain of reality – the one eternal and infinite divine substance and its attributes – that itself is neither mental nor physical. The relation between this domain and its aspects is what Frege (1892) refers to as sense, as the mode in which the psychophysically neutral – the 'third realm' in the parlance of Frege (1918) – presents itself mentally and physically. This third realm is the source and origin of all affects and thoughts of the mental and all behaviour of physical bodies.

This third kind of knowledge by intuition is capable of yielding meaningful insights – beyond causal implications or logical inferences – that are outside of what is achievable by reason alone, the second kind of knowledge. A sudden aha-experience, an instance of immediate understanding a great idea is something that cannot be logically derived from observed facts or from rules of inference. Yet it can open up entirely new directions of thinking, and expand our views on nature and how to act in it in meaningful ways.[5]

Along these lines, it is no accident that the Norwegian philosopher Arne Naess (1977) returns to Spinoza's kinds of knowledge when he proposes a deep ecology that does not merely intend to cure symptoms on the surface but focuses on the causes underlying the disease. The traditional strategy of trying to stabilize unstable situations using tools that actually were instrumental to produce those instabilities in the first place must be recognized as questionable, sometimes perhaps even detrimental. As the scientist and humanist Albert Einstein cautioned us more than seventy years ago, we won't be able to resolve the significant problems of our times at the same level of awareness we were at when we created them.

The re-orientation that Einstein, an admirer of Spinoza's philosophy, requests is precisely the move towards knowledge of the third kind. In the present days it is often regarded an indication for an astute modern mind to reject the idea that meaning can be found in life and, conversely, can be given to life. I suggest we rethink this. There is reason to believe that there is a greater reality, ignored by science and engineering, which may lead to an advanced kind of enlightenment – one that Spinoza was profoundly faithful about. It must be regretted that such faith is often overlooked and sometimes even culpably ridiculed as of today.

This brings me to the question raised at the beginning of this essay. Why did Spinoza choose the strangely rigorous and almost mathematical form for his *Ethics*? One answer is that he wanted to present his arguments in a way that looked as indubitable as possible. Another one may be that quasi-mathematical treatises were not uncommon in his time. But still the beginning of the *Ethics* with its positing of God among the very first definitions appears somewhat 'out of the blue' without any pretext. Here is a tentative speculation about such a pretext.

Spinoza was the son of Sephardic Jews forced into Catholicism by the Inquisition but then were expelled from their homeland in Portugal and emigrated to Amsterdam. He was educated in the Hebrew spirit, and at the age of fifteen he was anticipated as a future leading light of the synagogue. But his independent thinking brought him into serious and increasing conflict with Jewish scholars, so that he was finally ostracized from the community at age twenty-four. After surviving an assassination attack by a fanatic Jew, he left Amsterdam and separated from his family. He never converted to Catholicism – his free thinking, combining prudence and boldness, made it impossible for him to be a member of any religious tribe anymore.

For his economic security he learned to grind optical glasses and became very skilled and well-known with this craft. As a Dutch citizen he participated in the public and political life of the state and drafted partly incomplete treatises on political systems and ideas. Spinoza's biography at this point does not present a life full of turmoil and trouble. Although he entertained rich correspondence with contemporaries (especially after 1660) and contributed a clear-cut position to political discussions, he lived a humble and quiet life, and he experienced the self-confidence in solitude that is not uncharacteristic for free spirits.

Jaspers (1966) counts Spinoza as one of the great philosophers in world history. His list of those metaphysicians who think from foundational origins comprises Anaximander, Heraclitus, Parmenides, Plotinus, Anselm, Laozi and Nagarjuna, next to Spinoza. Here is what Jaspers (1966, chap. III: On Spinoza, translation by the author) writes about the origin of Spinoza's insights:

> His conceptual comprehensive vision is there with one stroke and with full perfection. To the question of how this may have happened there is only one answer: through a revelatory awareness of God, which ... became the one and everything that mattered to him.

This metaphysical total vision connects the divine with all of nature, including humans together in communion with other humans and with nature as a whole. Spinoza thus links theology and politics in a unique and unprecedented way, neither an esoteric mystic nor a cold rationalist. Its source and origin are what Jaspers (1967) characterizes as a 'philosophical revelation', not a religious experience directed towards mystical union.[6] This is the point where his *Ethics* derives from: a deep experience of knowledge of the third kind gives him the premise to start with.

Remember the four stages of creative work addressed in the preceding section. The key stage in which an illuminating insight happens is stage three, but the exact content of the experience, its detailed meaning, is usually too comprehensive to be communicable to others right away. It is introspective and holistic in its totality. With Jaspers we may speculate that Spinoza had a revelatory experience of such an extent that it massively overwhelmed his capacities of reason, the second kind of knowledge. Clearly, an experience like this cannot be written down or otherwise recorded the next day.

Spinoza does what all mathematicians typically do with their insights, just as the examples of Helmholtz and Dirac in Section 3.2 express: he tries to reconstruct it rationally to make it available to discursive thinking, as in stage four of the creative process, so that everyone with sufficient skills can follow and understand. This explains why he begins with God and finally arrives at the third kind of knowledge, which in fact was his personal point of departure in the first place. The premise that the revelation gave him was the unquestionable certitude of the eternal and infinite impersonal divine substance from which everything in the *Ethics* is developed, point by point with quasi-mathematical beauty and precision. *Sub specie aeternitatis*, and as far from anthropocentric as possible.[7]

5 What would it be like to be educated?

This question is the title of a graduation speech that the philosopher Peter Bieri delivered at the Pedagogical College Bern in 2005 – an extended version was later published by Bieri (2017). Bieri's text collects a set of remarkably obvious and yet easy to overlook principles and guidelines for education, at all levels from preschool to universities and to all of life in general. Why is this theme of education so all-important? Because education is decisive for the salutary growth and prosperity – in other words for the flourishing – of the younger members of humankind who will be in charge of its future development when those who are responsible for its present state are long dead.[8]

In this section I aim to relate Spinoza's kinds of knowledge to different kinds of education and its goals that are apparently less conspicuous than the extraordinary insights of scientists and philosophers addressed in Sections 3 and 4 above. Here I want to advocate that the educational dimension of knowledge is of paramount significance in a more general sense that leads us back to the human potential and its realization in a truly fulfilling life for everyone, as outlined in the introductory remarks to this essay. As Dahlbeck (2016) has argued, it may not be totally wrong to presume that this is close to the pedagogical motif and target of Spinoza himself in his *Ethics*.

The English term 'education' covers two meanings that the German language, for instance, distinguishes as 'Bildung' and 'Ausbildung'. The latter addresses the acquisition of facts and skills of using them whose certification may be conditional for a career in some selected profession. The success of this kind of education can usually be verified by examinations and assessments. In Spinoza's terms, such education would be a combination of knowledge of the first and second kind, hopefully with an emphasis on the latter. However, is this all we mean when we speak of education? Would we perceive someone who knows all the facts (and lets everyone know that he knows them) as an educated person?

Education as 'Bildung' is more comprehensive and, most notably, deeper than education as 'Ausbildung'. A well-known rule of thumb in science says that nature gives an answer to every question we pose to it (in an experiment, say), but it does not prescribe which questions to ask. In this picture, factual knowledge of the first and second kind is the route towards answering questions, not towards raising them. Yet the quality of all research depends decisively on good questions that open up interesting options to be investigated. Insight and understanding do not arise just from known facts, as the quote by Pauli in Section 4 aptly emphasizes. Raising a good question is a true measure of being human.

Half a century prior to Spinoza's *Ethics*, Francis Bacon coined the slogan that 'knowledge is power' (*scientia potestas est*) in his *Novum Organon* of 1620. This may have been a groundbreaking insight at the beginning age of enlightenment. Today, however, Bacon's view has gradually slipped into a form of knowledge that expresses a social-political gradient between those who are in power (often without much legitimation) and those who are not (often without their own fault). Knowledge for power is factual knowledge that one can possess at one's disposal and that can be acquired mechanically, with no deep insight or true understanding. A quote by the great American architect Frank Lloyd Wright (well known to be far from short of self-esteem) illustrates the opposite when he talks about (himself as) 'a man who understands what others only know about' (Wright 1994: 157).

At variance with Bacon's *potestas*, Spinoza uses the notion of *potentia* in the *Ethics*. What he means by that is the potential of agency, opening a scope for action, rather than the power to govern and realize goals. Negative affects are perceived as decreases of agency and lead to inadequate knowledge: noise and confusion. Positive affects are perceived as increases of agency and lead to adequate knowledge: prudence and wisdom. He who hates loses understanding, he who loves gains understanding. The viewpoint *sub specie aeternitatis* yields the highest enjoyment due to infinite *potentia*, with no goals remaining to be realized, hence no *potestas*. This state may be seen as the elation of revelation in the third form of knowledge.

In this spirit, Spinoza's third kind of knowledge ought to be seen as knowledge for orientation rather than for power. It provides a compass for judgement and action that cannot be derived from facts alone, a compass that gives our lives direction. The German 'Bildungsroman' is a literary genre reflecting this by exploring the coming-of-age of the protagonist, usually from a troubled childhood towards moral and psychological maturity. Goethe's *Wilhelm Meister's Apprenticeship* is often regarded as one of the first novels within this genre, other examples are *Great Expectations* by Dickens or *A Portrait of the Artist as a Young Man* by Joyce.

The question of 'what it would be like to be educated' highlights a topic that seems greatly underappreciated or plainly misunderstood in current educational policy concerning curricula, the training of teachers and many more factors contributing to the quality of schooling. And it is a question that many opinion leaders of today's digital intelligence, who like to think of themselves as the one and only redeemers towards a better future, surely will find utterly anachronistic. Yet it is of paramount importance for a human condition that is truly meaningful for us and our fellow citizens on this globe. Meaningful in both respects: finding meaning in our lives and bestowing meaning upon our lives.

While education as 'Ausbildung' is something one can receive from others, education as 'Bildung' is something one needs to cultivate for oneself. While education as 'Ausbildung' has the goal for us to master a number of skills to perform, education as 'Bildung' has the goal for us to become someone, to shape our mindset towards self-determination and liberation from barriers needlessly preventing us from flourishing. Education as 'Bildung' helps us mobilize our human potential and facilitates the great Greek poet Pindar's succinct demand to 'become who you are'!

6 Epilogue

When I was sitting at my desk a few months ago and thought about the invitation by Andrej Zwitter to contribute to the present volume, an email came in from Bob Moody with a 700-page manuscript about 'The Pattern of Change' attached to it (Moody and Deng 2024). It deals with the intricate relationship of patterns with symmetries and invariances, mainly from a mathematical and scientific perspective (including his own pioneering work on Lie algebras and their generalization in Kac-Moody algebras). I had met Bob a few years earlier at Esalen Institute, the birthplace of the idea of the human potential, where we had inspiring exchange on non-commutativity in physics and other areas.

Thrilled by Bob's work, I immediately felt the mixture of curiosity and intimidation that is typical if confronted with a text of such comprehensive extent – and hesitation too, because in my experience a math book can't be read front-to-end anyway. So, I first scanned it 'diagonally', as they say. What soon caught my eyes were occasional and unexpected references to Jan Zwicky, a Canadian philosopher with whose work I was, until now, unfamiliar.[9] Bob quotes her on the understanding of patterns as gestalts, as an arising awareness of a coherence that constitutes a pattern as a whole. This awareness is manifest even before the pattern gets decomposed into parts which it doesn't consist of in the first place.

Not surprisingly, it occurred to me that this is precisely how Dean Rickles and I address the decomposition of aspects in the metaphysical position of dual-aspect monism pioneered by Spinoza (Atmanspacher and Rickles 2022). I turned to Jan's book *The Experience of Meaning* and found in it a vast source of examples from Gestalt theory for her thesis about how the insightful understanding of meaning happens in what she calls imagination (Zwicky 2019). She nowhere refers to Spinoza though, likely because his pejorative equation of the term imagination with inadequate ideas is very different from hers.

After some more reading and thinking it was evident that Jan's term imagination must actually be conceived along the lines of Spinoza's intuitive insight, rather than his own take on imagination. Spinoza's highest knowledge, ultimately amounting to *amor dei intellectualis*, shines through when she says in her aphoristically designed 'Imagination and the Good Life' (Zwicky 2015) that 'ontological attention is a form of love' (aphorism 50). It did not take long, and I realized that some more aphorisms from her work are, in a way, the best possible conclusion to the present essay:

> Imagination ... as sensitivity to ontological resonance is the most direct route to the good life. ... It shows us that response is required.
>
> (52)

> But of course it is possible to understand that we must change our lives, and then fail to do so. The price is always a loss of integrity, an absence of interior attunement. This, then, is why the exercise of moral imagination requires courage: we could, in any instance, be risking everything.
>
> (53)

> If we do not cultivate the imagination as an organ of ontological insight, if we derogate it, we cripple ourselves as ethical beings. We close the door on our single best chance to espouse the world; to see things for what they are; to testify to that espousal through what we then become.
>
> (66)

Notes

1   The notion of the human potential has been coined by Michael Murphy, founder of the Esalen Institute (Big Sur, California). It has become known as a distinguishing characteristic of humanistic approaches to psychology and philosophy, attached with names such as Abraham Maslow, Carl Rogers, Fritz Perls, Viktor Frankl, George Leonard, Eugene Gendlin and many others.

2   I should say right away that 'happiness', another term often used in relation to flourishing, raises some ambivalent-poignant resonance with me, very similar to what Dakwar (2024) expresses in the first pages of his recommendable book. It is not purely accidental that the concert hall at Leipzig's Gewandhaus shows the inscription *res severa verum gaudium* as the guiding motto of its orchestra: true joy is a serious matter – very different from having fun or being happy.

3   At Spinoza's time, logic was broadly construed according to the Aristotelian system, which is close to what is known as Boolean logic today. Two basic rules of Boolean logic are the law of the excluded middle (truth values are either true or false) and the law of non-contradiction (no paradoxes: what is true cannot be false and *vice versa*).

4   For instance there is no way to explain the breathtaking insights of the Indian mathematician Ramanujan by cognitive-science models as discussed by Knoblich and Öllinger (2009). Ramanujan himself credited his theorems to inspirations that came upon him from the deity of Namagiri. Ramanujan's Notebooks, published in five volumes by Bruce Berndt between 1985 and 2005, contain about 4000 results, most of them correct, none of them (characteristic for him) presented with proofs. For an overview see https://en.wikipedia.org/wiki/Bruce_C._Berndt (accessed 26 October 2025).

5   A recent collection of articles by Zwitter and Dome (2023: 14) explores 'how meaning and meanings can help shape and direct the soft and the hard sciences towards a common goal solving complex problems and provide solutions that leave no one behind'.

6   Jaspers highlights the philosophical notion of revelation as an insight that has the capacity for transformative effects on the human condition. By contrast, mystical experiences leave us with an ineffability that defies thinking and communicating, base operations of the human condition.

7   Spinoza uses the phrase *sub specie aeternitatis* ('from the viewpoint of eternity') in a massively dominant way in the later parts of the *Ethics*. Its point is to emphasize as clearly as possible that human beings won't be able to flourish just for themselves, anthropocentrically, but only in harmony with all of nature. Naess (1977) picked up on this insight in his foundational contribution 'Spinoza and ecology' towards what is now known as deep ecology, as opposed to the shallow ecology dominating action schemes that seem to be preferred today.

8   While the theme of education enjoys much excitement about computer-assisted teaching and learning today, there are important critical voices too. Manfred Spitzer's provocative book on *Digital Dementia* (Spitzer 2014) sketches a number of psychological and biological malfunctions and impairments that humans, in particular young ones, acquire due to excessive usage of digital tools, not only in computer games and so-called 'social media' (where the term 'social' couldn't be

more inadequate) but also in school education. See also Manwell et al. (2022) for a recent review.

9   The name Zwicky is reminiscent of the astronomer Fritz Zwicky, who emigrated to the United States in 1925 from the small Swiss mountain village Mollis, only a few kilometres away from where I live. Fritz Zwicky is mainly known for his work in astronomy. His six-volume compilation of galaxies and clusters of galaxies in the Zwicky catalogue at CalTech in the 1960s has been used for much seminal work on the distribution of galaxies in the universe. In later conversation with Jan, it turned out that Mollis is indeed the origin of her family as well.

Part Three

Context and Culture – Flourishing as Being

7

The Arc of Hope

Richard D. Hecht

1 Introduction

The title of this chapter echoes a single sentence from the last Sunday sermon given by Dr Martin Luther King in Washington's National Cathedral on 31 March 1968, as he was launching the Poor People's Campaign just a few days before his assassination. He said, 'We shall overcome because the arc of the moral universe is long but it bends toward justice' (King 1991). This sentence is now so well-known that if we hear only the first words 'arc of the moral universe', we can easily complete it with 'it bends toward justice'. However, when the sentence is taken from its homiletical context, it may appear that King was suggesting an inevitable process that would culminate in justice. To the contrary, King was telling us that there is a long historical process of sacrifice and struggle against racism. The universe does not just bend towards justice on its own; it must be bent.[1]

As he drew his sermon to its conclusion, King noted that there would be difficult days ahead in the struggle for justice and peace, but he was unwilling to yield to the politics of despair. And, here is the critical introduction to this one sentence that has become so well-known. He said,

> I'm going to maintain hope as we come to Washington in this campaign. The cards are stacked against us. This time we will really confront a Goliath. God grant that we will be that David of truth set out against the Goliath of injustice, the Goliath of neglect, the Goliath of refusing to deal with the problems, and go on with the determination to make America the truly great America that it is called to be.

I wish to thank Andrea White, Professor of English Emerita, California State University, Dominguez Hills and Richard Cohen, PhD, for their helpful comments to an earlier draft of this paper.

It is hope that will give Dr King and the thousands who have marched with him in the past and will now march again the confidence to bend the universe towards justice. And as the peroration overturns the present situation King told the audience:

> If the inexpressible cruelties of slavery couldn't stop us, the opposition that we now face will surely fail. We are going to win our freedom because both the sacred heritage of our nation and the eternal will of the almighty God are embodied in our echoing demands ... We shall overcome because the arc of the moral universe is long, but it bends toward justice.

And, he returns to the theme of hope as he closes the sermon.

> With this faith we will be able to hew out of the mountain of despair the stone of hope. With this faith we will be able to transform the jangling discords of our nation into a beautiful symphony of brotherhood.
>
> (King 1991)

Hope then bookends 'the arc of the moral universe is long, but it bends toward justice' and Dr King drew a parallel between 'hope' and 'faith'. The human actor, and for Dr King, it was the generations upon generations of civil rights activists who bent the arc of moral universe. It is hope that provides agency and without it, in the religious worldview of Dr King, time and history would just go on and on. Hope generates the social and personal, the collective and individual grounds for human flourishing.[2]

This chapter explores some of the possible meanings of hope as it might contribute to a more fulsome understanding of human flourishing. Dean Rickles' chapter in this book laments the restrictive misrepresentation of science that requires it to be an anti-humanitarian worldview. Rickles writes that the scientific method would seem to require that the important things be measurable, repeatable and represented in abstractions. And yet the things that are most representative of humanity, and he lists 'Purpose, Meaning, Empathy, Joy and Love' and we would add 'hope', are not objectifiable in this way. Any effort to advance human flourishing must seek to preserve and enhance those things that initially appear to be unscientific. Worldviews are the ontological architecture that makes up lived experience. Similarly, Andrej Zwitter argues in his chapter that the classical virtues of faith, hope and love that were reformulated by Emmanuel Kant and we have inherited are at one and the same time both metaphysical and practical, are essential to human flourishing. There is thus both an ontology of hope and metaphysics of hope that stand at the heart of this discussion.

We begin with the work of Gabriel Marcel who is often described as a Christian existentialist. Hope, in his existentialism, is an essential element that contributes to the achievement of communion with the other. We then consider how hope gives access to transcendence in situations of extreme violence and trauma. We consider this meaning of hope through Viktor Frankl's immensely popular *Man's Search for Meaning*, first published in 1946, that led to the creation of what Frankl called logotherapy, a psychoanalytic therapy that could reconstruct an individual's sense of meaning. We then consider Ernst Bloch's monumental study *Das Prinzip Hoffnung, The Principle of Hope* and how what he called the 'not-yet-conscious', is reflected in cultural production. Finally, we consider a work of art, the Mexican muralist Diego Rivera's *Pan-American Unity* mural created for the Golden Gate International Exhibition in 1940 and discuss how it exhibits of Bloch's argument that art manifests hope. We will argue that hope is a way of seeing the world or a way of ascribing order, a way of discerning patterns, relationships and visualizations. We will suggest in conclusion that hope is one of the central ideational forces involved in the production and reproduction of the cultural and social worlds. Human flourishing draws its strengths or energies from hope.

2 'Things never change without hope'

Jane Fonda, the celebrated actress and political activist, sent out on her climate PAC blog on 29 April 2024 a memo with a banner that read, 'Things never change without hope'. Fonda's use of the word hope attests to there being no shortage in the use and citation of the term hope. Here, hope is connected to action and we are to understand that the climate crisis we face can be altered only by hope. This is the very same usage that Dr King used in his sermon.

Hope seems to suggest a desire or wish for change of some sort, or a wish for something to happen. Hope may suggest or be a reminder of the human capacity to see beyond the present. Another formulation of the meaning of hope might be that it is about the not-yet. Continuing at this most general level, hope is about things that can be or about something that could happen. It seems to be a word that directs our thoughts or our consciousness to what is in the realm of the potential or of possibilities. Hope might also be understood as a term that is parallel, but not a cognate, of the words faith or optimism.

Hope takes many forms. First, hope may suggest a temporal change or transformation, again in the future. A person, for example, might hope for the future to be better. Second, hope may suggest a spatial transformation in the future. It could be something as simple as hoping to visit New York City in the future. Or, it could be a much more substantial transformation as what is suggested by Jane Fonda's blog of 29 April. Third, hope may suggest an individualized or reflexive change. For example, one could hope for a successful medical treatment for a particularly aggressive form of cancer.

But, hope is much more complex than a tripartite division by time, space or reflexivity. In 1945 and in the wake of the Second World War, the French Catholic existentialist Gabriel Marcel published a series of essays under the title *Homo Viator: Prolégomènes à une métaphysique de l'espérance*. In using the term *homo viator*, he immediately drew on the tradition of medieval Christian thought in which the term was understood as a metaphor for the human as a wayfarer, a sojourner, or a pilgrim or wanderer between two worlds, or a term for the spiritual alienation or separation of man and God.[3] In his 'Preface' to the volume's essays, he begins with a fundamental distinction. Hope, Marcel suggests, is the availability of a soul or an individual, to enter into communion or a transcendent act with an 'other' or 'others'. He writes:

> This means that in the first place hope is only possible on the level of *us*, or we might say of the *agape*, and this it does not exist on the level of the solitary *ego*, self-hypnotised and concentrating exclusively on individual aims. Thus it also implies that we must not confuse hope and ambition, for they are not of the same spiritual dimension.
>
> (Marcel 2010: 4)

Here, he uses the Christian theological term *agape* or love to describe that hope is a communal experience. Second, hope only is found when the soul manages to liberate itself from the demarcation between factual knowledge and what it wishes or desires. Hope, he writes, means 'an act by which this line of demarcation is obliterated or denied'. He then notes that we cannot help to see that the soul and hope are the closest of connections – hope is to the soul what breathing is to a living organism. Without hope the soul would languish and become no more than a function that should be the object of psychology that cannot do more than register its location or absence. It is the soul that is the traveller that means 'being on the way' (*en route*). It is the soul that is the *homo viator* and thus hope is the motivation that provides the energy or perhaps the life-force to the soul (Marcel 2010: 5).

There is a danger that as we think of hope as the animating force of the soul and being, we make hope into a natural faculty. Does it depend on us or is it the result of an innate disposition, or a pure grace, or the result of supernatural help? He curiously says that we must recognize it is 'both true and false to say that it [i.e. hope] depends on us whether we hope or not'. Hope is offered to us, but we can refuse it, in the same way that we can refuse love when it is offered. He then suggests that hope might be considered a virtue and like other virtues represents the particularization of an interior force (Marcel 2010: 57). He comes to what he calls 'the intelligible core' of hope at the end of his discussion. What characterizes the core of hope is the movement by which it challenges the evidence upon which humans claim to challenge it.

He then offers the following as the definition of hope: 'hope is essentially the availability of a soul which has entered intimately enough into the experience of communion to accomplish in the teeth of will and knowledge the transcendent act – the act of establishing the vital regeneration of which this experience affords both the pledge and first-fruits' (Marcel 2010: 61). Hope is the act of communion, he suggests, against will and knowledge, that might preclude the possibility of communion. And hope then generates the pledge to do something in the future and also what is generated immediately from the communion. We can summarize this in the following way: hope allows an individual to reject despair and gives one the strength or courage to continue to create one's self in the presence of the other. We see in Marcel's understanding of hope a rejection of the Kantian tradition that argued that the third critique of reason (or what may I hope for?) whether speculative or practical, suggests that hope is accessible through aesthetics. For Marcel, what I may hope for is an existential question.

In addition to the hope that offers a temporal transformation, the hope that offers a spatial transformation, and the hope that offers individual or reflexive change, Marcel offers a fourth structure – the hope that establishes (or even, re-establishes) a communion with the other. At several points in the essay, he challenges the reader, saying that he wants to separate his analysis from theology. But the language that he uses is drawn from Catholic religious thought. This is important and should not disqualify the universality of his analysis. He is suggesting that hope gives access to transcendence and when this is considered, we can also say that hope gives access to the sacred.

3 Hope and meaning: Viktor Frankl's *Man's Search for Meaning*

One of the early attempts to confront the trauma of survivors of the Holocaust was Viktor Frankl's *Man's Search for Meaning* (1959), first published in German under the title *A Psychologist Experiences the Concentration Camp* (1946). Frankl was imprisoned in Theresienstadt (where he lost his father), Auschwitz (where his brother and mother were murdered; his wife died in Bergen-Belsen), and Kaufering III and Türkheim. Frankl was born in Vienna in 1905 and when he died in 1997, *Man's Search for Meaning* had sold ten million copies in twenty-four languages. Frankl studied with Freud and Alfred Adler, and before his imprisonment, worked in Steinhoff Psychiatric Hospital and Rothschild Hospital in Vienna.

A Psychologist Experiences the Concentration Camp was among the very first survivor accounts from the Holocaust. Frankl tells the reader that he was an ordinary prisoner, not a psychiatrist or a doctor in the camps, until near the end of the war. He begins by noting that prisoners first experienced what psychiatry describes as 'delusion of reprieve' in which the condemned person believes that he or she will be reprieved at the last moment. 'We, too', Frankl writes, 'clung to shreds of hope and believed to the last moment that it would not be so bad' (Frankl 2006: 10). For Frankl, hope is medicalized. It is a delusion. The prisoner passed from what Frankl calls the first phase of response to the camps, this delusion, to the second, 'relative apathy' in which he achieved a type of 'emotional death'. The prisoner no longer averted his eyes from the brutalization and death all around him. His emotions and feelings were blunted and he watched unmoved (Frankl 2006: 21). In this second stage, the prisoner's inner life was intensified in the midst of the emptiness, desolation and spiritual poverty of his existence in the camp, by escaping to the past. The imagination played with the past and transformed even the smallest and unimportant events into memories of great emotive power. Their world and their existence seemed very distant and the spirit reached out for them longingly: 'In my mind I took bus rides, unlocked the front door of my apartment, answered my telephone, switched on the electric lights' (Frankl 2006: 39). The prisoners experienced these with intense emotions, often moving them to tears.

All the values that a prisoner had were systematically destroyed. 'Under the influence', he writes,

> of a world which no longer recognized the value of human life and human dignity, which had robbed a person of his will and had made him an object to be exterminated (having planned, however, to make full use of him first – to the last ounce of his physical resources) – under this influence the personal ego finally suffered a loss of values. If the man in the concentration camp did not struggle against this in a last effort to save his self-respect, he lost the feeling of being an individual, a being with a mind, with inner freedom and personal

value. He thought of himself then as only part of an enormous mass of people; his existence descended to the level of animal life.

(Frankl 2006: 50)

Four decades after Frankl, some therapists working with soldiers who suffered PTSD as a result of their combat duty in Vietnam, described it as the 'undoing of character' (see, e.g. Shay 1994).

The third stage of reaction to the camp was the result of liberation. He describes this experience as 'depersonalization' in which everything appeared as unreal, unlikely as in a dream. The prisoner could not believe that liberation was true. 'How often in the past years', Frankl describes,

> had we been deceived by dreams! We dreamt that the day of liberation had come, that we had been set free, had returned home, greeted our friends, embraced our wives, sat down at the table and started to tell of all the things that we had gone through – even of how we had often seen the day of liberation in our dreams.
>
> (Frankl 2006: 88)

Of course, how the prisoners thought about their liberation and their return to their homes and families, the towns from which they had been taken, all of that was not as they had dreamed.

Frankl does not have a language to describe the horror of the camps. It is perhaps too early for there to be such a language. The most famous contribution that Frankl makes is not his description of the horrors of the camp, but how one survived the horrors. He points to the prisoner who has lost faith in his future. He describes how a prisoner who has lost his belief in the future, has also lost his spiritual hold. That prisoner declines and becomes subject to mental and physical decay. It should be underscored here that Frankl is only speaking of the prisoners who were selected to work in various tasks in the camps. These were not the huge number of prisoners who were murdered in the gas chambers. Here, he is speaking about the prisoners who refused to continue their daily ordered work. Those prisoners refused 'entreaties, no blows, no threats had any effect ....[they] just lay there, hardly moving. If this crisis was brought about by an illness, ... [refused] to do anything to help himself. He simply gave up. There he remained, lying in his own excreta, and nothing bothered him any more' (Frankl 2006: 74). Courage and hope had been lost and the only response, Frankl argues, is to change fundamentally one's attitude towards life. Frankl suggests that this change is that despairing people must understand that 'it did not really matter what we expected from life, but rather what life expected from us. We needed to stop asking about the meaning of life, and instead to think of ourselves as those who were being questioned by life ... ' (Frankl 2006: 76–7). The solution was for the individual to be helped to reclaim the meaning of his life. He called this 'logotherapy', coming from the Greek word *logos* that suggests 'meaning'. Frankl claimed that logotherapy, unlike Freudian and Adlerian depth psychology, centred upon the meaning of human existence and the human search for that meaning (Frankl 2006: 98–9). Meaning restored hope in those who had experienced the trauma of the death camps.

4 Ernst Bloch and 'Wishful Images' of hope

Ernst Bloch (1885-1977) was one of the most interesting and perhaps influential figures of the first half of the twentieth century. Jürgen Habermas described Bloch as 'a Marxist Romantic', a kind of soft Marxist who was particularly interested in utopian thought. He wrote 'Bloch's materialism remains speculative. His dialectics of the enlightenment transcends dialectics and reaches a doctrine of potentiality ... Bloch's thought processes are derived from the development of a pregnancy of the world which is generally assured rather than attempting to free us from the societal immobility of existing contradictions. Philosophy of nature here turns into the nature of his philosophy' (Habermas 1969: 324–5).

Bloch was born in Ludwigshafen, Germany and as a young man rejected his family's Judaism, but remained interested in Judaism as a system of powerful symbols and ideas that could be used to renew culture and society. He saw in Judaism the foundations of his utopianism and socialism. As a young man he studied philosophy, physics, music and art. As a socialist and pacifist, he spent much of the First World War in exile in Switzerland where he wrote his first important book, *Geist der Utopie* (1918), *Spirit of Utopia*. He fled Germany a second time in 1933 with the rise of the Nazis. During this second exile, he wrote two more important books, *Traces* (1933) and *Heritage of Our Times* (1935). In 1938, Bloch immigrated to Boston where he worked on his compendious *Das Prinzip Hoffnung* (*The Principle of Hope*), published in 1959. Bloch moved to East Germany shortly after it was founded in 1949, where he took a position teaching philosophy at the University of Leipzig. He later took a visiting professorial position at the University of Tübingen where he remained until his death.

Bloch's influence and significance extended into what became known as the School of Hope, one of the most important European theological traditions to emerge during the Cold War. Walter Capps traced the intellectual connection between Bloch, who Capps described as the 'philosophical catalyst' through the Catholic thinker Johannes B. Metz, and then the Protestant theologian Jürgen Moltmann (Capps 1972: xviii).[4] The School of Hope did not gain the traction in the United States that it had in Europe, in large part, because of the controversies involving the 'Death of God Theology' (Altizer 1966). The School of Hope did not have the theological power of either Karl Barth or Paul Tillich. The explosive growth of Protestant Evangelicals in the United States during the Cold War at precisely the same time that 'the School of Hope' was rising contributed to the inability of its theologians to gain much of a foothold. Perhaps it was the intense present-sense among Evangelical thinkers that provided a potent counter-point to the future orientation of 'the School of Hope'.

Bloch was influenced by discussions within the Frankfurt School, especially in regard to its explanation of the rise of Nazism. Many within the Frankfurt School understood that Nazism emerged from a desire for an authoritative figure. In *Heritage of Our Times*, Bloch argued that fascism was a distortion of a religious movement that attracted supporters through anachronistic kitsch and quasi-utopian ideas about the wonders of a future Reich. Fascism was, as a result, a paradox both ancient and modern, a rough and incomplete synthesis of hostility to anti-capitalism and support for capitalism. For Bloch, and also Walter Benjamin, fascism was a cultural synthesis that contained both anti-capitalist and utopian aspects. The Frankfurt School failed to emphasize in its broadly psychological analysis of fascism what Benjamin called the 'aestheticization of politics' (Wolin 1994: xxix). Benjamin and Bloch were thus particularly interested in how the Nazis used myths, symbols, parades and demonstrations to mobilize support.

One thing that clearly separated him from other Marxists was the significance he assigned to religion. Marx, of course, had rejected religion in 1843 and 1844 in his critique of Hegel and in his response to Bruno Bauer's proposal for Jewish emancipation as one of the major factors that stimulates the atavistic conflict between classes. But Bloch was interested in religion as a source of the 'wishful images of the fulfilled moment'. We might think of cultural phenomena, as Bloch did, as traces (*Spuren*) and signposts (*Wegweiser*) of hope. Indeed, Bloch was experimenting with very short, written descriptions, some of which were only a single sentence that captured the essential identity or importance of something, someone or sometime, as well as very short stories (Bloch 2006). Walter Benjamin was using the same kind of writing, for example, in his *One Way Street* (1928), *A Berlin Childhood around 1900* (1933), and his *Arcades Project* (1927-1940).

And, certainly the religious traditions of world history were those signposts. In the third volume of *The Principle of Hope*, for example, he titles chapter 53 as 'Growing Human Commitment to Religious Mystery, to Astral Myth, Exodus, Kingdom; Atheism and the Utopia of the Kingdom' and he provides a seamless account of the mythological traditions of ancient Greek and Roman religions, Confucius, Lao Tzu, Moses and the God of the Exodus, Zoroaster, Mani, Hindu ideas about Nirvana, the Buddha, Jesus and his apocalyptic kingdom, the serpent in the Garden of Eden, the three wishful mysteries of Judaism and Christianity, and he concludes that the submission to the will of Allah and the life of the Prophet Muhammed are all the signpost of hope. He argues that there are false forms of religion that he refers to as the religion of the 'opium-priests', here using the Marxist tradition of religion as an opiate. He writes, for example:

> Together with the triviality and clichéd conventionality with which Job's comforters, those prototypes of all opium-priests, dispense their kind of trust in God. The God of exodus is different in nature, in the prophets he proved his hostility to lords and opium. But he is above all not static in nature, like all pagan gods until then. For the Yahweh, the God of Moses, right at the beginning, gives a definition of himself, one which over and over again is breadth-taking, which makes all statics futile: "God said unto Moses, I will be who I will be." (Exodus 3,14)
>
> (Bloch 1995: 1235–6)

The name that God reveals to Moses is a perfect expression of what Bloch was pointing to, the future and hope.

At the centre of *The Principle of Hope* is an examination of what he calls the 'utopian consciousness'. The book might be described as a complete phenomenology of the utopian conscious that he weaves through all three volumes of *The Principle of Hope*, beginning in the first volume with his development of the idea of the 'not-yet-conscious' [*Noch-Nicht-Bewussten*], the anticipatory dimension of human consciousness (Bloch 1959: v.1: 10). The 'not-yet-conscious' is always oriented to the future and is central to human thought. He distinguishes it from what he calls the 'no-longer-conscious' that is always about the past. Human thought is always oscillating between these two forms of consciousness. The 'not-yet-conscious' is at the heart of everything utopian and he demonstrates this by examining the 'wishful images' in fairy tales, in popular fiction, in the accounts of travellers and travelling, in theatre, music, dance and cinema. In the second volume, he explores the outlines of a better world that he finds in utopian systems advanced by progressive thinkers in the fields of medicine, painting, opera, poetry and philosophy through an encyclopaedic account of utopian thought from the Greeks to the mid-twentieth century. The third volume then provides his prescription for how humans can reach their proper 'homeland' where social justice is coupled with an openness to change and to the future. In short, his prescription for how humans are to flourish.

What makes Bloch's vast consideration of utopian thought unique and important for discussions of human flourishing is that he is not interested in utopia as a social phenomenon. Utopia is important because of its aesthetic dimensions that reveal what he calls the 'not-yet-conscious'. Works of art, whether in the form of representational or abstract art and sculpture, literature and music all achieve their status in their breaking with convention and the *status quo*. And, indeed it is the break with convention that runs through artistic work that materializes the 'not-yet-conscious' (Zipes 2019).[5] He writes that utopian consciousness:

> wants to look far into the distance, but ultimately only in order to penetrate the darkness so near it of the just lived moment, in which everything that is both drives and is hidden from itself. In other words: we need the most powerful telescope, that of polished utopian consciousness, in order to penetrate precisely the nearest nearness. Namely, the most immediate immediacy, in which the core of self-location and being-here still lies, in which at the same time the whole knot of the world-secret is to be found.
>
> (Bloch 1995: 12)

He uses the metaphor of the telescope whose lenses are the utopian consciousness. But, it is a curious telescope because instead of looking for that which is distant, it takes us to that which is nearest. That nearness is the 'not-yet-conscious', that cleaves to the present and to the anticipation of 'Not-Yet-Become'. He rejects the psychologism that reduces 'not-yet-conscious' to what he calls the 'cellar' of consciousness, but at the 'front' of consciousness. And, he adds that he is describing a psychological process of approaching what is anticipated, and here he underscores that the anticipatory 'operates in the field of hope'. It is not only an emotion, but is purposeful and a directive act of cognition.

Bloch considered his focus on hope and the realm of the 'not-yet' would contribute to the reclamation of Marxism. In his introduction to *The Principle of Hope*, he described 'the architecture of hope' as standing on what humans see or understand in dreams and on a new Earth. He writes 'Becoming Happy was always what was sought after in the dreams of a better life, and only Marxism can initiate it. This provides fresh access to creative Marxism, even pedagogically and in terms of content, and from new premises, of a subjective and objective kind' (Bloch 1995: 17). Bloch suggests that the economic transformation or revolution that Marx envisioned could only be achieved through hope.

5 Hope in Diego Rivera's *Pan-American Unity*

Bloch understood that we are enmeshed in a vast field of markers and signs of the future and hope. Indeed, one of the most powerful dimensions of Bloch's *The Principle of Hope* is the dense and seamless fabric of examples he brings, not the least of which is from the arts. Consider how Bloch might have seen 'the principle of hope' in a single mural by Diego Rivera. Diego Rivera (1886-1957) first visited San Francisco with his wife Frida Kahlo (1907-1954) in 1930 or 1931. Rivera completed three commissioned frescoes; *Allegory of California* (1930-1931), located in the stairway of the City Club of San Francisco, a small mural titled *Still Life and Blossoming Almond Trees* (1931) originally in a private residence in Northern California, and *The Making of a Fresco Showing the Building of a City* (1931) in the student gallery of California School of Fine Art (later renamed The San Francisco Art Institute). These murals led to other commissions and murals in Detroit and New York.[6] As he was beginning the work on the first mural, he spoke at the Legion of Honor and urged American artists not to look towards Europe for their inspiration, but to the Americas to create and develop their own style of art and architecture, what he called 'the new American culture' (Castro 2022: 116). Claire F. Fox draws our attention to what Rivera understood as the true source of inspiration for American artists in a performative essay or script with many voices that Rivera published at the conclusion of his residency. One of them was the announcer for the 'Pan-American Continental Radio Corporation' who tells the listeners to reject European artistic traditions and themes and declare their aesthetic independence. He called to 'Inhabitants of America! … Your antiques are not to be found in Rome. They are to be found in Mexico. We have there a tradition of painting, of sculpture, of architecture … '. He dared American artists to have the courage to pursue their art forms independently of artistic styles, borrowed from Europe (Fox 2022: 224).

A decade later, when Rivera returned to California to complete a much more ambitious mural project, his ideas of 'Yankee imperialism' had changed as a result of the rise of Fascism in Europe, the Spanish Civil War, the Hitler-Stalin Pact of August 1939, and the Soviet invasion of Finland. In Spring 1940, Rivera agreed to contribute a north-south thematic mural for the Golden Gate International

Exposition. He wrote to Timothy Pflueger who had been instrumental in his first residency in San Francisco and who was now the architect designing the new library of San Francisco Junior College (now the City College of San Francisco). Rivera wrote, 'For years I have felt that the real art of the Americas mostly came as a result of the fusion of the machinism[7] and new creative power of the north with the tradition rooted in the soil of the south, the Toltecs, Tarascans, Mayas, Incas, etc., and would like to choose that as the subject of my mural'.[8] The subject matter of this new mural would draw its inspiration from the very same sources that Rivera had challenged American artists with a decade before.

While the organizers of the exposition titled his mural *Pan-American Unity*, Rivera called it *Unión de la Expresión Artistica del Norte y de Sur de este Continente* (*The Marriage of Artistic Expression of the North and the South on this Continent*). This title immediately suggests the collaboration of artistic traditions, rather than some unifying social, political or economic ideology of union. The mural would be painted in the Palace of Fine Arts on Treasure Island in the San Francisco Bay. The mural would be made up of fresco on plaster on five steel-framed cement panels (each panel was made up of two panels, the upper and lower registers), and would be 22 feet high and 74 feet wide when finished.[9] After the exposition, the mural would be relocated to the City College of San Francisco.[10]

Rivera described the first panel on the far left of the mural as 'The Creative Genius of the South Growing from Religious Fervor and a Native Talent for Plastic Expression'. Here, Rivera depicted Pre-Columbian Mesoamerica. In the upper register, he represents Tenochtitlan with its temples and the temple complex of Teotihuacan. Beyond them he painted the mountains that surrounded the ancient Mexico City. Below the Tenochtitlan temples he painted Quetzalcóatl teaching the Aztec ruling council. Yaqui deer dancers of Northern Mexico are portrayed, then a number of craftsmen and artists at work representing the many cultures of the Olmec, Maya, Toltec, Mixtec and Aztec peoples. At the opposite end of the mural, the fifth panel depicts, as he wrote:

> Just as the plastic tradition of the South penetrated into the North, the creative mechanical power of the North enriched life in the South, I depicted the greatness of the North in such engineering achievements as Shasta Dam, oil derricks ... and portraits of such geniuses as Ford, Morse, and Fulton, the last two of whom were artists as well as inventors. The creative force of the United States and the emancipation of women were symbolized by a woman artist, a woman architect, and a sculptress.
>
> (Rivera and Marsh 1960: 195)

In addition, he depicts gold mining techniques from panning for gold, the sluice box and then hydraulic mining. Henry Ford is depicted examining a fuel pump in front of the automobile assembly line.

The second and fourth panels depict events from the past and the present. In the second, he paints the city of San Francisco. In the lower register of the panel he depicts contemporary Mexican artists and craftsmen, and Rivera himself painting a portion of the mural that depicts the liberators and democrats of both the Southern and Northern hemispheres – among them Simón Bolivar, Miguel Hidalgo y Castilla, the father of Mexican independence, George Washington, Thomas Jefferson, Lincoln, and John Brown. On the left border of the panel, he depicts the peasant artist Mardonio Magaña (1868-1947) who Rivera called 'the greatest contemporary Mexican sculptor' carving the head of Quetzalcóatl protruding into the centre of the panel.[11] High above the city and the Bay Bridge he painted the National Diving Champion in 1939, Helen Crlenkovich (1921-1955) in one of her most extraordinary swan dives, over a bank of fog drifting in to cover the city. He described this as '[T]he conquest of time and space was symbolized by a woman diving and the Golden Gate Bridge spanning San Francisco Bay. A Quetzalcóatl figure personified the continuity of Mexico's ancient culture' (Rivera and Marsh 1960: 194). The Golden Gate Bridge was completed and opened in 1937, the Bay Bridge the year before, and here Rivera perhaps confuses the names of the two bridges.

The fourth panel is connected to the second by another depiction of Crlenkovich in her dive now high above Treasure Island where the exposition was held. In the lower register, the three dictators, Stalin with the bloodied ice pick used to assassinate Leon Trotsky, Hitler and Mussolini rising up in a poison-gas cloud and cinematic images of Chaplin's *The Great Dictator* (1940) that was screened at the exposition, and Edward G. Robinson's *Confessions of a Nazi Spy* (1939). Fox draws our attention to Rivera's description of the left of the panel from his *My Art, My Life*. She writes 'Rivera reveals that U.S. power will defeat the Nazis in the form of a bulging ... Caucasian-pink forearm and fist draped in a U.S. flag, which crushes in its grip a puny wrist emblazoned with a swastika tattoo ....the hand symbolizing [and quoting Rivera] "the American conscience reacting to the threat against freedom, in the love of which the history of Mexico and the United States were united"' (Fox 2022: 231–3). Fox also notes the imbalance between the left and right panels of the mural. In panels 1 and 2, he portrayed Toltec stone carvers and Mardonio Magaño carving the head of Quetzalcóatl, but she believes there is no comparable figure on the right. There are only two indigenous figures – a laborer operating a lathe and the racist image of a "cigar store Indian" being painted by a white artist. The absence of brown bodies on the right side of the mural evokes an unrepresented history of genocide and dispossession of Native peoples coterminous with U.S. westward expansion. On the left side of the mural, collectivity predominates over individuality, while the artist-inventors on the right are depicted posing with their own creations.

(Fox 2022: 233)

The viewer might try to understand the mural as a 'history' or the linear development of the Americas from the pre-Columbian period to the twentieth century, as if the mural has an historical *telos*. But this is clearly not what Rivera intended. The orientation of the mural is to the central panel, number 3,[12] that depicts the hybridization of the North and South.

In the background of the upper register are the intertwined images on the right of an auto-plant stamping machine that he had used in the Detroit Institute of Arts mural in 1933 and on the left two images of the Aztec Goddess Coatlicue. First, as the Mistress of Death and Earth with her outstretched right hand showing jade-stone calluses the result of the necessity and preciousness of working the soil. Second, and in the centre, as a human head, half skull and half covered with flesh, expressing the dualisms of death and life and past and present.[13] Rivera wrote that the genius of the South was its

> religious ardor and its gift of plastic expression and the genius of the North (the United States), its gift for mechanical expression. Symbolizing this union – and focal point of the whole composition – was the colossal Goddess of Life, half Indian, half machine. She would be to the American civilization of my vision what Quetzalcóatl, the great mother of Mexico, was to the Aztec people.
>
> (Rivera and Marsh 1960: 194)

In front of the machine-goddess, Rivera depicted Dudley C. Carter (1891-1993), a Canadian artist, who he met while working on *Pan-American Unity* and is working on a huge wooden sculpture of a Big Horn Mountain Ram that became the mascot of the City College. Carter and Rivera had become friends while working at the exhibition. Carter had become interested in the artwork of Native Americans on the northwest coast and sculpted using ancient tools like hand-axes. Rivera believed that Carter and his sculptures reflected the inspiration of the new American art and culture that Rivera saw and was the focal point of this mural. He also included a portrait of Frida Kahlo. 'I also painted a portrait of my wife Frida', he wrote, 'a Mexican artist of European extraction, looking to the Native traditions for her inspiration. Kahlo represented the vitality of these traditions in the South as [Dudley] Carter represented their penetration into the North' (Rivera and Marsh 1960: 194). Rivera's intention was then to re-centre or re-focus the viewer's attention on the centre panel. The North and the South were being integrated literally before the viewer's eyes in what is no less than an alchemical mural. Pan-American unity was being created as the viewer stood before the mural.

Many have noted that Rivera's politics, that had been so divisive and so confrontational, have been muted in the mural. Park understands that the key to what is a dramatic change in Rivera's work is his vision of Pan-Americanism that he now maps not on a vision of hemispheric tension and conflict but what Park calls 'a utopian version of San Francisco'. This 'local' understanding of Pan-Americanism and the multiple perspectives from which to contemplate it allows Rivera to offer a more usable template by which to understand hemispheric interactions rather than the unified and ideological visions of what was the mainstream version of Pan-Americanism (Park 2014, loc. 4346). Park drew from the work of Anthony W. Lee who argued that the new element in the mural was 'place' in which the exposition site itself becomes a space for a unifying vision, although the vision has two components, the dream from above of the benefactors and the practice from below where individuals inhabit and negotiate the fairgrounds (Lee 1999: 211). But, the new location of Pan-Americanism in the city of San Francisco and in the ambivalent view of the exhibition space, Rivera also asserted a new vision, a new hope, for a new *mestizaje*, not the new synthesis of the Indigenous people of Mexico and the Europeans who colonized Mexico, but the Goddess of Life in the centre panel, half Indian and half machine, half mechanical expression and half plastic or artistic expression. Diego Rivera's *Pan-American Unity* is a complete expression of Bloch's 'not-yet-conscious' and hope.

6 Conclusion: the Republic of Hope

Drew Gilpin Faust's *This Republic of Suffering: Death and the American Civil War* is arguably one of the most powerful studies demonstrating the power of ideas and symbols to make sense of extraordinary societal events, like the Civil War. She sets out the reasons why the Civil War matters to us in contemporary America – 'it ended slavery and helped to define the meanings of freedom, citizenship, and equality. It established a newly centralized nation-state and launched it on a trajectory of economic expansion and world influence'. But, the overpowering experience of those who lived through the Civil War was the presence of death. She took the title of her book from words written by Frederick Law Olmstead (1822-1903) who was a pioneer in American landscape architecture, a social reformer and made many contributions to public health. Indeed, he used the term 'republic of suffering' to describe the wounded and dying who arrived at Union hospital ships. As she wrote '[A] war about union, citizenship, freedom, and human dignity required the government attend to the needs of those who had died in its service ... Death created the modern American union – not just by ensuring national survival, but by shaping enduring national structures and commitments' (Faust 2008: xiii–xiv).

It is also possible to think about a 'republic of hope', a community that understands that hope is an essential element for a human society to flourish. Hope is always about the future. Hope is a call to bring into being something that is not yet present or is not complete. Hope cannot be quantified. Hope has no materiality. There is no index to measure it. And yet, it is impossible to conceive of human life without it. Hope is then a 'breeder-reactor' for the elements that compose human flourishing. If we think of ways of defining that which is essential to the human, we of course, know of *homo faber* or *homo ludens*. It is impossible to consider the human without hope, and thus the human as *homo spei* is as significant as what the human makes or the role of play and creativity in the human social world.

Hope is the human force that is intended to negate or overcome the present or the limitations of the present. In this negation, hope gives access to a transcendent order of meaning, a different scale of reality. Dr King's argument is that the arc of the moral universe bends towards justice through the intervention of hope. It is hope that moves those who will bend the arc. It is a way of seeing the world as drawn by Gabriel Marcel, a world in which hope is necessary for communion with others. Hope is also a way of ascribing order as in Viktor Frankl's logotherapy, an order of meaning that allows the individual to triumph over trauma. And lastly, hope is an expression of patterns, of relationships and of visualizations. Hope is a part of every utopian idea, system and pattern so that Bloch could argue hope lays at the very heart of civilization. Indeed, thinking about how a human society might flourish already is an example or a manifestation of Bloch's 'not-yet-conscious'. If we grant to Bloch's idea of visualizations, representations, like Diego Rivera's *Pan-American Unity*, we see that hope is also about the production and reproduction of culture.

Notes

1. On the fiftieth anniversary of the 1963 March on Washington (28 August 2013), President Barack Obama spoke from the steps of the Lincoln Memorial where Dr King had delivered his 'I Have a Dream' speech, and said, 'The March on Washington teaches us that we are not trapped by the mistakes of history, that we are the masters of our fate. The arc of the universe may bend toward justice, but it doesn't bend on its own'.
2. President Joe Biden is fond of quoting the Irish poet Seamus Heaney's *The Cure at Troy: A Version of Sophocles' Philoctetes* (1961), most recently when he spoke to the nation from the oval office in the White House about ending his presidential campaign 24 July 2024, on the relationship between hope and history: '*History says don't hope//On this side of the grave.//*But then, once in a lifetime//The longed-for tidal wave//Of justice can rise up,//And hope and history rhyme'. But this is very different from the formulation of Dr King. For Heaney, the rhyming of hope and history is completely arbitrary, maybe once in a lifetime. King's statement suggests that hope is the motivation for those who bend the arc of history.
3. On the theological and philosophical importance of *homo viator*, see the classic lecture by Gerhard B. Ladner (1967).
4. Capps wrote or edited three books on the School of Hope (Capps 1970; 1972; 1976).
5. Jack Zipes provides one of the very best short biographies of Bloch, and the role of artistic creativity in the realization of the 'not-yet-conscious' in his collection of Bloch's essays.
6. For discussion of the Detroit and New York murals, see Dickerman and Indyck-López (2011); Downs (1999); Rivera and Lozano (2008); Rosenthal (2015). One of the very best studies of Rivera's California murals is Sackman (1996).
7. Here Rivera may reflect an American artistic tradition documented in the exhibition and catalogue (Acker et al. 2018).
8. Cited in the City College of San Francisco, 'The Diego Rivera Mural Project': https://riveramural.org (accessed 9 October 2025). There are two other very important websites for the mural: https://www.sfmoma.org/exhibition/pan-american-unity (accessed 9 October 2025) and https://exhibits.stanford.edu/rivera/feature/the-mural-in-san-francisco</URI (accessed 9 October 2025).
9. For a high-resolution image of Diego Rivera's *Pan-American Unity*, 1940 visit: 'The Diego Rivera Mural Project': https://riveramural.org (accessed 9 October 2025).
10. The mural was moved to the San Francisco Museum of Modern Art and was exhibited from 28 June 2001 to 21 January 2024. There is a documentary film on how the mural was moved from this storage location to the San Francisco Museum

of Modern Art, 'Diego Rivera: Moving a Masterpiece', NBC Bay Area: https://www.youtube.com/watch?v=r6K2YTyN4jg (accessed 9 October 2025).
11 Cited in the City College of San Francisco, 'The Diego Rivera Mural Project': https://riveramural.org (accessed 9 October 2025).
12 For a high-resolution image of Diego Rivera's *Pan-American Unity*, 1940 (Plate 2) visit: 'The Diego Rivera Mural Project': https://riveramural.org (accessed 9 October 2025).
13 Cited in the City College of San Francisco, 'The Diego Rivera Mural Project'.

8

Folktales, Stories and Traditions – Conceptual Indigenous Wisdom and Knowledge for Human Flourishing

Wakanyi Hoffman

1 Introduction

At the heart of Indigenous thinking lies the continuous and evolving art of storytelling, where cultural wisdom is woven with the fine threads of tradition and community. This embodies the essence of human flourishing. To illustrate this, here is a version of a popular African folktale about the Hare and the Lion, which encapsulates the philosophy of ubuntu – a moral conception of collective human flourishing that depends on the cooperation of all beings for mutual success. In this version concerning a drought, the Tanzanian folktale 'The Hare and the Lion' demonstrates how anthropomorphized animals (a common element in African oral storytelling) collaborate to find a lasting solution to a severe drought. Each animal contributes labour to this communal effort, embodying the principles of cooperation and mutual responsibility. However, the hare abstains from participating in the collective work, choosing instead to exploit the community's resources without contributing. To address this breach of communal ethics, the animals devise a plan to confront the hare, ultimately leading to his punishment. This narrative highlights the values central to the ubuntu philosophy, emphasizing communal interdependence and the importance of contributing to the collective well-being. The tale serves as a moral lesson on the significance of collaboration and the social consequences of individualism within African Indigenous Knowledge Systems.

This tale serves as an introduction to this chapter, which explores folktales, stories and traditions that carry Indigenous wisdom and knowledge to promote human flourishing within the broader discourse of ecological flourishing. The argument invites readers to consider how knowledge systems and cultural narratives broaden a diverse understanding of community, resilience and the interconnectedness of all life. The argument further promotes using culturally inclusive narratives to transcend individual limitations, fostering a holistic vision of humanity where flourishing is a shared journey deeply rooted in the consciousness of humanity's collective existence.

This chapter explores how Indigenous cultures converge on the theme of interconnectedness, where all life forms – humans, plants, animals, land masses and the cosmos – exist as a unified, interdependent community in the same universe. By examining Indigenous folktales, stories and traditions, common themes and teachings regarding the origins of human flourishing emerge. Drawing from diverse Indigenous cultures, the richness of these narratives will be highlighted within the context of the shared Indigenous wisdom and values that inform them. It is contended that these narratives offer invaluable insights and guidance for navigating the complex challenges of our time and envisioning new paths towards a more just, sustainable and fulfilling future for all.

To this end, an expanded section will be dedicated to the principle of ubuntu, delving into how this African philosophy of 'human-beingness' embodies a collective human consciousness that supports human flourishing. Evidence will be drawn from academic and non-academic sources, with notable works from African philosophers such as J. S. Mbiti and Mogobe Ramose juxtaposed against contemporary thought on the concept of personhood, including ideas from popular culture that reflect the concept of ubuntu. Further exploration will delve into how Indigenous wisdom can be integrated into discussions surrounding sustainable development, contending that it offers fresh avenues towards a holistic and inclusive approach to the challenges of collective human flourishing. This section will examine the fundamental principles and values that underpin Indigenous worldviews, considering their implications for sustainable development. Additionally, it will highlight Indigenous-led initiatives and indicators that already contribute to realizing the Sustainable Development Goals (SDGs) while simultaneously challenging dominant development paradigms.

Finally, the chapter will explore the barriers and opportunities associated with incorporating Indigenous wisdom into the global sustainable development agenda, thus building a pathway for presenting recommendations tailored to policymakers, researchers and practitioners.

The chapter concludes with a call to celebrate the diversity of human expression and to recognize that within the tapestry of tales lies a roadmap to a shared humanity that continues to shape each life on Earth. Therefore, to truly flourish, stories matter; it is through storytelling that the oldest civilizations have persisted in harsh geographical conditions, relying not on material wealth but on the richness of strong social ties that emphasize the importance of living in harmony with nature and other life forms on Earth.

2 Storytelling as an Indigenous pathway for collective human flourishing

Storytelling in Indigenous cultures has long been a fundamental mechanism for preserving and transmitting knowledge. These narratives, embedded in folktales, songs and cultural rituals, are profound vessels that encapsulate Indigenous worldviews, values and epistemologies. Scholars such as Archibald (2008) and Kovach (2010) highlight the critical role of storytelling in maintaining the intricate tapestry of Indigenous cultural heritage, emphasizing that these stories serve functions beyond mere entertainment. Instead, they act as conduits for wisdom, ensuring the resilience and continuity of cultural identities through generations.

Central to many Indigenous narratives is the understanding of life as a web of interconnected relationships that supports harmonious existence with the natural environment, fellow beings and spiritual realms. This perspective is extensively explored by Kimmerer (2013), who articulates how Indigenous cultures prioritize a respectful and reciprocal relationship with all forms of life. These narratives foster a worldview in which human flourishing is inextricably linked to the well-being of the Earth and all its inhabitants.

As explored throughout this chapter, the convergence of Indigenous perspectives across diverse cultures significantly emphasizes the shared principle of interconnectedness. This presents a compelling case for the value of these narratives in contemporary society. These stories offer a lens through which we can view the complexities of human relationships with the environment, providing insights that are increasingly relevant to today's discussions on sustainability and equity.

However, it is essential to interpret these stories in light of their cultural specificity and the contexts from which they emerge. The insights derived from Indigenous narratives can provide guidelines for sustainable living, yet they must be respected as part of a broader cultural and philosophical framework that goes beyond simplistic interpretation.

Thus, understanding the role of storytelling in Indigenous cultures requires a commitment to openness to diverse ways of knowing or 'seeing through other eyes', as Andreotti (2008) suggests. This commitment will help one begin to comprehend wisdom-keeping as it relates to sustainable living. Engaging with these narratives allows exploration of pathways to human flourishing, offering a wide range of cultural expressions that unlock more possibilities for collective human flourishing.

Scholars and practitioners are encouraged to reflect on how these lessons can be thoughtfully integrated into contemporary discourse, recognizing the complexity and depth inherent in Indigenous knowledge systems. Thus, the journey towards understanding the role of storytelling in collective human flourishing continues, ever-inviting and ever-evolving. Yet, as articulated in a Native American philosophy explained in detail below, the source of life remains the same. Therefore, human flourishing depends on intercultural flourishing that acknowledges the interconnectedness of all life on Earth.

3 Life comes from it

3.1 Interconnectedness as an inter-Indigenous belief system

One of the central themes that emerges from Indigenous folktales and stories is the recognition of the interconnectedness and interdependence of all life governed by one natural law. This belief system is well captured in the Navajo Nation's conceptions of *beehaz'aanii,* or One Law, which extends to all beings, including human beings, non-human beings and all other entities in life. The concept of the One Law is expressed in the saying, 'Life comes from It', or 'Life comes from *beehaz'aanii*' (Yazzie 1994).

Numerous other Indigenous creation stories depict a world where all humans and non-humans are born from the same source and share a common essence (Cajete and Bear 2000; Vine Jr. 2001). For example, in the *Anishinaabe* creation story, the first human is lowered onto the back of a great turtle, which represents the Earth. The human is instructed to walk upon the land in a sacred manner, honouring the interconnectedness of all life (Benton-Banai 1988).

Similarly, amongst the *Mijikenda* community indigenous to the Indian Ocean coastal land in Kenya, there is a belief amongst the nine clans that make up one community that each clan is a part of the whole, without which the Mijikenda people would cease to exist. In his experience working and living with the Mijikenda people, posthumanist economist Will Ruddick (Ruddick and Peronne 2024) has established a primordial social and economic system of 'commitment pooling' still in practice within this community in which the human life is treated as an element of the larger element that is one life form and individually as active agents who can choose to contribute skills, knowledge and wisdom that keeps the systems flourishing as one life form from which all forms of life thrive. In an essay tracking Indigenous remediation protocols for healing and restoring the ecological networks that encompass the flourishing of human beings, he asserts that:

> Humanity's quest for healing and restoration is not confined to the boundaries of the individual body or mind. It extends to our society at large. What happens when we explore the interconnectedness of various remediation (healing) protocols? Perhaps there is a pathway to recovery and growth, drawing parallels between the regenerative processes of natural ecosystems, cognitive functions and social structures.

This understanding of a non-material conception of systemic interconnectedness is reflected in many Indigenous languages and concepts, such as *Mitakuye Oyasin* in Lakota, which means 'all my relations' and acknowledges the kinship between humans and all other beings (Little Bear 2000). Similarly, the Andean concept of *Pacha*, which refers to the interconnected realms of the cosmos, the Earth and the underworld, highlights the interdependence and reciprocity between all life forms (Stone 2024).

This recognition of interconnectedness has profound implications for how Indigenous peoples understand and practice sustainability. Rather than viewing the natural world as a resource to be exploited, Indigenous folktales and traditions emphasize the importance of maintaining harmonious and reciprocal relationships with the land, water, plants and animals. For example, in the *Tlingit* story of *The Boy Who Lived with the Seals*, a young boy is taken to live with the seals and learns the significance of respecting and honouring the beings that sustain human life (Dauenhauer and Dauenhauer 1987).

Similarly, in the Māori story of *Rata and the Tree,* a man who cuts down a sacred tree without permission is taught a lesson about the consequences of failing to respect the natural world (Taonui, 2006). Trees are also sacred among the *Kikuyu* people, one of the oldest Indigenous peoples in Kenya. They are used in rituals that reinforce the community's connection to the land, which makes cutting down a tree a ceremonial act requiring the blessings of *Ngai*, or God, the highest deity worshipped by the Kikuyu people, often under a fig tree and facing Mt Kenya, where the community is settled (Acquaviva 2019). Similarly, within the Swahili people who are indigenous to the East African Indian Ocean coast, poetry was a way of preserving ecological knowledge. In an extensively researched work tracing the interconnected nature of human beings and the environment, Graziella Acquaviva (2019) finds the following:

> In Euphrase Kezilahabi's poetry, he brings to the surface memories that are apparently "animistic" as in "Ngoma ya Kimya" ('The Dance of Silence'), "Lakini labda miti hii michache yakumbuka" ('But perhaps these few trees do remember;' Harrow 2015). In this poem, trees are used as symbolic images of continuity – the poet puts his trust in trees that incorporate the spirits of the ancestors and the wisdom of the elders – in contrast to images of change and destruction: "Kilichobaki ni uwanja uliokauka" ('What remained was a dry space;').

The poet's 'dreamlike' vision can be explained as a metaphorical manifestation of inner chaos and human alienation from the land, highlighting the relationship between human beings and their environment as central to promoting human flourishing. Thus, speaking of trees as persons gives a place to the psyche's propensity to personify and define what or who is felt as a necessary mode of understanding the world and being in it (Tidball 2014).

3.2 Common themes in Indigenous stories

The power of words and names is a common theme in Indigenous folktales and traditions that helps shape reality and bring things into being. In many Indigenous cultures, words are understood not simply as abstract representations of ideas but as living entities that have the power to create, heal and transform (Abram 1996). For example, in the *Diné* (Navajo) story of the 'Hero Twins', the twins use the power of words and naming to defeat the monsters that threaten their people (Zolbrod 1984).

This understanding of the power of words is reflected in many Indigenous naming practices, which often involve careful consideration of the meanings and associations of names and how they can influence a person's character and destiny. For example, in the *Anishinaabe* tradition, children are given names that reflect their unique gifts and potential and connect them to their ancestral lineages and responsibilities (Doerfler 2015). Similarly, the *Kikuyu* people of Kenya also traditionally nicknamed different generations of children based on their most common character traits, which show what a particular group was tasked to accomplish within their lifetime (Kenyatta 1938).

The power of words is also evident in many Indigenous storytelling practices, which use language, rhythm and repetition to create a sense of enchantment and convey essential teachings and values (Archibald 2008). For example, in the *Haudenosaunee* (Iroquois) story of the 'Sky Woman', the repetition of the phrase 'and she fell' creates a sense of inevitability and highlights the transformative power of the woman's descent to Earth (Mann 2000).

A second theme from Indigenous folktales and traditions is the importance of maintaining balance and reciprocity in all aspects of life. Many Indigenous cultures understand the world as a complex web of relationships and interactions in which every action has consequences, and every gift requires a reciprocal response (Cajete and Bear 2000). This understanding is reflected in concepts such as 'Ayni' in *Quechua*, which refers to the principle of reciprocity and energy exchange between humans and the natural world (Allen 2002).

The importance of balance and reciprocity is often illustrated in Indigenous folktales and stories through interactions between humans and other beings. For example, in the *Cree* story of 'Wîsahkecâhk and the Geese', a trickster figure learns the significance of sharing and reciprocity when he attempts to hoard all the geese for himself (Erdoes and Ortiz 1999). Similarly, in the *Hopi* story of 'Boy and the Eagle', a young boy discovers the importance of respecting the balance between humans and animals when he takes an eagle chick from its nest (Courlander 1971).

The importance of balance and reciprocity is also reflected in various Indigenous gift-giving and exchange practices, strengthening social bonds and maintaining the flow of energy and resources within a community. For instance, during the Potlatch ceremonies of the Northwest Coast peoples, gifts are given not only to redistribute wealth but also to acknowledge and honour the interconnectedness of all beings (Jonaitis 1991).

A third theme common in Indigenous folktales and traditions is the role of tricksters and transformers in shaping the world and imparting important lessons about life. Tricksters are often depicted as mischievous and unpredictable non-human beings, such as the hare mentioned in the introductory story of 'The Hare and the Lion.' They typically challenge the established order and expose the follies and weaknesses of humans (Hyde 1998). At the same time, they are also recognized as influential creators and transformers who can also use their wit and cunning to foster positive change (Archibald 2008). These tricksters are part of a web of relations that, from an Indigenous perspective, are essential for establishing bonds of interconnectedness between human beings and the cosmos, ensuring that balance and harmony are maintained in all aspects of this singular life from which all life originates.

Examples of tricksters and transformers appear in many Indigenous cultures, such as the Raven in Northwest Coast traditions, the Coyote in Great Plains and Southwest traditions, and *Anansi* the Spider in West African and Caribbean traditions (Erdoes and Ortiz 1999). These figures often play a central role in creating stories and cultural and spiritual transformation tales. For instance, in the *Haida* story 'Raven Steals the Light', Raven transforms into a pine needle and is swallowed by the chief's daughter, who keeps the sun, moon and stars in a box, enabling him to steal the light and bring it to the world (Reid and Bringhurst 1996).

The lessons taught by tricksters and transformers often highlight the importance of humility, adaptability and the ability to see beyond surface appearances. For instance, in the *Lakota* story of 'Iktomi and the Ducks', the trickster *Iktomi* is outsmarted by a group of ducks after he attempts to deceive them, imparting a lesson about the dangers of arrogance and deception. Likewise, in the Anishinaabe story of 'Nanabozho and the Maple Trees', the trickster *Nanabozho* learns the importance of moderation and respect for nature after trying to take all the maple syrup for himself (Johnston 1976).

In African mythology, trickster characters often emerge in anthropomorphic forms, embodying the essence of humanity and providing profound insights into the collective efforts required for human flourishing. These trickster figures, such as *Anansi* in West African folklore and Hare in Bantu-speaking traditions, challenge social norms and disrupt established hierarchies, reminding us that everyone plays a vital role in creating a thriving society.

Various scholars contend that trickster characters symbolize the transformative potential of human agency and the necessity for communal cooperation in achieving human flourishing. In her book, *African Philosophy: Myth and Reality*, Pauline Hountondji (1996) asserts that tricksters function as moral guides, teaching individuals the significance of reciprocity, mutual respect and collaboration. Similarly, in his seminal book, *African Philosophy Through Ubuntu*, Mogobe B. Ramose (1999) investigates how trickster figures embody ubuntu principles of communal 'personhood' by challenging oppressive systems and fostering inclusive, collective well-being. Through their narratives, trickster characters promote a mindset that embraces the interconnectedness of individual actions and acknowledges that the pursuit of human flourishing requires a collaborative effort, demanding the active participation and contributions of all members of society.

Further academic research underscores the importance of recognizing and valuing Indigenous knowledge systems. For instance, in his book, *Sacred Ecology*, Fikret Berkes (2012) illustrates how Indigenous knowledge contributes to ecosystem management and biodiversity conservation. Similarly, Robin Kimmerer's (2013) work in *Braiding Sweetgrass: Indigenous Wisdom, Scientific Knowledge, and the Teachings of Plants* highlights the ecological wisdom embedded in Indigenous stories and advocates for acknowledging Indigenous knowledge as a valuable resource for sustainability.

Thus, storytelling serves as a powerful tool for capturing and preserving Indigenous wisdom, which is essential for sustainable human flourishing and ecological sustainability. As elaborated, Indigenous narratives, folktales and oral traditions encapsulate profound insights into the reciprocal relationships between humans, nature and the spiritual realm. These stories transmit knowledge, wisdom and guidance across generations, embodying the collective experiences and values of Indigenous cultures. These references underscore the necessity of granting Indigenous wisdom and knowledge the same credibility as other forms of knowledge, recognizing its role in shaping sustainable practices and fostering a harmonious relationship between humans and the environment, as will be argued below.

3.3 Braiding Indigenous science into sustainable development practices

In light of escalating ecological crises and social inequalities, there is an increasing acknowledgement that prevailing development models fail to produce sustainable and equitable outcomes for everyone. The United Nations' 2030 Agenda for Sustainable Development, featuring its seventeen Sustainable Development Goals, seeks to confront these challenges by offering a global framework for action. Nonetheless, the SDGs have faced criticism for inadequately recognizing Indigenous peoples' knowledge, priorities and rights, and who have historically led sustainable development efforts in their own territories (Garnett et al. 2018; Yap and Watene 2019).

Indigenous peoples represent a rich diversity of cultures, languages and knowledge systems that have evolved over thousands of years through a close relationship with their ancestral lands and waters. Despite centuries of colonization, dispossession and marginalization, many Indigenous communities have preserved their unique worldviews and practices, which offer valuable insights and solutions for sustainable development (Berkes

2012; Mistry and Berardi 2016). As stated by Victoria Tauli-Corpuz, former UN Special Rapporteur on the Rights of Indigenous Peoples (UN Permanent Forum on Indigenous Issues 2018), 'Indigenous peoples have a very active role to play in finding solutions to the problems of climate change'. This statement acknowledges the importance of Indigenous scholars' contributions towards identifying sustainable practices that have been tried, tested and proven to promote human flourishing in harmony with the evolving planet.

As previously established, at the heart of Indigenous worldviews lies a profound recognition of the interconnectedness and interdependence of all living beings, including humans, animals, plants and the land itself. This holistic understanding is reflected in concepts such as *Sumak Kawsay* or *Buen Vivir* in the Andes, which emphasize living in harmony with nature and community; *Mino-Bimaadiziwin* or 'the good life' among the Anishinaabe people of North America, which involves balancing physical, emotional, mental and spiritual well-being; and Ubuntu in Southern Africa, which highlights the shared humanity and reciprocity that binds all people together (Gudynas 2011; LaDuke 2005; Museka and Madondo 2012).

These worldviews challenge the dominant development paradigm as unlimited economic growth, extractivism and individual accumulation, and instead prioritize values such as reciprocity, sufficiency, solidarity and respect for all life. As expressed by the *Kiowa* writer N. Scott Momaday (1976: 80):

> We Americans must come again to a moral comprehension of the earth and air. We must live according to the principle of a land ethic. The alternative is that we shall not live at all.

Such a land ethic, rooted in Indigenous knowledge and practices, has profound implications for how we approach sustainable development. Instead of viewing nature merely as a resource to be exploited for human benefit, Indigenous worldviews acknowledge the natural world's intrinsic value and agency, as well as humans' responsibility to maintain reciprocal relationships with the environment. This perspective is reflected in practices such as sustainable harvesting, rotational farming and ceremonial offerings to the land, which have allowed many Indigenous communities to thrive for generations without depleting their resource base (Berkes 2012; Kimmerer 2013).

Indigenous worldviews also highlight the significance of collective well-being and decision-making, contrasting with the individualistic and top-down approaches that frequently dominate development policy. Numerous Indigenous societies have established sophisticated governance systems rooted in principles of consensus, subsidiarity and intergenerational responsibility, ensuring that the needs and voices of all community members are acknowledged and respected. These systems are often governed by customary laws and protocols that have evolved over centuries of place-based learning and adaptation, prioritizing the long-term sustainability and resilience of the community over short-term gains (Granderson 2017).

Incorporation of these Indigenous principles and practices into the SDGs could reorient the global development agenda towards a more holistic, regenerative and life-sustaining approach. As argued by the Indigenous scholar Jeff Corntassel (2008), sustainable self-determination for Indigenous peoples involves 'reconstructing the cultural, political and economic conditions that support Indigenous nations' capacity to transform their visions of a good life into reality'. This is a counterargument to a widely held idea that promotes Indigenous knowledge as ancient and irrelevant for future development.

Around the world, Indigenous communities are already leading the way in implementing sustainable development initiatives grounded in their knowledge, values and priorities. These initiatives offer valuable models and lessons for the broader development community and demonstrate the power of Indigenous-led solutions to achieve the SDGs.

One example is the Indigenous Peoples' and Community Conserved Territories and Areas (ICCAs), which are lands and waters governed and managed by Indigenous peoples according to their customary laws and practices. ICCAs cover an estimated 22 per cent of the world's land surface and are home to 80 per cent of the world's biodiversity, making them crucial sites for the conservation and sustainable use of natural resources (ICCA Consortium 2021). In many cases, ICCAs prove to be more effective at protecting biodiversity and carbon stocks than state-managed protected areas, while also supporting Indigenous communities' livelihoods, cultural identities and rights (Garnett et al. 2018).

For instance, in the Bolivian Amazon, the *Tacana* people have established an Indigenous territory that spans two million hectares of rainforest, managed according to their traditional knowledge and governance systems. The Tacana have developed a variety of sustainable livelihood activities, including agroforestry, ecotourism and the harvesting of non-timber forest products, which generate income for the community while conserving the forest and its biodiversity. They have also created their own indicators for monitoring the health and sustainability of their territory, which encompass measures such as the abundance of key wildlife species, maintenance of sacred sites and cultural practices, and participation of youth in traditional knowledge transmission (Fernández-Llamazares et al. 2020).

Another example is the Māori-led *Whānau Ora* initiative in New Zealand, which aims to enhance the health and well-being of Māori families and communities through a holistic, culturally grounded approach. Whānau Ora recognizes that individual well-being is linked to the well-being of the larger family and community. Addressing social determinants such as education, housing and employment is essential for improving health outcomes. The initiative is guided by Māori values such as *whanaungatanga* (kinship and connectedness), *manaakitanga* (caring and hospitality), and *rangatiratanga* (self-determination and leadership), and it emphasizes the importance of building upon the strengths and aspirations of Māori families (Baker, Pipi and Cassidy 2015).

Whānau Ora has established its own set of outcome measures, grounded in Māori understandings of well-being and success. These measures include indicators such as the quality of relationships within families, the transmission of Māori language and culture to younger generations, the ability to access traditional foods and medicines, and the level of participation in Māori community events and decision-making processes. By emphasizing Māori knowledge and priorities, Whānau Ora presents a strong model for Indigenous-led development that aligns with and promotes the SDGs, particularly those related to health, education and social inclusion.

Another groundbreaking Indigenous-led initiative is Grassroots Economics, a pioneering organization based in Kilifi, Kenya, the homeland of the Mijikenda Indigenous people of coastal Kenya. This project showcases the potential of Indigenous-led initiatives to promote sustainable development by integrating local knowledge, values and priorities. By concentrating on community-driven economic systems, such as implementing Community Inclusion Currencies, Grassroots Economics empowers local communities to build economic resilience while honouring Indigenous wisdom. These currencies serve as a medium of exchange that supports regional trade, encourages the circulation of goods and services, and strengthens economic stability in marginalized areas. By grounding its operations in the principles of ubuntu, Grassroots Economics establishes an inclusive platform where monetary transactions are not merely about exchange but about reinforcing communal bonds and improving collective well-being. This approach aligns with the principles of Indigenous knowledge systems, which emphasize interconnectedness and shared prosperity.

In *Designs for the Pluriverse*, Escobar (2018b) critiques conventional development models rooted in Western, individualistic and market-driven ideologies. He advocates for alternative approaches grounded in relational ontologies, emphasizing the interconnectedness of communities, nature and knowledge systems. These perspectives align with Indigenous worldviews that prioritize collective well-being and harmonious relationships with the environment. Such frameworks often favour holistic and relational approaches to development, which sharply contrast with conventional paradigms emphasizing individualistic economic growth. Grassroots Economics illustrates how culturally sensitive and locally contextualized initiatives can significantly contribute to achieving several SDGs, including poverty alleviation (SDG 1), decent work and economic growth (SDG 8) and reduced inequalities (SDG 10).

The success of Grassroots Economics underscores the power of Indigenous-led solutions in shaping effective and sustainable development models. As a testament to the efficacy of such initiatives, the organization demonstrates how grassroots movements can effectively leverage local knowledge systems to address systemic challenges and build resilient, adaptive communities. This, in turn, provides valuable insights for the broader development community, suggesting that integrating Indigenous perspectives can lead to more sustainable and inclusive forms of global development. Meanwhile, Fernández–Llamazares et al. (2020) highlight how such initiatives not only harness the inherent strengths of community-based approaches to human flourishing but also enhance social cohesion by aligning with cultural practices and collective values, thus offering an innovative paradigm for sustainable development.

From a sustainability perspective, Indigenous wisdom presents a crucial pathway for reimagining and transforming the sustainable development agenda into a more holistic, equitable and life-affirming future. By learning from and uplifting Indigenous worldviews, knowledge systems and practices, we can transcend the limitations of dominant development paradigms and move towards a vision of shared prosperity and abundance for all. This necessitates a fundamental shift in collective cultural values, relationships and ways of knowing and being in the world. It demands a recognition of interdependence among species and with the living Earth, as well as a commitment to creating conditions for all life to thrive. As expressed by the Anishinaabe scholar Deborah McGregor (2018):

> Sustainability is not just about the environment or the economy, it is about our relationships with each other, the natural world, the spiritual realm, our ancestors and our descendants. It is about the ways in which we come to know and relate to the world around us, and the ethics and values that guide our actions.

This sentiment is broadly reflected in African ubuntu philosophy. This concept suggests that human agency and development are deeply interconnected with the lives of others. Therefore, it encompasses relationships with other life forms for the ongoing collective flourishing of both human and non-human life.

The following section will examine how the concept of ubuntu embodies an ontological African consciousness that underpins human awareness and fosters human flourishing. This understanding of the interconnectedness and shared humanity of all people is deeply rooted in African folktales, proverbs and traditions, which serve as vehicles for transmitting the wisdom and values of ubuntu across generations.

4 Ubuntu: an ontological African consciousness for sustainable human flourishing

Ubuntu is a fundamental concept in African philosophy that encapsulates interconnectedness, shared humanity and reciprocity, forming the foundation of human consciousness and flourishing. The term 'ubuntu' is derived from the Nguni phrase 'umuntu ngumuntu ngabantu', which translates to 'a person is a person through other persons'. This concept is deeply rooted in African folktales, proverbs and traditions, which serve as vehicles for transmitting the wisdom and values of ubuntu across generations (Gyekye 1997). Ubuntu is not merely a social philosophy or ethical framework but a fundamental understanding of the nature of reality and the place of humans within it. As the Kenyan scholar John Mbiti (2010) famously stated, 'I am because we are, and since we are, therefore I am'. This statement encapsulates the ontological understanding of ubuntu, asserting that human existence is inherently relational and individual identity is inextricably linked to the community.

This understanding is reflected in many African folktales and proverbs that emphasize social harmony, collective responsibility and mutual aid. For example, the *Akan* proverb 'Onipa nua ne onipa' ('A human being's brother is another human being') highlights the shared humanity that binds all people together (Gyekye 1996). Similarly, the *Igbo* proverb 'Igwe bụ ike' ('There is strength in community') underscores the power and resilience derived from social solidarity (Achebe 1958).

The ontological nature of ubuntu is also evident in African conceptions of personhood, which differ markedly from Western individualistic notions. In many African cultures, personhood is not an inherent quality of individuals but rather something acquired through social relationships and participation in community life (Menkiti 1984). As the Ghanaian philosopher Kwasi Wiredu explains (1996): 'A person is not just a certain biological entity with a certain psycho-physical endowment, but, more essentially, a being with a certain status in a network of social relationships'.

This understanding of personhood has profound implications for conceptualizing human flourishing and the good life. From an ubuntu perspective, human flourishing is not a matter of individual achievement or self-actualization but rather a collective endeavour that involves cultivating harmonious relationships and fulfilling one's social responsibilities (Metz 2011). As the South African philosopher Mogobe Ramose (1999) puts it, 'To be a human being is to affirm one's humanity by recognizing the humanity of others and, on that basis, establish humane relations with them'.

4.1 Ubuntu and human flourishing

The concept of ubuntu is reflected in many African folktales and traditions that emphasize the importance of generosity, compassion and mutual aid in promoting community well-being. For example, in one of the Zulu folktales about 'The Hare and the Lion', one version tells the story of a hare who saves a lion from a trap and is later rewarded for his kindness when the lion shares his food with the hare during a time of scarcity (Canonici 1996). This tale highlights the reciprocal nature of ubuntu, in which acts of kindness and generosity are viewed as investments in the collective good. Similarly, the *Yoruba* tradition of 'eniyan l'aso mi' ('my garment is human beings') emphasizes the significance of social relationships and mutual support in fostering human flourishing (Akiwowo 1983). This tradition asserts that one's true wealth and success are measured not by material possessions but by the strength and quality of one's relationships with others.

The ubuntu understanding of human flourishing also emphasizes the importance of moral character and cultivating virtues such as empathy, forgiveness and reconciliation. As the late South African philosopher Desmond Tutu explains:

> A person with ubuntu is open and available to others, affirming of others, does not feel threatened that others are able and good, for he or she has a proper self-assurance that comes from knowing that he or she belongs in a greater whole and is diminished when others are humiliated or diminished when others are tortured or oppressed, or treated as if they were less than who they are.
>
> (Tutu 1999: 31)

This emphasis on moral character and virtues is evident in many African folktales and proverbs that serve as moral guides for individuals and communities. For example, the Akan proverb 'Woforo dua pa a, na yepia wo' ('When you climb a good tree, you are given a push') highlights the importance of virtuous behaviour and the support that comes from adhering to moral values (Gyekye 1996: 71). Similarly, the Igbo proverb 'Onye kpọ ọbara, ọbara enyeghị ya aka' ('He who invites trouble will not be helped by others') cautions against anti-social behaviour and the repercussions of failing to fulfil one's social responsibilities (Achebe 1958).

From an ubuntu perspective, human flourishing is a collective endeavour that involves cultivating harmonious relationships, fulfilling social responsibilities, and developing moral character and virtues. This understanding challenges individualistic and materialistic views of the good life, offering a more holistic and relational vision of human well-being. As the Ghanaian philosopher Kwame Gyekye argues (1997: 38), 'The African conception of personhood is grounded in the belief that the human being is an inherently communal being embedded in a context of social relationships and interdependence'. This conception of personhood has important implications for understanding and promoting human flourishing in Africa and beyond.

Thus, ubuntu contrasts with the dominant Western philosophical foundations that have shaped the frameworks of our contemporary global society, offering a distinct perspective on the nature of existence and human interconnectedness. Descartes' influential assertion, 'I think, therefore I am', exemplifies the focal point from which this divergence arises.

In a light-hearted vein, one might jest that Descartes inadvertently veered the discourse away from the act of thinking towards a preoccupation with individual existence. By fixating on the ruminations of the self, Descartes failed to fully acknowledge the multifaceted requirements of being, which underpin the ubuntu mindset. Ubuntu philosophy is predicated upon the notion that genuine being necessitates a profound engagement with the interconnected web of interpersonal relationships. The maxim 'I am because we are' elucidates the essence of ubuntu, signifying that an individual's identity is intrinsically intertwined with the collective tapestry of humanity. The very notion of humanity derives its significance from its inherently pluralistic and collective nature. Consequently, an individual can only ascertain their humanness through the presence of a broader humanity.

By embracing the wisdom and values of ubuntu, we can cultivate a more inclusive, compassionate and sustainable vision of human well-being that acknowledges our shared humanity and responsibility to one another. As th e South African proverb states, 'Umuntu ngumuntu ngabantu' ('A person is a person through other persons'). This captures the essence of ubuntu, serving as a powerful reminder of the interconnectedness and interdependence at the core of human experience.

By embracing a relational understanding of sustainability and centering the leadership and wisdom of Indigenous peoples worldwide, who share similar worldviews despite occupying different terrains, we can chart a new course for global development that honours the interconnectedness of all life and the dignity and potential of every human being. This is the transformative vision of ubuntu's principles and values, which also requires intellectual recognition and understanding if we are to create a more just, sustainable and flourishing world for all.

5 Conclusion

Indigenous wisdom and knowledge in folktales, stories and traditions offer a richness of insights and teachings about the nature of reality, the meaning of the good life and the origins of human flourishing. These narratives reflect a profound understanding of the interconnectedness of all life, the power of words and names, the significance of balance and reciprocity, and the transformative roles of tricksters and other beings in shaping the world and imparting essential lessons.

At their core, these narratives highlight a fundamentally different approach to understanding and relating to the world compared to what predominates in modern Western societies. Instead of viewing humans as separate from and superior to the natural world, Indigenous folktales and traditions stress the interdependence and kinship among all life forms. Rather than emphasizing individualism and competition, they focus on cooperation, reciprocity and the collective good. Additionally, instead of aiming to control and exploit the world, they underscore the significance of living in harmony with the rhythms and cycles of nature.

These teachings provide deeper implications for how we understand and practice sustainability amid the complex challenges of our time. They suggest that true sustainability entails not just reducing our ecological footprint or finding more efficient ways to use resources but fundamentally transforming our relationships with the natural world and with one another. This requires a shift from a mindset of domination and exploitation to one of respect, reciprocity and reverence for the sacred in all things.

One of the critical lessons that emerges from Indigenous folktales and traditions is the importance of deeply listening to the world around us and learning from the wisdom of other beings. Many Indigenous stories feature animals, plants and other non-human beings as teachers and guides, reminding us that we have much to learn from the natural world (Kimmerer 2013). By paying attention to the messages and lessons embedded in the land, water, sky, and the beings that inhabit them, we can cultivate a deeper sense of connection and belonging, discovering our unique gifts and roles in the web of life.

Another critical lesson emphasizes the importance of honouring the power of stories and the role of storytelling in transmitting knowledge, values and wisdom across generations. Indigenous folktales and traditions are not merely entertainment or curiosities but a vital part of the cultural fabric that binds communities together and provides a sense of identity, purpose and resilience (Archibald 2008). By engaging with these stories and the teachings they encompass, we can reconnect with deeper sources of meaning and purpose in our lives, finding inspiration and guidance for navigating the challenges of our time.

As the story of the Hare and the Lion illustrates, Indigenous folktales, stories and traditions offer more than mere wisdom; they provide guiding frameworks for human flourishing that are rooted in cooperation, resilience and the power of community. Much like the animals' recognition that collective strength and mutual support are the only ways to survive harsh droughts, Indigenous narratives emphasize that true well-being arises not from individual dominance or strength alone but through a shared commitment to reciprocity, balance and unity. In their relational worldview, flourishing is an endeavour in which the strengths of each member, whether human, animal or natural element, contribute to the whole, echoing the core principles of ubuntu – 'I am because we are'.

This story underscores a worldview that stands in stark contrast to the individualistic focus of many modern societies. Rather than promoting individual prowess as the hare once did, these Indigenous narratives invite us to rethink our relationships with the Earth and each other as interdependent and sacred, where kinship extends to all forms of life. This worldview sharply contrasts with dominant paradigms in Western societies, which prioritize competition, resource extraction and the accumulation of power. Instead, Indigenous traditions, much like the animals' collective approach, provide a model of cooperative survival that values mutual support and harmonious existence with nature. In this context, sustainable development becomes a matter of conserving resources and fostering a deeper transformation towards relational and ecologically rooted approaches to life (Kimmerer 2013; McGregor 2018).

The story also teaches us that thriving in adversity requires wisdom that modernity often overlooks. By heeding the Lion's call for unity, the savannah's animals could harness the collective strength necessary for survival – a lesson reflected in Indigenous knowledge systems, where animals, plants and even natural forces are revered as teachers and guides. Like the Hare and Lion story, these tales invite us to listen to the land and its inhabitants with an openness to their wisdom. The significance of storytelling here is transformative, reminding us that narratives serve not only to entertain and connect but also to transmit values and unite communities through shared purpose and identity (Archibald, 2008).

Yet, as the challenges faced by Indigenous communities today remind us, realizing this vision requires addressing the ongoing marginalization and dispossession that Indigenous peoples experience worldwide. The loss of their lands, exclusion from policy discussions and the undervaluation of Indigenous knowledge present obstacles similar to the drought that parched the savannah – struggles that undermine collective resilience. Overcoming these obstacles necessitates an inclusive shift in power dynamics, enabling Indigenous voices to lead decisions impacting their communities and lands while honouring their rights as caretakers of invaluable ecological and cultural wisdom.

Despite these barriers, there is also the promise of meaningful collaboration. Just as the creatures of the savannah united under the Lion's vision, Indigenous-led initiatives in sustainable development – such as traditional ecosystem management practices like fire stewardship, agroforestry and community-based conservation – offer practical solutions aligned with global sustainability goals. Indigenous languages and cultural heritage are also indispensable in protecting biocultural diversity, echoing the importance of ubuntu as a guiding principle for resilience and ecological stewardship.

Thus, when considered as a pedagogical approach to understanding the interconnection between humans and the Earth, the story of the Hare and the Lion emphasizes that human flourishing is deeply intertwined with our relationships with each other and the natural world. Just like the creatures of the savannah that found strength in unity, Indigenous narratives encourage us to reconnect with our kinship with all living beings and to recognize our responsibilities as stewards. These traditions remind us that flourishing cannot be achieved through individual ambition or wealth, but rather through fostering respectful, reciprocal relationships with the Earth. LaDuke (2005) states, 'We have forgotten our relatives and our responsibilities', and through this forgetfulness, we have weakened the web that sustains us all.

The vision of the animals in the savannah encourages us to recognize humanity's interconnectedness, honour our shared responsibilities and discover unique roles within the collective flourishing of life. By absorbing the wisdom embedded in these Indigenous stories, we are reminded that a sustainable and just future requires us to suspend our individual pursuits of wealth, which, after all, depend on others' collective skills and efforts. We must embrace the spirit of collaboration and mutual support that has sustained the savannah, learning that our strength lies not in isolated power but in the unity that comes from understanding, *I am because of who we all are.*

Confucian Language of Flourishing: Reading the Classic Daxue

Victoria Sukhomlinova

1 Introduction

The Book of Odes says, 'The imperial domain of a thousand li is where the people stay [zhi]'. The Book of Odes also says, 'The twittering yellow bird rests [zhi] on a thickly wooded mount'. And Confucius said, 'When the bird rests, it knows where to rest. Should a human being be unequal to a bird?'

(Chan 1963: 88, Daxue, chapter three)

Imagine a society where everyone practices self-cultivation, shows proper behaviour and enjoys good governance. Where everyone knows just exactly what to do, at what moment and under what circumstances. In such a society, human flourishing manifests as a harmonious interplay of individual growth and collective well-being, where each person's actions ripple outwards, nurturing their own development while simultaneously uplifting the community. According to Fei Xiaotong, a pioneering Chinese anthropologist, this is exactly how traditional Chinese society functioned – shaped by Confucian principles (Fei 2017: 45).[1]

But let us take a step back. A significant body of work has emerged to expose and challenge epistemic extractivism (Alcoff 2022; Grosfoguel 2011). In that spirit, I would argue that if we aim to genuinely study the Confucian understanding of flourishing – without romanticizing or flattening it – we will not gain any fresh insights by simply extracting yet another Confucian 'approach' to flourishing, self-cultivation, governance and so on.

But what exactly is sought in non-Western approaches? The answer may lie in things that are not necessarily Western in origin, but that have come to exist primarily within the dominant Western worldview – one that makes sense of the world by placing something at the centre of the argument, proceeding from it, and circling back to it, treating it as the main character. Works in decolonial studies, such as Gayatri Spivak's (1988), and research in non-Western and comparative philosophy (Hall and Ames 1987) have highlighted this point, arguing that Western epistemologies, overall, are less receptive to notions of interrelation and processuality than their non-Western counterparts.

From this perspective, the urge to explore non-Western approaches to flourishing may, at its core, remain deeply Western-centric – if not subtly appropriative. Instead, I propose taking a different path, a longer one. Rather than seeking a Confucian approach to flourishing, I suggest that we explore the Confucian worldview, manifested in its 'language of flourishing'. Scholars of non-Western philosophical methodologies have been working in this direction: for instance, similar to what I refer to as 'language,' Jana Rošker calls a 'framework of reference' (Rošker 2021: 9–10). While the notion of approach works with what is already clearly thematized within a field, the notion of language moves beyond straightforward thematization. This way, flourishing remains unnamed, undefined, not thematized, yet still perceptible.

The starting point to learn the Confucian language of flourishing would be to choose a classical manuscript for thinking with and through. After all, the study of primary text sources is the major knowledge practice within Confucianism (Elman 1984; Makeham 2012). Two Chinese terms – and more accurate ones – for what we commonly call 'Confucianism' are *ruxue* (the scholarly tradition) and *jingxue* (the study of the classics). Confucianism *is* the reading and the re-reading of classical texts, as well as the recording of the history of these readings and re-readings.

The first classic that a Confucian from pre-modern China would start studying would be a text called *Daxue*, or 'The Great Learning'. It is a part – and the first one – of the Four Books (*si shu*) of Confucianism. These Four Books were texts selected during the Song dynasty (960–1279) by the philosopher Zhu Xi to serve as the foundation of Confucian thought, and since then have been studied in a particular order: *Daxue*, *Zhongyong* (The Doctrine of the Mean), *Lunyu* (The Analects), and *Mengzi* (The Mencius). Reading and re-reading them – usually to the point of full memorization – was and still is essential to developing a Confucian worldview (Shen 2014).

I suggest following Zhu Xi's idea and focusing on *Daxue* as an entrance point both into the Confucian worldview and its language of flourishing. *Daxue* – with its structure and its effect on the reader – gives a perfect idea of what a Confucian text is. It also happens to be the text that addresses the matter of flourishing, which it describes as an 'abiding within the highest possible good'.

This idea, however, would be difficult to grasp without attending to the text's structure and its effect – that is, without seeing it as an instance of Confucian language.

Finally, *Daxue* is not just the Confucian text that teaches us about flourishing – it's also, to take Donna Haraway's expression, 'the people of this text' (Haraway 2016: 16–20)[2] and how they flourished – often in non-innocent ways – with and through it. I will therefore not only explain how the text conveys its core message, I will also shed light on the history of *Daxue*, of which I consider myself a humble part.

2 Daxue – a text

Daxue consists of 1,750 characters, which is not much – it takes only seventeen minutes to recite at a heartbeat pace. It is one of the shortest Chinese classics, a 'little Classic' in the words of Wing-Tsit Chan (Chan 1963: 84).[3] It contains eleven sections – the 'canonic core' (*jing*), likely written by Confucius' disciple Zengzi (505–436 BCE?), and ten 'expansion chapters' (*zhuan*).[4] These chapters, written by later generations of Confucian thinkers, try to deepen and clarify each point in the canonic core.[5]

The text of the canonic core starts with the following sixteen characters (Chan 1963: 86):

> The Way to the Great Leaning is through the manifestation of one's clear character (*ming ming de*), the love for the people (*qin min*), and the abiding within the highest good (*zhi yu zhi shan*).

Seemingly ignoring the first two, the canonic core then proceeds to explain why it is important to abide/to remain within the highest good (Chan 1963: 86):

> Only after knowing what to abide in can one be calm. Only after having been calm can one be tranquil. Only after having achieved tranquillity can one have peaceful repose. Only after having peaceful repose can one begin to deliberate. Only after deliberation can the end be attained. Things have their roots and branches. Affairs have their beginnings and their ends. To know what is first and what is last will lead one near the Way.

Translating *zhi* as 'abiding/remaining within' is central to this interpretation. Character *zhi* 至 signifies a natural limit. The phrase 'abiding within the highest good', therefore, has this meaning of remaining within one's locality, however wide the knowledge of the 'roots and branches' is. The story about a bird, told by

Confucius and recounted in the epigraph here, further explains this point: can a human being be like a bird which is only aware of its immediate surroundings and proceeds from it?

Daxue shows how goodness is a journey where a human being cannot relocate themselves to a branch when they are at the root, or vice versa. It emphasizes that the form of goodness they practice right now might not be the highest in an absolute sense, but it is the highest that is possible within a particular context.

Then follow the actual eight steps, each corresponding to a form of the 'highest possible' goodness. These forms progress from the most internal to the most external. The steps are: from the 'investigation of things' *ge wu* to the 'extension of knowledge' *zhi zhi*, then – to the 'sincerity of the will' *cheng yi*, then – to the 'rectification of the mind' *zheng xin*, then – to the 'cultivation of the personal life' *xiu shen*, 'regulation of the family' *qi jia*, 'ordering the nation' *zhi guo*, and 'bringing peace to the world' *ping tianxia*. If a person manages to figure out the *investigation of things* and transitions to the *extension of knowledge*, this person still operates on the deepest internal level. By contrast, if they start to see how *ordering the nation* brings about the *world peace* and consciously contribute to it, they operate on the highest external level.

The canonic core finishes with reiterating that the manifestation of goodness can only flow in one direction – from the most internal/present/intimate to the most external/distant/public (Chan 1963: 87):

> From the Son of Heaven down to the common people, all must regard cultivation of the personal life as the root or foundation. There is never a case when the root is in disorder and yet the branches are in order. There has never been a case when what is treated with great importance becomes a matter of slight importance or what is treated with slight importance becomes a matter of great importance.

The internal and the external, the fundamental and the secondary, the first and the last, must be clearly distinguished. Only this way, three general expressions of goodness are followed. According to Wing-Tsit Chan, this concept of proper arrangement and proper transitioning is one of the few in the Confucian corpus whose meaning has never been disputed by later Confucian philosophers (Chan 1963: 85). The exact meaning of concepts like 'investigation of things' or 'sincerity of the will' has often been debated. Interpretations of *ge wu*, for instance, have appeared consistently over the past half-century – among them D. C. Lau's 1967 study on Ki-Wu (Lau 1967) and Wang Huaiyu's (2007) research. However, what was never questioned is the need to transition from one concept-stage to the other, and to do so in a specific rhythm and order. At every point, a human being should know where to abide.

3 Daxue – a history

Daxue has a history far beyond the Four Books (Johnston and Wang 2012: 1–19). Before Zhu Xi made it an independent manuscript in the eleventh century CE, *Daxue* served one of the chapters of *Liji* (Book of Rites) – maybe in another form and without citations of Confucius, but still recognizable. The Book of Rites is an earlier Chinese classic that existed long before Confucius's lifetime and was later compiled and edited by him and his disciples in a certain manner (Chik 2021). This Book describes the social norms, rituals and ceremonies that shaped proper conduct and moral order in ancient Chinese society. For instance, the chapters of *Liji* neighbouring to *Daxue* addressed concrete ceremonial prescriptions, such as how to enact a ceremony of capping or how to organize mourning.

When a text is part of a ceremonial corpus, its form and its enactment are as important as its content. If we frame *Daxue* that way, we can see that the purpose of reading and chanting it helps set the pace of thought and reach calm transitions from one concept to another, visualizing the transitions apart from the concepts themselves: from the 'regulation of the family' we move to 'ordering the nation', from 'ordering the nation', we move to 'bringing peace to the world'. Or, when 'bringing peace to the world', there is no chance of not simultaneously aiming at the 'regulation of the family', whatever that would mean. It is an intellectual exercise which takes the focus off the exact meanings of concepts. It feels as though the reader is invited not so much to precisely understand, but simply to *carry it along*.[6]

Studies on ancient Chinese manuscripts have shown that textual variants are not merely errors or corruptions, but often reflect deliberate adaptations that contribute to a text's evolving identity over time (Richter 2013; Shaughnessy 2006). In this sense, Zhu Xi's compilation of the Four Books reshaped how each was read and practiced.

During the cultural essentialization under the Song dynasty (960–1279) and the territorial expansion and imperial build-up under the Yuan, Ming and Qing dynasties (1279–1911), the rulers of China were tempted to appropriate the Four Books and to employ parts of them whenever it justified their – mainly reactionist – policies (Elman 2013: 14–15). And it is easier to reinvent a term, concept, norm or any other non-material cultural artifact when it is disentangled from its original context and history and subjected to focused analysis. It has been proved that Zhu Xi practiced this approach with parts of *Lunyu* (Perelomov 1998: 110),[7] and the same happened to parts of *Daxue*.

As I mentioned earlier, Confucian scholars have long debated – and still do – the precise meaning of certain concepts in *Daxue*, the 'glaring gaps' in the text (Plaks 2014: 143). In the case of *ge wu* ('investigation of things') one could ask: what does it mean – to investigate things? What is its practical blueprint? The corresponding expansion chapter of *Daxue* (chapter five), i.e. the text itself, explained it very poorly. It simply put *ge wu* and its neighbouring concept 'extension of knowledge' *zhi zhi* together in the heading and backed it with one single explanatory line: 'It means to understand the fundamental core'.

How mesmerizing it must have felt for the Confucian imperial advisors to project one's personal history and web of contexts – or one's policy advice to the rulers of the empire – onto this concept and explicate it accordingly. But how vastly it must have differed from the pre-imperial practice of *carrying ge wu along*.

Under the Ming and Qing dynasties (1368–1912), knowing *Daxue* and other classics by heart and being able to translate them into policy decisions was the basic job requirement for a public servant. This knowledge and ability were subject to imperial civil examinations – a rigorous merit-based system used from 1371 to 1905 to select government officials. Four Books, Five Classics, iconical Chinese poetry of Tang and Song dynasties, and further canons and poems were to be memorized through repetitious recitation (Blitstein 2021).

Therefore, *Daxue* would most certainly be the first thing that a Chinese in 1371–1905 China, who aspired to a statesman's career, would lay eyes upon when they started preparing for these examinations. While technically the memorization through repetitious recitation resembled chanting, the spirit of the exercise changed completely. I suggest that, on the normative level, the foundational arrow of *Daxue* 'internal first, external next' no longer existed. The *ordering of nation* came as the first priority to those who embarked on the path of Confucian scholarship.

To aid the studies and to ensure the cultural unification, the authorities installed Confucian symbols everywhere: from calligraphy pieces to musical instruments with inscriptions of classical poems on them, not to mention the system of Confucian temples built over the empire. The text of the 'canonical core' of *Daxue* was, for instance, inscribed in elegant calligraphy on the wall of *minglun tang* (The Hall of Illustrious Talks) at the Confucius Temple in Tainan City. The temple was built in 1665, so the Hall served as the main venue for intellectual talks among the Confucian literati in Taiwan for more than three hundred years.

Local civil examinations were held every two years, provincial and nationwide – once in three years. The competition was insane – only five percent of the examinees could eventually succeed in becoming officials.

Yet are we certain that only those who stubbornly set their minds on taking the imperial exams every two years – and followed through – were the ones who entered The Hall of Illustrious Talks in Tainan and similar spaces across the empire? Who were the others? Could the inscriptions on the walls, along with the broader study of the texts, have undermined the dynasty's cultural monopoly over the classics in unexpected ways?

> One of the unintended consequences of the civil examinations was the creation of legions of classically literate men (and women), who used their linguistic talents for a variety of nonofficial purposes, from physicians to pettifoggers, ritual specialists, and lineage agents; from fiction writers, playwrights, printers, and bookmen to examination essay teachers; from girls, maids, and courtesans competing for spouses and patrons to mothers educating their sons.
>
> (Elman 2013: 5)

A Neo-Confucian thinker Wang Yangming (1472–1529) initially aspired to succeed in the imperial civil examinations but faced several failures. His avid and receptive mind, however, no doubt along with his resentfulness towards the imperial bureaucracy, led him to famously reframing Confucianism. Wang was annoyed by having to apply Zhu Xi's 'empirical' method of Confucian inquiry: it was the 'extension of knowledge' *zhi zhi* preceding the 'sincerity of the will' *cheng yi* that struck Wang as inappropriate. So, he rearranged respective passages in *Daxue*, built on that basis a few powerful Confucian concepts of his own, such as *liang zhi* ('moral knowledge' / 'intuitive comprehension') and *zhi xing he yi* ('the unity of knowledge and action'), attracted disciples into his private academy, and happened to inspire many of the contemporary Confucians (Dong 2019).

The author of the brilliant collection of Chinese supernatural tales, *Strange Stories from a Chinese Studio* (*Liaozhai Zhiyi*), Pu Songling (1640–1715) also struggled with the imperial examination system. Due to his exam failures, Pu Songling lived most of his life in relative obscurity, working as a private tutor for wealthy families to support himself. This position, though low in status, afforded him the time to write and gather the materials that would eventually form *Liaozhai Zhiyi*. The collection features ghosts, fox spirits, scholars and common folk encountering strange and otherworldly events. Despite the fantastical elements, the stories frequently offer subtle social commentary, criticizing the corruption, hypocrisy and injustices of the time. But most importantly, in his vernacular writing (as opposed to the higher classical writing) he infused expressions from the body of classics, comically torn from their poetic contexts and applied to mundane affairs. On the one hand, what Pu Songling did was destigmatize the non-thoughtful and unintentional usages of classical expressions. But on the other hand, he created a system of signals to his fellow classically educated literati, associating the classics with an allure of underground culture.

On the descriptive level, the foundational principle of *Daxue* – 'internal first, external next' – remained present. For both Wang Yangming and Pu Songling, preparation for the imperial exams may have served as a starting point, but it ultimately became secondary. Each of them did the *ge wu* in their own way and created nations (*guo*) of their own. Some formed a 'nation' of scholars devoted to 'moral knowledge' or 'the unity of knowledge and action' (among contemporaries, Stephen Angle (2006), Harvey Lederman (2022), Artem Kobsev (2002), and others). Others shaped a 'nation' of educated lovers of humour and superstition, as shown in Judith Zeitlin's study (Zeitlin 1993).

4 Daxue – my story

In 2012, I came to Peking University for an exchange year and took Professor Zhao Yanfeng's course *Basic Facts about China*, which proved to be as distanced from basic as it could. According to Professor Zhao, the best way to interact with Chinese culture was to engage with it – spatially and timely – in a way that Confucian scholars would choose when they assumed that they were acting in accordance with their culture.[8] The way was quite clear – to start memorizing the Confucian Classics, out of which *Daxue* was the first one. Then I could proceed to Five Classics, Daoist texts, to collections of Tang poetry and to other ancient Chinese manuscripts that are being continuously excavated and revisited up till the present day.[9] Professor Zhao gave me one hundred gigabytes of audio recitations of the texts, for alternatively memorizing by listening, and said: 'This would be enough for your lifetime'.

What I did accomplish over the ten years after that was memorizing *Daxue*, partly – *Zhongyong*, partly – *Lunyu* and *Laozi*, as well as selective Tang poems. I was curious to observe what kind of understanding would grow in me if I learned this Confucian language.

When I was memorizing *Daxue* back in 2013 and then *Zhongyong* in 2017, there was one concrete prescription that I was given by Professor Zhao and my other supervisor Professor Wang Caigui: never look in the commentaries yet, as well as into translations and other explications of the meanings inside the texts, either in English or even in contemporary Chinese. The texts are written in *guwen* – the classical Chinese language, and apart from growing attuned to the sound of it and refining my own articulations, I did not feel that I benefitted a lot. Some expressions offered their decodings easily, but many did not. And still, I never looked at the commentaries.

In 2016, I went back to Beijing for another exchange programme and was invited by Professor Zhao Yanfeng to visit Wenli Shuyuan, the academy of classical Chinese education located in the village of Zhuli near Wenzhou city of which Professor Wang Caigui was the director. Wenli Shuyuan is one of the communities in China that are reviving the study of the Four Books and further classics among young Chinese and foreign students of school and university age.

Clearly, the imperial civil examinations no longer exist: following the political upheavals of the twentieth century, China has become a modernized technocratic state operating through Western institutional mechanisms. But the Confucian education in the academies like Wenli Shuyuan follows precisely the imperial model. Younger children there memorized the texts through collective recitation. Around the age of thirteen, they would take an exam to demonstrate their knowledge of the canons which meant that they had to recite a text by heart. Once they pass and only after this, they begin studying the well-known commentaries by medieval Confucian scholars and interpreting the text's overall meaning in their own words.

After I stayed at Wenli Shuyuan, I was invited to share my reflections at a large forum. When I was preparing the speech, I felt that I could use one particular phrase from *Daxue* to describe the intention that I deduced from the founding of Wenli Shuyuan and the similar private schools China-wide.[10]

The phrase went as follows:

If one pursues this with sincerity, even if they never reach their goal / hit the point, they are never too far away from it.[11]

The phrase, when I used it, was met with applause. After my speech (not before it!), I consulted *Daxue* for its translation and commentaries. The phrase comes from the expansion chapter nine of *Daxue*, headed *[From] 'regulation of the family' qi jia [to] the 'ordering the nation' zhi guo*. It happens to be a continuation/an explanation of another phrase – a quotation from another manuscript the *Book of Documents* 'Treat them as infants'. James Legge's English commentary says:

> In the Announcement to Kang, it is said, "Act as if you were watching over an infant." If a mother is really anxious about it, though she may not hit exactly the wants of her infant, she will not be far from doing so.[12]
>
> (Legge 1893)

Therefore, not being fully aware of the meaning of the phrase, I used it spontaneously, but it happened to land in a very good way in that particular setting. Gathered at the forum were mostly Chinese parents whose children attended or were to attend Wenli Shuyuan. They were educated and wealthy people. They ventured into signing their children for a Confucian academy despite many odds. Instead of going to a modern school and university and further competing for a successful career with conventional tools (macro-management, academic specialization, economic productivity), their children were going to study the ancient manuscripts and engage in artificially reconstructed cultural practices (Makeham 2008).[13]

I am not claiming that I managed to respond to these concerns. But what I did was enact an ancient Chinese ceremony of chanting and recitation, showing that the impulse can come from a foreigner. That proved to that particular audience that classical Chinese wisdom can transcend the restrictions of modernism and progressivism and assume its role within a contemporary setup.

5 Conclusion

It is important to let a non-Western worldview speak in its own language. No matter how exotic and thus illuminating the portrayals of a flourishing society may be – as seen at the beginning of this chapter – a Confucian scholar from pre-modern China might express it this way: 'When the bird rests, it knows where to rest. Should a human being be unequal to a bird?'.

At the same time, rather than being served on a plate ready for extraction and application elsewhere, this phrase is deeply contextualized within its text – the *Daxue*. And not only within the content, but also within the specific rules and practices surrounding its reading and transmission.

In Confucianism, classical texts are like large living, evolving organisms rather than static writings. They are spaces to inhabit, places to return to in search of meaning. I have shown how being with even a single text – through reading and re-reading, through connecting with its other people – can be a deeply satisfying, lifelong commitment for a Confucian.

In the longer journey I have taken you on, we carried the various layers of *Daxue* along with us and revealed how all sorts of bridges have been built between its people in ways that have been diverse, inconsistent, often strange. But this is precisely what *Daxue* affirms: the flourishing that is not a static achievement but a situated, relational process, marked by attentiveness to one's place, timing, and capacity – an attentiveness that frees one from abstraction while binding one to the concreteness of humanity.

Notes

1. The metaphor of water ripples used by Fei Xiaotong compares a person in traditional Chinese society to a stone thrown into water: the self is the centre, and one's goal is to create ever-expanding circles of strong relationships. These ripples represent layers of connectedness – from immediate family to distant community members – with no strict boundary between 'insiders' and 'outsiders'. A person takes responsibility for all within their network, arranging them by closeness, though these positions can shift depending on the individual's capacity to maintain ties. This flexible, relational view of identity also reflects the Confucian ideal of *ren* (仁), which emphasizes interconnectedness over individual autonomy.
2. The idea of addressing objects of study not with an anthropocentric view, but rather through the lens of their horizontal relationships with 'their people' (including researchers) is discussed extensively in Donna Haraway's work 'Staying with the Trouble'. My idea to include classical texts in the category of objects with certain relationships with its peoples is inspired by the work of Alexander Dobrokhotov, a Russian historian of philosophy of culture. According to Dobrokhotov, inanimate cultural artifacts – a manuscript, a ritual, a myth – are inclined to generate worlds of their own. He summarizes this by asking on behalf of a cultural object: 'What should the world look like for me to be part of it?' (Dobrokhotov 2008: 8–10).
3. Wing-Tsit Chan (1901–1994) was one of the first translators of Chinese thought to Western audiences.
4. Translations by A. Plaks (2014).
5. It is important to note that the original authorship of *Daxue* remains unidentified to this day. When names such as Confucius or Zengzi are mentioned, they are usually preceded by qualifiers like 'likely'. Scholars explain that the text must have had multiple authors and evolved over the span of centuries.
6. In my conversations with another author in this volume, Karma Ura, we discussed rituals in Buddhism and Confucianism. Karma agreed that understanding and meaningfulness may accompany a ceremony or ritual – but they don't have to. Sometimes, it's more important simply to take part. When one wants to express

something but has no clear words or personal form to do so, the ritual speaks on their behalf. They enact something – and that might be enough for them.

7   Passage 2.16 in *Lunyu* says: '*Zi yue: gong hu yi duan, si hai ye yi*'. This translates as 'The Master said: To attack alternative opinions is harmful indeed!' Character gong has previously been only used in contexts implying battling and therefore interpreted as 'attacking'. Zhu Xi, however, replaced it by 'studying', arguing in his commentary that studying is 'cognitively attacking' the object of your study. Which was well-suited to the internally oppressive nature of imperial politics. That resulted in the further misreading of the passage – and suppression of dissident thoughts – which had stretched until the nineteenth century: James Legge's English translation of *Lunyu*, for instance, draws exactly on this misreading, saying 'The Master said: The study of strange doctrines is injurious indeed!'. Available online: https://ctext.org/analects/wei-zheng (accessed 9 October 2025).

8   These are not the exact words that Professor Zhao used to explain her point, but are rather the lens that I have assumed later to reflect on it. The exact words come from the work *The Interpretation of Cultures* (1974) by an American anthropologist Clifford Geertz (Geertz and Darnton 2017). Geertz argued that there are no actual agents of a particular culture and that even 'the natives' are rather its observers and interpreters. Geertz writes: 'what we [ethnographers] call our data are really our own constructions of other people's constructions of what they and their compatriots are up to' (Geertz and Darnton 2017: 10). I take Geertz's idea as my main theoretical lens whenever I approach Chinese cultural material.

9   For example, the Daoist manuscript 'Wenzi', allegedly written more than 2000 years ago by a disciple of Laozi himself, was for centuries considered a later forgery. But in 1973 it was excavated, inscribed on the bamboo strips, from a 55 BCE tomb.

10 There are more studies which analyse thoroughly the case of Wenli Academy and Wang Caigui's *dujing* education movement, with both its strengths and controversies (Wang and Gao 2023).

11 Translation by the author.

12 Available online: https://ctext.org/liji/da-xue (accessed 9 October 2025).

13 For more than a hundred years after the collapse of the Qing empire in 1912, Confucianism has been subjected to harsh critiques from leaders of China's revolutionary movement. It has therefore gradually become, in John Makeham's words, a 'lost soul', an intellectual tradition without a major culture in which it could function as the core.

10

Towards a Feminist and Radical Democratic Theory of Human Flourishing

Maggie O'Neill

1 Introduction

Over the past thirty-three years, my research has centred on working with marginalized groups – including sex workers, migrant women and survivors of sexual violence – within the framework of 'violent dominance hierarchies'. I have explored these issues in depth through a combination of ethnographic, participatory, biographical and arts-based research methods, often in collaboration with artists and utilizing mobile methods. This chapter draws upon this body of work to propose three interconnected pathways that can move us beyond normative understandings of gender equality towards a radical democratic imaginary of human flourishing. This vision seeks to dismantle violent dominance hierarchies by engaging theoretically with feminist concepts of social justice through socially engaged research in and with communities, aimed at fostering our collective social futures as human flourishing.

First, this chapter argues that we need to acknowledge the significant shifts in contemporary understanding of sex, gender and sexualities, informed by diverse feminist perspectives including post-feminism, Black and intersectional feminism, masculinity studies, and LGBTQI+ activism and scholarship. This first pathway emphasizes the importance of gender studies in challenging normative concepts of gender equality, particularly through the critical potential of intersectionality to address racialized, gendered and class-based harms. However, it is crucial to critically assess how these concepts are often co-opted and neutralized by dominant hegemonies. Hence it is important to ask how can our work as academics, researchers and artists 'play a critical role when every critical gesture is quickly recuperated and neutralized by dominant hegemony' and how does neutralization work (Mouffe 2019; O'Neill and Perivolaris 2020)?

Second, the chapter calls for inter- and transdisciplinary research that is theoretically driven and socially engaged (collaborating and co-producing research in and with marginalized communities and groups) and combines feminist critical theory with lived experiences (ethnographic, biographical sociology/narrative work) and the arts. It is particularly important that we learn and draw on the contributions of Black feminists and scholars from the Global South. Biographical research is central to this approach, not merely because it reflects the storied nature of human lives but because it offers profound insights into societal structures. The integration of feminist theorizing and collaborative, biographical and arts-based methods, can create conditions conducive to human well-being and human flourishing, re-ethicizing social research and critically addressing social pathologies and 'violent dominance hierarchies'.

These two pathways converge to suggest a third: a critical, transdisciplinary feminist approach that advances a radical democratic imaginary for our collective social futures (Cornell 2016; Mouffe 2019; Smith 1998) by systematizing the different components in pathways one and two, that taken together make up a feminist and radical democratic foundation for human flourishing. This approach is exemplified in this chapter through socially engaged, participatory, ethno-mimetic feminist research with students and community-based agencies who together created a Feminist Walk of Cork, a city located in the south of Ireland.

2 Feminist genealogy in addressing violent dominance hierarchies

Moving beyond normative liberal understandings of gender equality towards a radical democratic imaginary of human flourishing requires a deep engagement with the genealogy of critical feminist thought, particularly the contributions of Black feminists. Despite significant shifts in our understanding of sex, gender and sexualities – driven by various feminist movements including post-feminism, Black and intersectional feminism, masculinity studies, and LGBTQI+ activism and scholarship – sociological research often remains entrenched in heteronormativity.

The transition from women's studies to gender studies reflects this shift, with gender studies emerging as a transdisciplinary field informed by sociology, literature, philosophy, history, science, politics and the arts. My research, situated at the intersections of sociology, women and gender studies, and criminology underscores the importance of participatory action research. By actively listening to marginalized voices, particularly through participatory and arts-based research with groups such as sex workers (O'Neill 2001) and migrant women (O'Neill 2018), I have explored the critical relationship between feminist critical theory, lived experience and praxis (the development of purposeful knowledge). This approach emphasizes the use of creative methods – like narrative and biographical research and the arts, exemplified in collaborations with Bea Giaquinto and Fahira Hasedzic using photography and life history research (O'Neill, Giaquinto and Hasedžic 2019), and theatre making with Erene Kaptani, Umut Erel and Tracey Reynolds (Erel et al. 2022; Kaptani et al. 2021; O'Neill and Roberts 2019), and film making with Jan Haaken (Haaken and O'Neill 2014). These methodologies, central to the feminist project, transcend traditional research methods, pushing the boundaries of ethnography, biographical sociology and art-making, which I term 'ethno-mimesis' (O'Neill 2001; O'Neill et al. 2002).

In my graduate seminar 'Feminist Epistemologies' at University College Cork, we explore these shifts through theoretical debates and cultural practices, focusing on 'sites of oppression' and 'sites of resistance'. The graduate seminar led to a feminist walk in Cork that was developed with students and community organizations, and serves as a theoretical and practical tool to write women into the city's topography, uncovering both their contributions and sites of resistance as well as the ongoing sites of oppression. This project embodies the continued relevance of intersectionality in feminist theory, particularly in addressing violent dominance hierarchies.

The walk not only maps these issues but also challenges the absence of women's contributions in public spaces, prompting critical reflection on intersectionality as a necessary lens for understanding and combating systemic violence.

We begin the module with a feminist walk in the city, celebrating women's contributions past and present while uncovering sites of oppression and resistance. Since 2021, the Feminist Walk of Cork project has mapped feminist issues and women's roles in the city's topography, addressing their absence by engaging with feminist theory and practice embedded in various organizations that are also women led. This walk, inspired by postgraduate students taking my MA Women's Studies module on *Participatory Action Research methods*, and MA Sociology module on *Feminist Epistemologies* explores intersectionality and its importance in confronting violent dominance hierarchies. Prompted by the question of whether women's contributions are visible in the topography of the city, we discovered a lack of monuments dedicated to women and set out to document women's impact through time, on the city's topography in a feminist walk.

Walking through the city and engaging with spaces, places and stories connected to women, social justice, sexual and social inequalities offers a convivial, critical and imaginative method to practice feminism on the move. Drawing inspiration from walking artists and decolonial sociologists who pioneered participatory research methods, our approach emphasizes three key aspects of critical feminist theory and feminist participatory research:

1. Collectivity: we work collaboratively, prioritizing teamwork, trust-building and convivial feminist research process and practice, fostering a subject–subject relationship throughout the research process.
2. Critical History Recovery: we research women and places using feminist intersectional theory and historical, personal, folk and archival materials, especially oral traditions, to support and value community knowledge.
3. Knowledge sharing: we ensure our research and walks are accessible and meaningful and can reach broader audiences beyond academia.

2.1 Intersectionality

In the graduate seminar, we focus on understanding intersectionality, a concept rooted in the activism and scholarship of Black feminists and writers like Maria Stewart, Sojourner Truth, Toni Morrison, Barbara Smith, Alice Walker, Angela Davis, Patricia Hill Collins, Kimberle Crenshaw, Nira Yuval Davis, Analouise Keating, Gloria Lopez and Akwugo Emejulu. Black feminist work challenges racial, sexual, class-based oppression, rejects stereotypes of Black women and seeks to 'forge self-definitions of self-reliance and independence' (Collins 2000: 3). This is discussed alongside standpoint feminisms[1] and Kimberle Crenshaw's work on intersectionality.

Crenshaw (1989) argues that sexism and racism cannot be understood or addressed separately, as they intersect in complex ways. Intersectionality examines how social structures, marked by power and violence, create vulnerabilities. It posits that social hierarchies and oppressions are inseparable and mutually modifying, making any unqualified reference to 'woman' as a standalone category problematic. For example, Black women's experiences of discrimination cannot be understood by separately analysing sex and race discrimination. Ignoring this complexity risks overlooking the experiences of those who are multiply marginalized.

Standpoint feminisms emphasize that knowledge is perspectival, recognizing multiple standpoints from which knowledge is produced. Sandra Harding teaches us that subjects of knowledge are embodied, socially located and also objects of knowledge (subject-object dichotomy), and the knowledge produced is multiple, heterogeneous, contradictory and not homogeneous. Dorothy Smith (1996) argued in *The Relations of Ruling: a Feminist Inquiry* that knowledge from multiple standpoints can be less partial and reflective of the conditions of ruling. Feminist standpoint theorists explore how knowledge can be situated yet still 'true', without claiming a universal experience. They stress the importance of reflexivity, relationality and acknowledging difference without undermining the critique of feminist politics.

Kimberle Crenshaw builds on work of Combahee River Collective, a collective of Black feminist activists who have been meeting together since 1974 and Cherríe Moraga and Gloria Anzaldúa (1981) in *This Bridge Called My Back,* linking intersectionality to both individual and collective identity and emphasizing the need to focus on social structures, power relations and violence against women of colour. Intersectionality, according to Crenshaw, is fundamentally about social justice.

Patricia Hill Collins (2000), a key figure in intersectionality, views it as field of study situated within power relations, an analytical strategy offering new angles, and a critical praxis informing social justice efforts. She argues that race, class, gender, sexuality are not only intersecting but mutually reinforcing, highlighting the connection between intellectual work and activism.

However, intersectionality has been critiqued for being appropriated by academic feminism, often narrowly attributed to Crenshaw's work alone. Hill Collins and Bilge (2020) argue that this erases the broader history of Black women's activism. Erel et al. (2010) note that its incorporation into neoliberal academia has depoliticized the concept. Emejulu (2022) warns that intersectionality, rooted in Black feminism, is a theory and politics of liberation that can be co-opted and colonized by White feminists, diluting its original intent and power.

Hence a key takeaway from our work in the graduate seminar as we map the feminist spaces, and sites of resistance and oppression in the city is to consider *whose knowledge is being produced for whom and for whom will it bring benefits.*

2.2 Neutralization and appropriation of intersectionality

In this context, we must ask how our work as academics, researchers and artists can 'play a critical role when every critical gesture is quickly recuperated and neutralized by dominant hegemony' and how does neutralization work (Mouffe 2019; O'Neill and Perivolaris 2020)? Ana Louise Keating (2012) along with Erel et al. (2010), highlight how references to intersectionality often 'signal' rather than actively pursue an alternative, collective, democratic social future. Keating (2012) notes that Cherríe Moraga and Gloria Anzaldúa's (1981) edited collection *This Bridge Called My Back* 'broke new ground and was instrumental in introducing intersectionality into mainstream feminist discourse'. However, as Jasbir Puar (2012: np) points out in her discussion of feminist and queer theories, 'Much like the language of diversity, the language of intersectionality, its very invocation, it seems, largely substitutes for intersectional analysis itself'. Those using 'intersectionality' mark, classify and divide; they stop with the labels and do not use this labelling process to generate new commonalities. The differences function like walls, not thresholds (Keating 2012: 37). Keating's answer is to focus upon the relational aspects of interconnectivity, and 'build on intersectionality's lessons by developing and enacting nuanced politics of interconnectivity' (Keating 2012: 38).

This analysis is crucial and will be revisited later through examples from the Feminist Walk of Cork. However, it is important first to understand how neutralization operates. Sociologically, we must recognize that our work is situated within the broader context of neoliberal advanced capitalism, where feminist research occurs within 'Violent Dominance Hierarchies'. To explain the neutralization of radical critiques like intersectionality, as theorized by Black feminists, and the persistence of these hierarchies despite resistance, we can consider Robert Jay Lifton's concept of 'malignant normality', which can lead to what Nicholson (2019) describes as 'moral injury'.

2.3 Malignant normality and moral injury

In *Malignant Normality and the Dilemma of Resistance*, Shierry Nicholsen (2021), a critical theorist and psychoanalytic psychotherapist, explores Robert Jay Lifton's concept of 'malignant normality' in conjunction with Adorno's Minima Moralia, particularly in the context of the Nazi holocaust and regime. Malignant normality is a 'social actuality' 'presented as normal, all encompassing, and unalterable', yet it fosters inhumanity (Lifton 2017: xv). It 'promotes suffering and it is normal because it is couched in accepted social norms and is all encompassing'. To illustrate, I will share two examples from our Feminist Walk of Cork.

2.3.1 Irish Travellers: racism as malignant normality

A stark example from our feminist walk in Cork highlights the racism faced by Travellers in Ireland, often described as the last socially acceptable form of racism in the country. Irish Travellers (*an lucht siúil – the walking people*) are an Indigenous, nomadic people who were given ethnic recognition by the Irish government in March 2017. Traveller culture and history is passed on through oral tradition, storytelling and traditional songs and music.

Research with Traveller communities documents significant inequalities in maternal health, access to education, higher education outcomes and their over-representation in the prison population. Irish Travellers are 'over-policed as suspects but under-policed as victims' (Joyce et al. 2022: 8). Forced to live in 'halting sites' in the worst areas, such as next to the council dumps, with limited facilities, Travellers also represent the hidden homeless. It is common for Travellers to be denied entry to restaurants despite having reservations and they often feel unwelcome in the city. The normalized, routine everyday racism Irish Travellers experience exemplifies the operation of malignant normality, where their human rights, cultural rights and opportunities to flourish, are systematically denied, and suffering is perpetuated.

Nicholson (2021) explains that malignant normality operates both inwardly and outwardly, referencing Adorno's *Minima Moralia* where he states, 'dwelling is now impossible'. Adorno reflects on his experience of living a 'damaged life' in *Minima Moralia* under the Nazi regime. Nicholson explores how malignant normality functions within the broader context of advanced capitalism. Nicholson argues that Adorno offers a faint hope of resistance against the all-encompassing nature of malignant normality. She notes that 'the surface of social relations and ordinary sociability then becomes the normality of those whose individuality has been eliminated, sucked up into collective madness, which under capitalism also takes the form of commercial relations in the marketplace' (Nicholsen 2021: 4). Quoting Lifton, Nicholson underscores that 'nothing does more to sustain malignant normality than its support from a large organization of professionals' (Nicholsen 2021). In the case of Travellers, this includes institutions like schools, teachers, welfare services, city councils and housing services, all of which play a role in perpetuating malignant normality.

In our discussions with the Traveller Visibility Group (TVG), we come to understand that the city represents many sites of oppression and also resistance. The Traveller Visibility Group was founded by three Traveller women and responded to 'malignant normality' by self-organizing and creating a city centre organization that offers support to Traveller communities, that includes providing a crèche for Traveller families, supporting social enterprises that offer employment, especially for those leaving prison via 'community employment schemes'. The organization offers domestic violence and addiction support. It also supports families in ensuring full access to education for their children and young people (Figure 10.1).

The three women who co-founded TVG set out to challenge and change the malignant normality of racism against Travellers. TVG is founded on the principles of community development as well as 'Nothing About Us Without Us'. TVG supports Travellers throughout the city and county and ensures Traveller voices are heard, and are represented in decision-making forums.

2.3.2 Sexual violence as malignant normality

Further along our walk, we connect with the Sexual Violence Centre Cork, which has been operating for over forty years. The centre provides services to all victims

Figure 10.1 Traveller rights are human rights. Image: Marcin Lewendowski. Feminist Walk of Cork 2, July 2024.

of domestic violence and coercive control and the women-led organization works closely with Traveller groups to address sexual inequalities, oppression, racism and violent dominance hierarchies. This collaboration highlights the intersectional approach to combating these deeply rooted societal issues.

The Sexual Violence Centre Cork officially opened in 1983 with two main aims: to work towards the elimination of sexual violence in society and to provide the highest quality of services to victims of sexual violence. Mary Crilly, who has led the work of the Centre for forty-four years, emphasized the importance of societal change, stating,

> As a society we need to have a zero tolerance for sexual violence. We need to keep talking. We need to keep challenging. We need to change the culture that tolerates sexual violence.

In July 2024, violence against women and girls (VAWG) was declared a national emergency in the UK by The National Police Chiefs' Council and College of Policing. The evidence cited is alarming: 3,000 crimes of VAWG are recorded each day, with at least one in twelve women becoming victims each year. However, the true extent of the issues is likely much higher, as most victims do not report to the police. The End Violence Against Women Coalition estimates that the actual number of victims and perpetrators is significantly higher, with one in twenty people estimated to be perpetrators of VAWG each year (National Police Chiefs' Council 2024).

Andrea Simon, Executive Director of the End Violence Against Women Coalition, stated:

> Violence against women and girls is one of the most pervasive human rights violations that we face. Recorded data will never completely capture the extent of this abuse, which we know remains highly underreported because as a society we still frequently treat victims as being responsible for their own abuse and keeping themselves safe, while many are denied routes to support and justice.
>
> (Jacques 2024)

Violence against women and girls has long been a central focus of feminist analysis, with research consistently showing that women who are victims of violence are often pathologized or held responsible (Grace et al. 2022).

The operation of 'malignant normality' in both the examples above, and the intersecting issues they reveal, is evident. Each case represents a 'social actuality' that is 'presented as normal, all encompassing, and unalterable' yet is fundamentally 'conducive to inhumanity' (Lifton 2017: xv). Malignant normality 'promotes suffering and it is normal because it is couched in accepted social norms and is all encompassing'. In both cases, the normalized racism against Irish Travellers and violence against women and girls, along with their complex interconnections, the focus often remains on the individuals rather than on the social conditions that give rise to, perpetuate and advance malignant normality and moral injury. This limited perspective fails to incorporate intersectional analysis into normative understandings and responses, allowing the status quo to persist.

The Feminist Walk raises awareness about these intersecting issues by:

(1) Understanding in situ and in practice: building on Hill Collins' work, the walk highlights that intersectionality as a field of study is situated within power relations.
(2) Providing an analytical and spatial strategy: the walk draws attention to and connects the intersections, making them visible in both analysis and physical spaces and places of the city.
(3) Delivering critical praxis: through feminist theoretical and methodological work, a mobile feminist pedagogy is developed and the walk informs social justice projects by collaborating with community groups and agencies to develop knowledge and understanding through critical recovery of history, supportive of community knowledge and oral history that enables knowledge sharing across the widest audiences.

The feminist intervention includes a feminist walk, a map and a website, offering research evidence to support the guided walk.[2]

2.3.3 What is the relationship to moral injury?

In her writing about moral injury, Shierry Nicholsen (2019) defines it as the 'betrayal of what's right by a person in legitimate authority, or by oneself in a high stakes situation'. Nicholsen explains that for victims, this betrayal severely damages their capacity to believe in a just world and morally trustworthy world. Furthermore, she argues that participating in wrongdoing – even if coerced – inflicts additional moral injury on both the victim and the self.

Nicholson, drawing on Jonathan Shay's work, emphasizes that moral injury is a distinct aspect of human-caused trauma, which might not be recognized as traumatic, especially when no immediate threat to life or physical injury is involved. Shay notes, 'While other aspects of trauma destroy the sense of safety, moral injury destroys the sense of trust-especially social trust'. As a therapist, Shierry Nicholsen emphasizes that acknowledging moral injury is crucial. The therapist needs to be a 'trustworthy listening audience' for the morally injured person's trauma narrative and the extent to which recovery from moral injury is possible, requires a restoration of social trust. Nicholson also explores that 'apology opens a transitional space'. Moreover, in thinking about large-scale injustices she asks what apology would look like in larger-scale social structures/injustices. For Shay (Shay 1994), 'healing from moral injury begins by trusting a single individual, it needs to continue in a trustworthy social setting, a trustworthy community'.

In considering larger-scale apologies, Nicholson examines examples such as truth and reconciliation efforts in South Africa and reparations, and she concludes with reflections on climate change and 'the certain sins of the future', highlighting how these could inflict moral injury on future generations. She states:

> Think of how we, the complicit beneficiaries of a fossil fuel dependent life style are simply by continuing to live as we do inflicting moral injury on our children and their children and grandchildren, who have a right to life on a liveable planet. If young people are protesting this is why.
>
> (Nicholson 2019:10)

To address the challenge of malignant normality and resist the ways in which our critical advances are co-opted and neutralized by dominant hegemony, we must adopt a multifaceted approach. This involves resisting the normalization of harmful practices, challenging the structures that perpetuate them and striving for transformative change. Moving towards the hopefulness of human flourishing, as defined by thinkers like Nussbaum and Zwitter et al. (see Introduction to this volume), requires embracing a radically re-politicized intersectionality, as advocated by Black feminists.

It is imperative that we create and protect spaces for conducting critical feminist interdisciplinary research. This research must prioritize listening to the voices of those with lived experience and work collaboratively with communities to drive meaningful change. By doing so, we can ensure that our efforts are rooted in the realities of those most affected and are capable of challenging and ultimately dismantling the oppressive systems that sustain malignant normality and violence dominance hierarchies.

3 Dismantling violent dominance hierarchies through critical praxis that informs social justice projects towards radical relationalism

To effectively dismantle violent dominance hierarchies, it is essential to integrate the concepts of intersectionality, malignant normality and moral injury into a cohesive strategy. This requires not only thinking with intersectional theory, as Keating (2012) advocates, but also advancing it through deeply socially engaged critical feminist research. Such research should involve biographical, narrative and participatory-co-productive methods, working ethno-mimetically (O'Neill 2024) at the intersection of the arts and social science.

By working within the 'hyphen' between feminist theory and feminist ethnographic, biographical and art-making practices, we open up a potential space (drawing on Winnicott 2005) for transformative possibilities and counter-hegemonic thinking – representing a critical theory in practice. Ethno-mimesis functions as a counter-hegemonic space, practice and process. As a theory and relational research practice, ethno-mimesis creates room for the voices of experience, using active listening and participatory arts-based methods to explore ways of seeing and relating that emphasize social conditions rather than focusing solely only on individual actions, performances or victimhood. This approach helps to resist and challenge the malignant normality that often results in the responsibilization and blaming of individuals, thereby fostering a more just and equitable understanding of social life.

3.1 Feminist Walk of Cork as exemplar of critical praxis towards human flourishing

In devising the Feminist Walk, we encountered intersectionality in high relief through the radical interconnections we uncovered, both materially and symbolically, as we explored various city spaces and landmarks in search of women's contributions to the city and their role in building safer and fairer communities. This socially engaged research process involved deep collaborations with women-led community organizations, highlighting the rich tapestry of women's work and influence in the city.

The map, walk and website that emerged from these creative collaborations served as important counter-hegemonic interventions. They drew attention to and critiqued malignant normality and moral injury while celebrating women's substantial contributions to the city's development and the creation of fairer, more sustainable and safer communities. These interventions not only celebrated human flourishing but also made visible the radical intersecting issues in women's lives and the profound connections within the city's topography, organizations and communities; all of which played a crucial role in making these interactions and contributions visible and understood.

The relational interconnections created by the Feminist Walk offer a powerful message in the context of neoliberalism and its ideologies. As Layton (2020) argues in her latest book, neoliberalism encourages people to disregard their history, downplay the importance of relationality and focus solely on the development of their sovereign individual selves. This mindset promotes a forward-looking, positive attitude, often at the expense of acknowledging the complexities of our past and the importance of social ties – for Layton, a perspective reflected in the popularity of positive psychology courses at Harvard during the 1990s and 2000s.

This raises crucial questions about social justice and its role in human flourishing. The Feminist Walk as a collaborative research project, aimed to highlight social justice and celebrate human flourishing not as isolated, individual achievements, but as collective and interconnected pursuits essential for our shared social futures. In the words of Keating (2012, 38), we made 'connections through differences', by embracing 'radical interrelatedness' and practicing active listening with 'raw openness'. This approach underscores the need for a better definition of social justice than what normative liberal equality theories typically offer, emphasizing the importance of collective well-being and interconnectedness in achieving true human flourishing. The following section seeks to outline a critical feminist and relational approach to social justice.

3.2 Social justice and human flourishing

Barbara Hudson's (2006) pioneering work on social justice is crucial for thinking beyond malignant normality and moral injury towards human flourishing; human flourishing as conceptualized by Zwitter et al. in the Introduction to this book. Hudson aims for a 'conceptualisation of justice that has the potential to escape being sexist and racist', and she observes the long time it has taken for racialized as well as sexualized harms to be acknowledged and taken seriously by the law (Hudson 2006: 30). She argues that this delay is due to the law reflecting the subjectivity of dominant, White, affluent adult males. Feminist and critical race scholars have provided many examples of how this male, White subjectivity pervades the law, highlighting the systemic biases that still exist within legal frameworks (Emejulu 2022; Hudson 2006; O'Neill and Laing 2018).

Hudson argues that the constructions of law and 'the liberal philosophies on which western law is based, reveal the closures of law and, therefore, the limits of justice that can be expected by marginalised Others' (2006, 31). She points out that women, especially women of colour, have faced greater challenges in securing bodily rights than in gaining property rights or rights in the public sphere. Justice, she suggests, is often only served when claimants are 'like white men in their mode of being in the world as well as their basic characteristics' (Hudson 2006: 31). Hudson highlights issues like ambiguous definitions of 'racial' crimes, inadequate protection for non-Christian religions, and racial disparities in sentencing as evidence of the entrenched whiteness of criminal law (Hudson 2006: 32). To move beyond this 'white man's justice', Hudson proposes that justice should incorporate three key principles: (1) discursiveness, (2) relationalism and (3) reflectiveness (Hudson 2006: 35–9).

3.2.1 Social justice as discursive

Hudson (2006) argues that those most excluded from justice should have privileged access to discourse, allowing them to present their claims in their own terms rather than conforming to dominant legal and political discourse. She stresses that telling their story in her own words is vital to reach intersubjective understanding. However, having a forum in which to speak does not necessarily ensure that one's claims to justice will be recognized. Discursiveness goes beyond just having 'space in proceedings to speak' but also being 'open to challenges to the identity of law and open to identity claims that are not based on similarity' (Hudson 2006: 35). In Ireland, Traveller women and Travellers, until recently, did not have a place in legal proceedings to voice their concerns. Even now, it is questionable whether their claims are genuinely listened to when expressed in their own terms, as they are often expected to conform to the dominant discourse.

3.2.2 Social justice as relational

Hudson suggests that identities are situationally and relationally contingent and shared. Feminists working on intersectionality, and I suggest, the participants in the feminist walk, can perceive and empathize with these situationally and relationally contingent and shared experiences, particularly those of marginalized communities like Travellers in the Irish context and victims of gender-based violence.

Methodologically the understanding of social justice as relational emerges through the socially engaged, participatory and collective methods used in feminist research and the feminist walk, which creates an epistemological and embodied awareness of relationalism, including an appreciation of intersectionality as theory, method and praxis. The research process and practice are collaborative, fostering subject-subject relationships, undertaking critical recovery of history using folk knowledge, oral traditions and archival research and sharing research knowledge in accessible ways and forms, such as through a feminist walk and walking as critical pedagogy.

Social justice as relational is exemplified by the example of the Sexual Violence Centre Cork and the work they undertake through a relational and intersectional lens. The founder of the centre offers free services to victims of all forms of sexual violence aged fourteen and over, addressing issues such as rape, child sexual abuse, sexual assault, sexual harassment, human trafficking, stalking, forced marriage, female genital mutilation and coercive control. The centre also functions as a community hub, fostering connections with other agencies, including Travellers. The SVCC has hosted exhibitions that examine the interconnections between slavery, modern-day slavery and violence

Figure 10.2 Ask consent. Oliver Plunket Street. Image: Feminist Walk of Cork 1.

dominance hierarchies and has co-organized forum theatre events to address sexual violence and gender-based inequalities and runs various campaigns such as #Ask Consent (see Figure 10.2, a photo taken on the walk highlighting the important message of consent in relationships, in SVCC banners situated along a main street in Cork, Oliver Plunket Street).

3.2.3 Social justice as reflective

Reflective justice, according to Hudson (2006), involves considering each case in terms of its subjectivities, harms, wrongs and contexts, and then measuring it against concepts like oppression, freedom, dignity and equality. She argues that gendered theories of justice based on fixed identities fail to consider the unique circumstances of individual cases, leading to a one-dimensional approach to justice. She emphasizes that relational and reflective principles are closely tied to discursiveness is to transcend the limitations of established justice and the fixed identities of dominant legal discourse.

For example, violent dominance hierarchies are much in evidence when it comes to crimes against women, sexual and violent crimes, for it is they, the victims, who are pathologized, as much research (e.g. Grace et al. 2022) and media analysis such as Jane Gilmore's book and campaign 'Fixed It' (Gilmore 2019) highlights. In response to these gender biased narratives based on 'fixed identities', Jane Gilmore launched the 'Fixed it' campaign in 2014 to correct media misrepresentation and advocate for accurate reporting of men's violence against women against the limitations of established justice and the fixed identities of dominant media and legal discourse.

> "Fixed It is a project I started in 2014. It was born of frustration with the then constant victim-blaming and erasure of perpetrators from media headlines about men's violence against women. At the time media headlines were full of "scorned lovers" who raped women they claimed to love, "good fathers" who killed their children, and adults who had "sexual relationships" with children they were abusing".
>
> (Gilmore 2019)

#FixedIt – Ireland was inspired by journalist Jane Gilmore's work and the reflective analysis of SVCC to the pathologizing of the victims of sexual violence in the mainstream media. SVCC developed a 'Fixed It-Ireland' Twitter campaign to challenge and change media representations of sexual violence by highlighting the importance of reflective justice. This campaign addresses the issues of victim-blaming and gender bias in mainstream media, which pathologizes women and/or shifts responsibility onto victims.

An example of this approach is transforming the headline 'Domestic violence victims forced to live in violent homes due to housing crisis' into 'Domestic violence victims forced to live with violent men due to systemic gendered economic disadvantage and decades of government prioritizing wealth creation over housing equity'.

The Sexual Violence Centre in Cork extends its efforts through its X/Twitter and Instagram campaign, where they 'fix' inappropriate and incorrect headlines, tagging the respective news source to ensure visibility of the corrected versions. This initiative is part of the Centre's broader commitment to challenging harmful narratives and advocating for reflective justice in the portrayal of sexual violence.

Two examples are as follows.

Irish Times	#Fixed It
'Gardai are investigating claims that a 15 year old girl was visiting hotel rooms in Dublin to have sex with men in exchange for cocaine'	'Gardai are investigating claims that a 15 year old girl was raped and exploited by men'

The Journal.ie	#Fixed It
'Man with 41 previous convictions who defiled teen girl has six-year sentence halved on appeal'	'Man with 41 previous convictions convicted of rape and sexual assaults has six-year sentence halved on appeal'

By incorporating the principles of reflective justice (bound up with discursive justice), it becomes possible to position oppression and inequality at the heart of justice considerations. For Hudson, this approach provides criteria for resolving conflicts and competing claims without requiring participants to mimic formal justice's ideal of impartiality. Instead, they can advocate for causes, promote respect for rights and integrity, and encourage the discursive recognition of wrongs and harms (Hudson 2006) as the Fixed It campaign does.

3.3 Resistance and recognition

The numerous acts of resistance documented in feminist research and scholarship are testimony to the vital role of resistance and hope. In answering the question of how to resist malignant normality, our experience with the feminist walk demonstrates that resistance is pervasive – in sexual violence campaigns, Traveller pride marches, research and writing, and in everyday actions and practices. As Nicholsen notes, 'The individual and reflection on individual experience becomes the crux of complicity in and resistance to malignant normality' (Nicholsen 2018: 12).

Resistance and hope are intertwined, for, as it is said, 'without hope truth would scarcely be thinkable'. This is closely linked to the future of critical theory and our collective social futures. Resistance requires moral commitment, which includes rational self-interest, human responses of sympathy and respect, and one's 'moral identity' as essential to self-respect.

Where does this lead us? As a feminist researcher, I emphasize the importance of understanding violent dominance hierarchies as patriarchal structures, practices and processes. Resisting and working to change these hierarchies involves grappling epistemologically and methodologically with malignant normality and moral injury through radical intersectional feminist research and practice, in participatory and collaborative ways. Implementing a critical, creative theory in practice, aimed at a radical democratic imaginary for our collective social futures and underpinned by Hudson's (2006) account of social justice, that echoes words attributed to Angela Davis (2014): 'You have to act as if it were possible to change the world. And you have to do it all the time'.

As feminists, academics, students and practitioners we sought, in the Feminist Walk, to create and sustain spaces for critical discourse, theory and praxis, using creative and arts-based methodologies that can often convey what words and theory alone cannot. The New Institute and the workshop where I delivered this paper are crucial for fostering a radical democratic imaginary. It is equally important to consider our methodologies, particularly the inclusion of biographical methods, given the focus upon storytelling at The New Institute's Human Flourishing group.

3.3.1 Storytelling – biographical sociology, recognition and meaning making

As we live our lives, we constantly engage in biographical work, constructing and reconstructing our 'selves' in response to social contexts, happenings, relationships. Biographical sociology encompasses a broad range of areas, including oral history, life story/history, family histories, narrative forms, literary biography, autobiography, biography-discourse and auto/ethnography. These methods highlight the dynamic nature of identity and social experience, capturing the interplay between individual lives and broader social forces. Hence, a biographical approach to understanding violence dominance hierarchies should not be seen merely as a research tool for 'it is primarily a social phenomenon that must be seen as a fundamental constituent of sociality' (see Chantfrault-Duchet, 1995: 212 in R. Miller 2000).

Critical feminist theory that is intersectional, premised on a feminist understanding of social justice (Hudson 2006) in addressing malignant normality and moral injury (Nicholson 2018a, 2018b) is underpinned by feminist methods that are socially engaged, collaborative, participatory and biographical. I want to reinforce the crucial need for biographical sociology today to connect individual lives with history, structure and culture. Wengraf (2001) argues that researching social processes and social change is essential to biographical sociology, with Fritz Schütze's exploration of suffering, disorderly social processes and biographical trajectories (and ruptures) adding an important interactionist and psychosocial interpretation (Schütze 1992). Herbert Blumer's 'sensitising concepts' and Anselm Strauss's stress on social interaction and lives in process are key contributions from the Chicago School of sociology, alongside Mead and Schütze's work in phenomenology and symbolic interactionism.

The combination of biographical and arts-based research, when conducted collaboratively and co-produced, can significantly contribute to fostering conditions that support human well-being and flourishing. This approach helps to re-ethicize social research by grounding it in real human experiences and creative expression, making it more responsive to the complexities of social life. By critically addressing social pathologies and violent dominance hierarchies,[3] this method promotes a deeper understanding of how individuals and communities are affected by and can resist oppressive structures. Ultimately, it offers a powerful means to engage with and transform the social conditions that undermine human flourishing.

4 Towards a radical democratic imaginary for human flourishing

Bringing together the two pathways outlined in sections one and two into a third, we might approach a radical democratic imaginary for human flourishing. Section 1 emphasizes the importance of a genealogy of gender studies, particularly the contributions of Black feminists, in challenging normative concepts of gender equality, particularly through the critical potential of intersectionality to address racialized, gendered and class-based harms. Section 2 focused on the importance of dismantling violent dominance hierarchies, through the concepts of intersectionality, malignant normality and moral injury into the narrative that forms the Feminist Walk of Cork. This requires not only thinking with intersectional theory, as Keating (2012) advocates, but also advancing it through deeply socially engaged critical feminist research. Such research should involve biographical, narrative and participatory-co-productive methods, working ethno-mimetically at the intersection of the arts and social science and marginalized communities. Barbara Hudson's work on social justice as discursive, relational and reflective (explored through Traveller and Sexual Violence Centre Cork's contributions to the Feminist Walk of Cork) is important here in critiquing, challenging and changing violence, dominance hierarchies towards a radical democratic imaginary for human flourishing.

A radical democratic imaginary for human flourishing allows us to move beyond entrenched binaries often found in gender equality debates. The Feminist Walk of Cork exemplifies this radical democratic imaginary by creating and sustaining spaces for feminist critique, understanding and dialogue. It upholds principles of social justice as discursive, relational and reflective, making visible the struggle for recognition, and includes stories of suffering, resistance and resilience.

In the walk, we engage with the rich contributions of Traveller women and their leadership in advancing Traveller rights (TVG and Traveller Women's Network). We also explore the relational and interconnections between women's experiences of the criminal justice system, sexual and intimate partner violence (Sexual Violence Centre Cork), sexual health and inequalities, and broader social issues like poverty and homelessness. The walk highlights the strengths-based work of these organizations, especially their efforts to combat racism, and documents the important contributions made by feminists past and present, situating history within the present.

As Means (2014: 123) argues, 'the power and universal capacity of the *demos* [the people] to engage and participate directly in political life is a basic presupposition held by those who advocate for some form of radical democracy today'. Participatory, biographical and arts-based research methods such as walking methods can create spaces for trust, sharing and recognition of each other's humanness, which in turn supports trust and confidence in working together.

In previous work, I have argued that arts-based participatory work can be transformative by developing trust as a 'relational good' (Archer 2015) when working with marginalized people. Drawing on Carl Rogers, it is important to recognize that 'human beings become increasingly trustworthy once they feel at a deep level that their subjective experience is both respected and progressively understood' (Thorne 1992: 26). However, building trust requires time, attention and a focus on the relationship between collaborators, facilitators, processes and practices, as well as the methods used by the researcher. Lynne Layton (2020) shows us that culturally sanctioned recognition – through approval, love and the conditions of social belonging – is the primary way norms are transmitted and internalized. While sociologists describe this as socialization, Layton emphasizes deeper, specific ways of relating that puts the emphasis on the social conditions rather than the individual.

In closing, I want to draw on Anne Marie Smith's engagement with Laclau and Mouffe's concept of radical democratic pluralism, as well as Drucilla Cornell's work on the imaginary domain, to suggest some next steps for advancing a transdisciplinary, feminist approach towards a radical democratic imaginary. This approach is aimed at fostering human flourishing and shaping our collective social futures, aligning closely with the goals of the Deep Institutional Change Project.[4]

The Feminist Walk creates a space for understanding radical intersectionality and it also raises awareness of how neutralization operates. How critical and radical concepts and contributions are neutralized – there were no monuments to women celebrating their contribution to the city and particularly their contributions to building fairer, safer communities for human flourishing. In Section 1 we asked how can our work as academics, researchers and artists 'play a critical role when every critical gesture is quickly recuperated and neutralized by dominant hegemony' (Mouffe 2019; O'Neill and Perivolaris 2020) and we answered through feminist theory, feminist methodologies and feminist praxis through generating critical pedagogy. The Feminist Walk enabled us to connect with Barbara Hudson's (2006) pioneering work on social justice as crucial for thinking beyond malignant normality and moral injury, beyond neutralization towards human flourishing.

To advance a radical democratic imaginary for human flourishing, and the work that this chapter has started, we now need to continue to:

(1) Collaborate Through Radical Intersectionality: Work collaboratively using a radical intersectionality lens to ensure 'The circulation, radicalisation and institutionalisation of democratic discourse' (Smith 1998: 7). This includes transdisciplinary collaboration and participatory methods in research.
(2) Recover Hidden Histories: Continue the critical recovery of hidden histories of the marginalized and oppressed groups through socially engaged, biographical work, storytelling and arts-based participatory research. This approach aims to develop purposeful knowledge that informs critical praxis.

(3) Preserve Space for Democratic Contestation: Ensure the preservation of spaces for democratic contestation against 'regulatory and governing authorities', 'techniques of neutralisation' and the forces of authoritarianism (Walters 2003: 11). This involves advocacy and challenging various forms of discrimination.
(4) Support Marginalized Groups: Through participatory research, ensure that marginalized groups have 'access to material resources necessary for self-development and meaningful participation in social, cultural, political and economic-decision-making' (Smith 1998: 31).
(5) Focus on Anti-Essentialism: Keep a central focus on anti-essentialism and resist all forms of domination and 'disciplinary normalisation' (Smith 1998: 35).
(6) Integrate Indigenous Knowledge and Decolonial Perspectives: Work collectively with Indigenous knowledge makers and apply a decolonial lens to our research and analysis, aiming to shape our collective social futures.

Notes

1   Feminist standpoint theory – emerged in the context of Marxist politics – experience – critique of capitalism and systemic oppressions.
2   https://www.feministwalkcork.ie (accessed 11 October 2025).
3   As part of understanding the wider social context, Adorno points to the individualizing and responsibilizing nature of advanced capitalism and the way it 'attributes to people everything that in fact is due to external conditions, so that in turn the conditions remain undisturbed'. This process can lead to alienation and 'people's alienation from democracy reflects the self-alienation of society' (Adorno 1984: 93). As Honneth (2009) identifies, the multiple approaches and work of those who connect with the Frankfurt School relate to the notion of social pathology and 'exploring the social causes of a pathology of human rationality' (Honneth 2009: viii).
4   The Deep Institutional Change Project for Sustainability and Human Development (DIIS) is led by Ian Hughes in collaboration with the Institute of Sustainability at University College Cork, in collaboration with Ariel Hernandez (The New Institute Hamburg and GIGA) and colleagues at UCC – Ed Byrne, Ger Mullally, Bob Grumiau and myself. It is underpinned by theories and concepts of human flourishing by Nussbaum and Sen and aims to lay the foundations for a critique and reimagining of the major social institutions in society – 'economics, democracy, religion, technology, gender and higher education'. This chapter connects to this project on the theme of gender.

The King's Philosophy and a Happiness Train: A Conceptual Overview of GNH

Karma Ura

1 Introduction: conditions for happiness

Gross National Happiness contrasts with other well-known concepts such as subjective well-being where life-satisfaction scores become central phenomena to be explained and by which nations are ranked (Helliwell et al. 2024) with normative connotation all ranking and rating carry (Mau 2019: 57). Such approaches analyse satisfaction dependent on individual's income, work, health, family and demographic characteristics, without considering larger contexts. Individuals strive to enhance their happiness by working on such life-satisfaction domains. Their life journey can be improved without being concerned with what happens to the society. They try to improve their narrative from their own first-person perspective, as their journey in their life unfolds on a programmed moving train. However, trying to do this alone is effective only to some extent. A more systemic approach to happiness should include the broader factors that go beyond individual practice to collective, policy making in organizations. When GNH is applied to the train, the direction and speed of the society can be altered.

2 Time

Like all mammals, irrespective of body size, people's lives can be measured in terms of the average number of heartbeats. End of life comes to all mammals,

---

I would like to express my gratitude for the extensive comments on the draft by Ian Hughes. I thank Andrej Zwitter, Ruth Chang and Sabine Alkire for stimulating directions. The article is a summary of a book project, which includes quantitative analysis, carried out at THE NEW INSTITUTE (TNI) in Hamburg. Thus, the primary debt goes to TNI for supporting a period of writing and drawing.

Figure 11.1 His Majesty Jigme Singye Wangchuck (reign 1972–2006), founder of GNH.

but it is met in different ways and different spans despite a similar number of heartbeats for all. Most lives consist of the historical average of one and a half billion heartbeats (West 2017).

Time as a condition of happiness is not comparable to any other condition. Other things cannot substitute for time, and it is inelastic. The passage of objective time is beyond anyone's decree or control. A billionaire cannot turn his day into thirty hours because it will be tomorrow for all including them. They cannot choose to sleep whole of this week and be awake whole of next week since they must live within the circadian rhythm lasting twenty-four hours and fifteen minutes. Mental and physical risks arise for those who deviate from this rhythm.

Moreover, subjective time differs from objective time within and between individuals. Judgement on objective time is often based on subjective time (Shipp and Jansen 2021). Although the days are of the same duration for all, the quality of experiences and happiness are not. Life span may be the same among two individuals; yet the quality of their health and happiness can differ widely.

How individuals use the same objective time is governed by state regulations, economic structure, cultural norms, individuals' preferences and chance. Money is not a constraint for most people in materially prosperous places; objective time has become the new constraint. Adequate objective time is a condition of happiness but there is often insufficient degree of autonomy over time by individuals. Ample objective time for adequate rest and restoration, socialization and sleep are significant issues.

Sleep of seven to nine hours for an adult should be prioritized over waking time. Waking time comes second in importance, strictly from the point of view of physical and mental health. Stress,[1] worry, night shift work, caffeine, alcohol, digital disturbance at bedtime and prolonged 'sunlessness' can contribute to sleep deficit (Walker 2017: 26–7, 341–2). Inadequate sleep, as opposed to insomnia, is a contributory cause of Alzheimer's, depression, chronic pain, cancer, diabetes, weight gain, obesity, immunity deficiency and a host of other physical and mental health issues (Walker 2017).

Objective time duration given to sleep may be the same between two individuals, but differing quality of their dreams and sleep makes a difference to the following day and after. Dreaming – both during the day and night – is a necessary activity for human beings, and an indicator of our internal world. Sleep expert Walker (2017: 207) characterizes nighttime dreaming as 'overnight therapy': (1) to 'nurse emotional and mental health' and (2) to enhance problem solving and creativity. But joyful daytime dreaming is equally an essential indicator of a positive mental world. To drift into the stillness of a lifting and restful daydream marks a high mental health. This too is threatened by screen time and productivity obsession that undermine positive daydreaming.

Leaders and organizations often focus excessively on the management of objective time as a strategic means towards the transformation of their organizations, at the expense of subjective time. However, individuals are often not present in the present moment: mind wandering, and time travelling are pervasive experiences. People spend more than half the day in subjective time (Shipp and Jansen 2021). Mind wandering incessantly to negative thoughts and ruminations undermine happiness.

The quality of subjective time is an analytically important condition for happiness, meaning and narratives structure of human lives. In affective terms, people feel happy when they are subjectively unaware of the passing of time (Levine 2006). Time feels non-existent at that point. Experiences of subjective time differ among individuals. That makes a difference to happiness.

3 Natural environment

The cosmos consists of the essence (beings) and the vessel (the physical world) according to the antiquated Buddhist spiritual cosmology. All beings live in a shared vessel, as it is referred to nature in the Abhidharma-kosa that was formulated much later after the Buddha lived. Both the physical world and the sentient beings arise and experience 'reality' according to their cumulated virtues and non-virtues. They are happy or unhappy, fortunate or unfortunate due to their unfolding intentions and actions, i.e. *karma* (Jampalyang 2018).[2] On this view, happiness is a broader temporal outcome of much broader collective conditions they have created. The Abhidharma views the ebb and flow of life on the planet as being driven by intentions and actions of beings.

But human beings do mistakenly believe themselves to be standing outside of nature instead of nature being a condition in which they develop (Nassar 2022).[3] Nature is falsely excluded from daily interaction with our senses, making people infer that human beings can achieve happiness by standing outside nature. Once human beings conceive themselves of being outside of nature, it can precipitate domination of nature. Both social domination and the domination of nature are shown to be interconnected. It is a problematic duality consisting of culture versus nature, and reason versus nature (Plumwood 2002).

The role of the immediate natural environment in happiness is muted in most happiness literature. Happiness is foregrounded in terms of income, work, personal health, family and so forth. Income is discussed in the context of economy, and the economy is discussed in relation to the supply of natural resources. Nature thus appears primarily as a distant store of resources, and only the consumption of resources matters for happiness. In the metaphor of a train journey, nature can look like a two-dimensional image seen through the windows of the train. Knowing something as an abstracted thing does incline people to think human beings are outside nature while the train is driven through the landscape it can destroy. They can be desensitized from any aesthetic and sensory experience of it having been confined too long to the compartments of the train, which is totally man-made and devoid of communion with nature.

Yet the beauty of the physical world is the most common everyday source of happiness for mankind. Human happiness is impacted daily by the natural environment. The richer the natural environment, greater their daily positive impact on happiness. Nature's aesthetics shape human sensory awareness and happiness profoundly. Not only bio-physical needs are met by nature, but an

Figure 11.2 An expression of an early view of the web of life. Ink on parchment by Karma Ura.

individual's happiness is raised by the beauty of flora and fauna, and its reflection in the arts.

As nature's aesthetic stimulus is largely free, it is taken for granted and has become an under-appreciated condition. Skills for aesthetic experiences are not cultivated especially among those alienated from nature. Beauty of forms, movements, sounds[4] and colours of nature enliven the mental and physical life of human beings both at conscious and unconscious levels. Underlying the beauty of the natural environment are fractals and scale invariance,[5] Fibonacci numbers and golden ratios (Dunlap 1997; West 2017). Natural forms, as nature's works of art, are beautiful at a visceral level because they manifest symmetry, economy and vibrant colours. Symmetry in nature is defined as 'harmony, balance, and proportion', while economy in nature is defined as nature's ability to produce 'an abundance of effects from very limited means' (Wilczek 2016).

Aesthetic experience affects simultaneously the sensuous, emotional and intellectual aspects of being.[6] It consists of 'direct experience of being in nature as well as indirect experience of nature through works of art' (Nassar 2022: 216-17). A deep perception of form, for example of plants,[7] is crucial in the realization of aesthetic experience and ecological understanding. They could be described as deep visualization rather than direct observation. It is only in systematic visualization that we can create and relive a coherent and integrated picture of nature, such as form and colours of a plant.[8]

Colours have inherent psychological qualities and evoke happiness or sadness. Goethe linked colours to emotional experiences. For instance, experiencing the yellowness of sunflowers was associated with liveliness and joy. For human beings, the exposure to the greenness of a forest, for example, has significant psychological effect. A lush green forest is associated with balance and calmness, and hence a direct perception or visualization of a forest can induce tranquillity and restfulness. Goethe also discovered the phenomena of colour, such as coloured shadows (Heitler 1998) that the eye naturally sees as the complementary colour to the colour which is viewed.

Finally, the role of fauna, both wild and domesticated as everyday sources of awe, pleasure, emotional support, socialization, companionship and learning is highly underestimated. Louv (2019: 137) conveys this point strongly with respect to wildlife:

> For millennia, humans have drawn solace and cultural and spiritual illumination from wildlife. Nonetheless, the academic world has expressed only glancing interest in the inclusion of wildlife in therapy directly or indirectly. The dangers of wildlife—as attackers, as vectors of disease—are more easily measured than their positive direct or indirect influences on human mental and physical health.

Seeing animals as part of the community of lives that influence each other's happiness and well-being could also contribute to bridging the divide between man and nature.

4 Living standard, market and ethics

The economy consists of physical goods that are priced and transacted in a market. The production and consumption of goods and services increase year by year. This increase in transactions is the main concept of progress at present. The measurement of this increase is equated with the increase in welfare of the population. It is essentially a false equivalence. Material output increase per year is assumed to be translated into increased positive mental and bodily experiences per year in a linear relationship. A crude example is certain length of road construction, which is assumed to be translated into positive mental and bodily experiences in a linear relationship. Another example is military expenditure increase per year. This is also assumed to be translated into positive mental and bodily experiences per year in a linear transformation. At a macroeconomic level, any increase in production and consumption of goods and services is largely equated with improvement in human welfare.[9] This is, in brief, the way progress is being measured through GDP. It is obvious that such a linear relationship is not coherent with reality.

GDP as an account of progress is problematic, although it has achieved a legendary status as a metric of progress. The amount of yearly materials extraction is too excessive to generate so little well-being (Abdallah, Hoffman and Akenji 2024).[10] Legitimizing GDP growth as an achievement legitimizes and condones excessive exploitation, which is a form of confiscation from the future generations, not to say other creatures, should they have any right. The continuing use of GDP as the only measure of human well-being and progress is illogical and unethical. As such indicators are considered problematic from the point of view of happiness: other values that represent well-being and happiness must be explored and adopted.

The assumption underlying a linear relationship between GDP and happiness is incoherent. But even if the assumption of linear relationship between materials output and human welfare were not false, the planet cannot go on supplying such colossal amounts of material for much longer. Ninety-two billion tons of materials are extracted every year (Kallis et al. 2020). 'About 28 tons per person per year, on an average' are extracted, but 'a sustainable level of material footprint … is about 8 tons per person' (Hickel 2021: 106–7).

Supposing global GDP is pushed to grow 3 per cent every year, global GDP will double every twenty-four years, unleashing devastating side effects on the welfare and happiness of future generations, caused by pollution and the emissions people leave behind. From the perspective of costing, pollution arises because someone shifts costs from themself to another person (Kallis 2018). Fossil fuel pollution arises because its users shift the cost of its use to others, including future generations. Trains that are driven by dirty fuel are an example of shifting pollution cost to non-users. Pollution of rivers and air from mining arises because the mine owners shift the cost of mining on to others who use the rivers and breathe the air.

Moore commented that 'the *condition* for large scale industrial production is Cheap Nature' (Moore 2015: 93), but 'Nature is finite. Capitalism is premised on the infinite' (Moore 2015: 87). He names the four main inputs that have fuelled capitalism as 'the Four Cheaps: food, labor-power, energy, and raw materials' (Moore 2015: 127). All these four major cheap inputs have been produced with 'the interlocking agencies of capital, science, and empire' (Moore 2015: 63). Surplus or profit is based primarily on the exploitation of the 'Four Cheaps', particularly labour. Both social domination and the domination of nature are interconnected (Plumwood 2002).

Under capitalism, marketization and commodification displace and colonize other values in society and compel them to be part of the monetary value (Kallis 2018). For instance, when essential services like education or healthcare become commodities primarily driven by profit motives – as big pharmaceutical corporations, private hospitals, elite schools and universities often do – access to health and education can no longer be based on needs or rights but on wealth. In many spheres of life, commercialization and marketization should have no role, but the penetration of the rationality of market has weakened social bonds and diminished the sense of community and shared responsibility (Sandel 2012). Sandel argues that many activities that have moral and ethical implications should not be subject to market forces, as doing so can lead to outcomes that are unjust or morally objectionable.

5 Community and commons

A community is an organically related group of individuals who have repercussions on the happiness and well-being of its members and is an essential condition for happiness. It encloses other relationships such as parental, romantic, kinship and friendship (Dunbar 2010; 2012). These relationships are crucial for happiness and can be deepened by commitment and caring, intimacy and sharing. Quality of contacts matter more than mere frequency of contacts

(Jamieson and Simpson 2013). Happiness is most often an intangible and unbidden gift by one person to another belonging to these circles of relationship.

The size of each circle of relationship varies, in the Bhutanese case, by spheres of contact such as sickness, money, emotional intimacy, illness and life-passage events. Bhutanese data show that belonging is not significantly different between in-migrants and non-migrants among citizens of the country, but that is because it is a small society with culture that is similar enough between communities. Data does not compare foreign-born versus domestic citizens: both in-migrants and non-migrants are citizens.

Members of a community share a sense of belonging, a primal emotion not only among human beings but among higher animals where cooperation and care among the group members are necessary for survival and flourishing. An individual who feels alone and isolated has no one to belong to and no one to fall on for support.[11] Living alone in old age, without any care by children has become a determinant of well-being. In a fraction of cases, the elderly die in isolation without having someone to carry out their funerals (Jamieson and Simpson 2013).

Bhutanese data confirm that the score of subjective well-being on the scale of 0–10 is lower for people living solo compared to multi-member households. Isolation deprives people of love and laughter and drives them into mental ill-health. Humour, which arises from recognizing incongruities specific to a culture, and triggers laughter is an unexpectedly powerful uniting factor (Dunbar 2012). Dunbar calls laughing communal laughter since human beings laugh socially or together.[12] Laughing together is an act of renewing a community of laughter which is bound by understanding a norm (Carroll 2017).

Underlying all close human relationships is a degree of unconditional loving kindness, a sentiment more universal than even a sense of justice. Every human being has been an infant, and all have received such unconditional loving and nurturing kindness from their parents, and therefore all can give. Maternal love, which is an aspect of loving kindness is the most important altruistic gift in the perpetuation of a species. Unconditional love and nurturance of a mother towards her infant is the perfection of generosity because the giver does not do it with any expectation of reward. Parents and alloparents' nurturance instinct is the basis of altruism towards biologically unrelated people (Ricard 2016).

Loving kindness is cultivatable through the practice of *tonglen* meditation, which is a form of mind training practice along with mindfulness practice (Ricard 2016). A practitioner first simulates in their mind by being grateful to all those who have been kind and generous to them. Feeling of generosity is correlated with feelings of compassion, contentment and forgiveness in Bhutanese data. Remembering all the acts of kindness is conducive to the next stage where they can imagine giving away whatever they cherish. This is followed by actual generosity. Generosity is an antidote to selfishness and greed that thanks community and humanity.

A community existed often prior to the nation in which it sits, and can have its own history and legal concepts (customs). Its existence is crucial because a child develops as a social being only by being in a community. Commons, especially land, as belonging neither to a government nor to private property but held in trusteeship by a community, is crucial to the concept of community. Customary rights, which entail legal pluralism rather than a single legal order of the state, are favourable to the continuation of community. Interactions over material and intangible commons in a community lead children to become fuller social beings. They shape individuals into social and political beings who forge decisions affecting the local community. It is more natural for individuals to embrace a community as an object of love and loyalty for its members, in addition to love and friendship among its members (Ludwig 2002).

6 State and governance

What kind of governance system do we need to flourish? Individuals can give each other happiness, but a government cannot. Yet a government can profoundly shape collective life and promote conditions of happiness more than individuals can. For instance, in relation to the train journey, a government formulates transport policies that influence on which tracks people and goods move. Policies and legislation affect almost every aspect of life, penetrating pervasively and deeply into the personal realm. The separation between private and public realms are artificial from the point of view of happiness.

An individual might assume that they are free to choose happiness, and it is situated in their emotions and biology. But they are ensnared in a socially constructed web of institutional networks of politics and economy and religion into which they are socialized, and which together regulate their behaviour and thoughts, and their relationships with other human beings.

It is in response to such institutional networks of power and dominance on how an individual's life has been structured that agency and freedom are necessary for happiness. If only the driver controls the direction and speed of the train, there is no agency, or the exercise of imagination, freedom, intellect and intention for those on the train. In the architecture of GNH, the governance domain incorporates these aspects. A person's agency for directing their life must be enlarged by increasing the space for freedom of thought, speech and practice. These are integral to both Buddhism and Western traditions of human rights. Buddha encouraged freedom of thought, speech and practice. These are necessary for individuals to test what is empirically true from a first-person perspective as well as to forge a third-person perspective on different collective paths.

Freedom and rights are necessary conditions for responsibility, co-creation and agency, and for Buddhist metaphysics of *karma*. From an axiomatic Buddhist point of view, the primal drive for happiness is common among all beings, and this central commonality is the basis of the shared right to be free from pain and suffering for both people and animals, since pain as a psychophysical phenomenon is experienced by both. This also leads to the argument for the concept of right to life reserved so far to human beings to be extended to other sentient beings.

Buddhist virtue ethics does not dwell on the architecture of governance or institutions. Its emphasis is far more on ethical training so that the best potential of human nature is realized. It is aimed radically at transformation of three afflictive states – ignorance, aggression or anger, and greed – into wisdom. Ignorance has a special meaning: that of an illusion of a permanent and independent self, leading to attachment. These three mental phenomena are described as the root causes of suffering and the obstacles to happiness and beyond.

Where a lack of these virtues and qualities persists in the character of individuals and their leaders, it will inevitably get reflected in the institutions and systems. A system of checks and balances, the rule of law, and the idea of conflict of interest and so forth are based on systemic or institutional performance but they cannot further systemic freedoms without qualities and virtues of individuals within the institutions. We need to shift once again from outside architecture to inner architecture, while valuing both in their rightful proportions.

7 Culture

A significant cultural and historical difference between the East and the West has been a culture of meditation and mindfulness that influences happiness in the East. Culture can be defined as entrenched, legitimized habits and goals.

In certain ethics, such as Buddhism, happiness and activities that support it have philosophical and ethical standing. In others, discourse on happiness and associated activities are muted, while greater prominence is given to the prosperity in terms of the material world. There is less thought given to how people sense and experience these material objects, and how attention, an ever-present fundamental consciousness while awake, is deployed. Attention is the link between the inner and the outer world.

Meditation can be loosely defined as the practice of bringing attention to the present moment of experience and observing that experience (Olendzki 2003). A wide variety of meditations has been grouped into three categories, namely, those that focus on improving attention; those that improve virtues such as loving kindness and compassion; and those that focus on insight going beyond the ordinary cognition to 'non-dual' awareness (Dahl, Lutz, and Davidson 2015; Davidson and Dahl 2017). Mindfulness meditation is not simply a question of brain state. It is part of behaviour and how we relate to others as an enacted phenomena in our social and economic world (Seth 2021).

A compulsive attention to rumination and distraction, along with spotty attention on the present moment, has been associated with activeness of the default network mode. In extreme cases, a hyperactive default network mode (involving the singulate cortex area of the brain) is associated with anxiety and depression. Distraction or the inability to control attention, i.e. mind wandering, is an existential challenge for most people, a major analysis from a Buddhist cognitive and psychological standpoint.

Mindfulness meditation that trains attention to focus on something while ignoring distractions (Davidson and Dahl 2017) builds a flexibility and control over attention, interest and feelings to enable an individual to feel relaxed and be mentally spacious, an inner condition for happiness. It prevents attention from drifting into the past and future in an endless succession of momentary thoughts and perceptions by building awareness towards equanimity, an ability to remain in neutral feeling, without reacting emotionally to any sensations and feelings (Anālayo 2018). But the key characteristic of mindfulness is not only to hold attention onto something but being aware of being in that state, which distinguishes it from being merely absorbed.

There is no global data on mindfulness meditation practice. GNH data shows that at any given day, 7 per cent of the population practice meditation as opposed to 62 per cent who engage in daily prayers. It would be preferable for the numbers to be switched, and hence meditation programmes have been instigated.

8 Emotions, psychological well-being and memory

Modern education has elevated skills and rationality over emotions. From the perspective of happiness, emotional states must be given significance, if not primacy, over the rational and causal thought process. GNH metrics and policy recognize happiness is constituted by positive emotions of calmness and joy, and absence of worry, fear and sadness. They are taken prominently into account in a conception of happiness. High frequency of negative emotions indicates a lack of happiness.

Emotions as mental states occupy a profound position in Buddhism.[13] Desire-attachment, greed and anger are the mental afflictions that represent core negative emotions. Nussbaum (2016: 14) notes that: 'the idea that anger is a central threat to decent human interactions runs through the Western philosophical tradition'. While she notes the role of anger in responding to injustices, her analysis also resonates with Buddhism on transforming anger into actions for personal and social welfare, through compassion and generosity of forgiveness (Nussbaum 2016). Negative emotions ought to be extracted and neutralized, for us to move forward. They can rise but then subside with mastery of non-judgemental awareness meditation (Rubin 2003). Among the six virtues of *Prajnaparamita*, the most relevant virtue for the psychological well-being domain of GNH is generosity. Compassion and generosity are at the heart of positive emotions that enable us to engage beneficially with others. Generosity is a concrete behavioural expression of compassion, testifying its demonstrative existence. Indeed, GNH metrics measure not only self-reported frequency of such emotions but also how frequently they engage with others through various forms of material and non-material contributions.

Worry, anger, fear, sadness and selfishness are major emotions confronting human beings. Their prevalence rate occurs in that descending order among Bhutanese populations. Fear is linked with anger. In the context of global situation, coercive leadership creates fear for political expediency either explicitly or implicitly among the people. Fear inhibits people from gaining agency. They cannot dispel the internal sources of fear or confront the external penetrators of fear. A fearful rule of a country fosters blind obedience and creates what has been called 'police society' (Peckham 2023). Fear 'as a tool for acquiring and retaining power' was formulated in a callous way in Machiavelli (2003). Equally, the role of fear in gaining oppressive power over the public features indirectly in Locke and Hobbes (Peckham 2023). In modern corporations, Faraci et al. (Funnell

2024) argues that evidence from corporate surveys in a few Western countries show that presence of fear among the employees in corporate workplace is a factor in the loss of productivity and dynamism, and a cause of unhappiness.

Despite huge economic, educational and health improvements, it is difficult to make societal progress in terms of positive emotions unless there are deeper strategies. Three national surveys carried out over the last seventeen years in Bhutan show this. The average frequency of various positive emotions and negative emotions has not changed significantly over the last seventeen years. Bhutanese GDP, for example, has doubled over the last seven years from Nu 152 billion in 2015 to Nu 301 billion in 2022. Over the same period, the percentage of the sample who felt angry a few times a day to once a day was 7.75 per cent in 2015 and 8.1 per cent in 2022. Similarly, the percentage of the sample who felt generosity once a day or a few times a day was 14.8 per cent in 2015 and 14.6 per cent in 2022. National averages are steady. It means that an increase in per capita income doesn't necessarily push the frequency of positive emotions up nor bring the average frequency of negative emotions down. But their distribution is sensitive to age, gender and region. In general, the frequency of anger, fear and worry declines with age but worry and fear rise again in old age.

Current emotional experiences are not the only factors for happiness over the long term. It is how emotional experiences cumulate as memory, i.e. through retrospective assessment (Breithaupt 2025). An individual appraises how far she has been able to close the gap towards an ideal life. An ideal life is not the same as aspirational life. An ideal life is something in harmony with deep ethical paradigm whereas an aspirational life is fulfilment in terms of contemporary standards. The ideal life is culturally and ethically shaped. In this sense, happiness is socially constructed for every individual.

Individuals seek justification in terms of the meaning of their goals and aspirations. They view their life from a first-person, subjective perspective or from an impersonal, third-person perspective (Nagel 1979). Our personal experiences have meaning for us, yet they may seem insignificant or without meaning from the third-person point of view of the wider objective universe. There is often a tension between the two views. The justifications in terms of meaning come to an end, and end in nowhere because there is no other higher reason to justify it, without regressing further and further. This can lead to a sense of absurdity and despair and crisis in meaning (Zwitter 2023a). The response to this existential crisis is different in philosophies of Camus and Nietzsche. From a Buddhist point of view, 'final causes are introduced only in the intentional thought and actions of living beings, and they go out of existence with them'

(Carpenter 2021: 188). Suffering is brought about by the confusion of the human condition of ignorance, and that matters hugely in getting rid of suffering. To think that life is meaningless is itself a form and condition of suffering.

9 Virtue ethics and being

Leadership is not independent of the substantive question of existence, and the substantive question of existence is inseparable from the idea of the good and the ideal life. Haybron clarifies one of the notions of good life. A theory of the good life should specify all the things that ultimately matter in life, whether they benefit the agent or not (Haybron 2008). The content of the ideal of a good life in Aristotelian ethics is happiness through virtue cultivation, while for Buddhism it is the reduction of suffering as an immediate goal and enlightenment[14] as a distant goal, through practice of six virtues or excellences (Aristotle 2009; Conze 1975; Wright 2009). They share many similarities, although virtue specifications and end goals are different. If it is counted in a detailed way, there are twelve virtues in Aristotelian analysis, but it is often summarized to four whose cultivation are necessary for happiness: justice, temperance, courage and practical wisdom.

Aristotelian and Buddhist virtues are not the only ones to have a conception of good life. The good life for Kant came about from actions of rational agents, which are carried out in accordance with a sense of duty for the moral law that he called the universal law of the Categorical Imperative. The good life, in the consequentialist view, is based neither on virtuous character nor in following fulfilment of moral laws legislated by human reason in a third-person perspective. It is centred on either aggregate maximization or average maximization of pleasures and happiness, technically known as utility, accrued from a first-person perspective. It is important to note this, as the concept of consequentialist pleasure and happiness differ from Buddhist ethics.

Buddhist ethical framework takes the relevant consequences of action not to be pleasure and pain in terms of introspectable experiences of actors (first-person), but suffering caused to others (second-person and third-person effect). The very basis of morality seems distinct from consequentialism or utilitarian. It is not how happy an agent feels that is the criterion but how much beneficial happiness others have felt by his action. In general, suffering can be reduced simultaneously for others and oneself if an individual guides her efforts by the six excellences (*paramita* in Sanskrit) or virtues. The six excellences or virtues are generosity, morality, toleration, energy or effort, concentration and wisdom.

They 'provide concrete guidance for the construction of character' (Wright 2009: 8).

The *Prajnaparamita sutra*, which discusses the six excellences or virtues, comes in various versions with differing length: 8,000 lines version (*brgyad stong pa*), 25,000 lines version (*gnyis khri*), and a 100,000 lines version (*bum*). Buddhist ethics for flourishing, which confronts the three pervasive human qualities, is the practice of the six excellences or virtues simultaneously. Those who practice the six paramita virtues are model beings in Buddhist ethical conception (Conze 1975; Wright 2009). So, Buddhist ethics is based on character-exemplars rather than rules. In relation to governance or management, those who hold offices therefore ought to embody them, as their character.

Liberal moral theory does not particularly emphasize virtues and dispositions because the question of character is left as a private matter. In liberal theory, liberty or right of the individual to be what she or he is must be protected, mostly as a negative freedom, subject to affordability of the state. In this sense liberal theory cannot advocate any character virtues (Garfield 2022). Such virtues as generosity and compassion, which are central to Buddhism, are dependent on occasional charitability in liberal theory. Rights and freedom can remain unfulfilled, existing merely as a formality. They cannot be fulfilled as a claim against someone else. However, if virtues such as compassion, kindness and generosity were cultivated, they would facilitate recognition of the rights. Garfield (2022: 125) argues that 'without a foundation in compassion that recognition facilitates, rights become pointless. And if there is an antecedent relation of compassion, rights are unnecessary'.

There has been a steady shift away from virtue ethics. This shift to ethics prevalent today has been taking place since the eighteenth century (McCloskey 2010), in conjunction with the rise of neo-liberal capitalism. Consequently, leadership positions are not necessarily filled by individuals possessing virtues. They can then easily be masked agents of hierarchy, power and wealth. In such virtueless space, there can be too many elections and far less democracy, and too many leaders and far less liberty.

Neo-liberal capitalism has risen rapidly since the late nineteenth century and spiked after the 1970s, bringing both innovation and prosperity on one hand as well as inequality, climate change and biodiversity collapse on the other. The global consequences of greed have escalated through capitalism in the last two centuries. Equally, anger and fear shape the stances of international relation as much as hope for external peace and inner happiness. The share of global military expenditure as a percentage of global GDP was 2.3 per cent, equivalent to US$

2,443 billion, in 2023. The amount that would be available if it were divided among each person in the world in 2023 would be US$306 per person (Tian et al. 2024). In comparison, the percentage of official development assistance (ODA) as a percentage of global GDP was 0.36 per cent, is equivalent to US$ 204 billion, in 2022 ('ODA Levels in 2022 – Preliminary Data Detailed Summary Note' 2023).

Virtue ethics including Buddhist ethics assign importance to individual qualities or virtues, instead of systems and institutions. Where a lack of virtues and qualities persists, it will inevitably get reflected in the institutions and systems that individuals inhabit. The dilemma posed by the need to have institutions while they cannot be held responsible has been a recurring theme in morality. The institutional insulation of crimes, and crimes committed in and through the offices and institutions people hold, have emerged as the greatest threat to freedoms and rights. At the widest level, colonialism and imperialist projects could fit into the nature of such institutional crimes. Nagel (1979) discusses the difficulty of assigning responsibility and culpability, for example, in the context of war crimes and atrocities. Responsibility cannot be shifted to a relational network. There is, then, a difficulty in holding an entire government or institution accountable for actions that may involve numerous individuals with varying roles and moral culpability. The greatest crimes have been arguably committed by public figures using official institutions. Institutions are vulnerable when those who control them have no virtues.

10 Official application of GNH indices[15]

There are five technically specific ways in which GNH indicators are being applied in the administration of the country.

Firstly, the GNH index and some of the domain indicators and sub-indicators are directly used as a benchmark in the five-year plan (FYP). Every FYP is guided by national targets and key results. In the current, thirteenth FYP, seventeen baselines or targets are drawn from GNH indicators such as sufficiency level in mental health, safety, community vitality, skills, political participation, fundamental rights, subjective happiness, values, assets, income, and housing. The composite GNH index is used as an overall national baseline.

Secondly, the GNH index is also used as weighted criterion in the allocation of budget among the twenty districts and four urban municipalities of Bhutan. GNH index of a district is given 65 per cent weight in the district budget

Figure 11.3 Idealized scenario of implementation of GNH, which varies from government to government.

distribution in the current FYP. Those districts with poor GNH index get more resources to implement relevant projects.

Thirdly, policies are formulated by subjecting them to vetting with GNH policy screening tools, which consist of twenty-two criteria drawn from GNH and implemented according to a well-defined process since 2008, revised further in 2015 (Cabinet Secretariat 2015). So far, fifteen out of twenty-two draft policies have been approved, with majority of them getting modified to some degree by the process of policy screening.

Fourthly, big state corporations like the Bank of Bhutan, Bhutan Telecom, Bhutan Power Corporation and all nine hydro-power plants of the Druk Green Power Corporation have been assessed according to GNH certification for business so that businesses in the government align with GNH values. GNH business certification is being implemented in Brazil by some private companies.

Lastly, GNH is taught as a subject in academia and schools, and civil service and public enterprises management training. GNH survey findings are communicated publicly through outreaches in the 205 gewogs, the lowest tier of administrative organization.

The GNH index for the country has improved from 0.743 in 2010, 0.756 in 2015 to 0.781 in 2022. The slow rate of change is a consequence of 135 variables, some of which fall back during broader movement forward.

11 Conclusion

To conclude, passengers in the train have limited success trying to maximize their individual happiness in their compartmentalized lives. Far more possibilities of collective happiness exist but they are bound up with leaders who exercise state powers. The conditions of happiness discussed here are fields of leadership's action. The virtues are modes of being, especially for leaders or the drivers of the train, in both private and public realms.

The GNH index and indicators provide both internal and external compass for the leaders and the passengers on their journey. A passenger is relative to the train, and the train is relative to the natural environment. All are relative to emotions and feelings – happiness – in the passenger's sense-making of the journey. A journey can be assessed in multiple ways. But one perennial way is first-person experience of it. No wonder you are asked at its end: hope your journey was pleasant, and you have happy memories of it. Happiness does not await at the end of the journey. The path is the goal.

Notes

1. Severe distress affected about 4 per cent of the nationally representative sample of Bhutan in 2022, having doubled perhaps by the Covid's impact. More women were affected than men.
2. See Tibetan commentary on Vasubhandu's Abidharma-kosa by *Chim Jam' dpal dbyang* (1245–1325) titled *Ornament of Abhidharma (mdzod 'grel mngon pa'i rgyan)*.
3. 'Living beings are not caused by their environments. Rather, their environments are the conditions in which they develop. To this Humboldt added: living beings are also the conditions in which environments develop. There is no relation of

external causality here, i.e., between two originally separate entities' (Nassar 2022: 210).

4   See Wilczek (2016). Beautiful tone follows the Pythagorean rule. Two tones sound harmonious when the ratio of the length of two strings with equal tension vary in 1:2; 2:3; 3:4 and so forth.

5   See Wilczek (2016). He characterizes invariance as 'features that are common to all representations ... valid from any perspective'.

6   The key to aesthetic experience of nature, according to Humboldt (Wulf 2016) and Goethe (Goethe 2009; Heitler 1998), is a mode of thinking called intuitive judgement. It involves going from a particular to the whole through imagination. Heitler (1998: 66) described such perception and archetypal images: 'Above all we must agree with Goethe that the "archetypal images" – the formative principles behind phenomena – are spiritual realities accessible to our cognition, and we must view them as a part of the spiritual content of nature'.

7   See Schmithausen (2009). Plants were categorized as one sense-faculty being as they have sense of touch. Thus, early Buddhist thought they had fundamental sentience.

8   An essential phenomenon, which emerges from intuitive judgement, is called archetypal phenomenon, which Goethe calls archetypal (holistic, essential) phenomena.

9   Such linear relationship is not assumed in microeconomic analysis of consumption. But at GDP or macroeconomic level, non-linear analysis is scarcely used, and 5 per cent growth is assumed always better than 4 per cent growth.

10 One index that tests this statement is the happy life index, which is equal to life span multiplied by self-reported well-being divided by carbon footprint.

11 Bhutanese data shows the pattern of trust and belonging levels differ between migrants and native residents in a community.

12 Dunbar (2012) says that a good belly laughter leads to release of endorphins such as oxytocin and dopamine.

13 The six main factors of emotional bewilderment or obscuration are part of the fifty-one mental formations in Abhidharma. They include (1) desire and (2) anger (*khong khro*).

14 See Kyabgon (2010). He discusses enlightenment as a matter of moving beyond subject-object duality towards Buddha nature on the subjective side and emptiness on the objective side.

15 This section is adapted from my lecture delivered in Sheldonian Theatre, University of Oxford, 8 January 2019, titled 'GNH: Development with Integrity' (Ura 2019).

Part Four

New Imaginaries of Flourishing

12

Planetary Thinking for Planetary Flourishing

Frederic Hanusch and Claus Leggewie

1 Planet-human relations

Planet-human relations take place in spatial, temporal and material dimensions that are planetary in extent and initiated by the planet or humans or both. Earthquakes, for example, can be caused by plate tectonic movement, by the building of reservoirs or by fracking in regions that are already geologically fragile. Similarly, the Earth's magnetospheric shield is perforated both by solar winds and nuclear tests, resulting in failures within our electricity and telecommunication infrastructures. Planet-human relations are thus a quasi-agent in their own right, even though we tend to only notice this in case of incidents. Generally, we can ascribe to them three features: Planet-human relations are *metabolic* in that they concern flows of matter between the planet and humanity, but without equating the two spheres and lapsing into material relativism. They are *recentring* since they remove humans from their special position, but without releasing them from responsibility. And they are *transversal* because they connect things and concepts like nature and culture, but without merging them.

Let us have a closer look at these three features of planet-human relations. First, they are *metabolic* inasmuch as the material exchange occurring within them (Fox and Alldred 2015) begins with our own bodies: about half of a human body's mass – along with all other matter in 'our own' Milky Way – actually descends from other galaxies (Anglés-Alcázar et al. 2017). Since we consist of the same stuff as the Earth, humans are fundamentally planetary creatures, 'walking, talking minerals', as Margulis and Sagan once put it (Margulis and Sagan 1999: 49; with recourse to Vernadsky 1998 [1926]). This kinship between humans and things was elaborated within 'vital materialism', a concept coined

---

Some arguments here draw on and extend material previously published in Frederic Hanusch, Claus Leggewie, and Erik Meyer, Planetar denken: Ein Einstieg (transcript Verlag, 2021).

by ecological theorist Jane Bennett (1987; 2001; 2002; 2010) which questions the traditional Western distinction between the modes of existence of humans, animals, plants and minerals. The old juxtaposition of active human subjectivity and passive materiality is dissolved in planetary thinking. From this perspective, even inorganic matter seems strangely alive and has agency – not in the sense of intentional action, of course, but in the sense that it causes effects. Recognizing that inorganic, organic and human 'activities' are linked and interdependent does not take away from the specific agencies of humans.

Once we concede this kind of agency to 'inanimate' nature, animals, etc., this will make us recognize humans as material beings, respect our kinship with non-humans, and realize that we are all united in a planetary community. The status of human agency will change when we acknowledge that our world is greatly influenced by all kinds of non-humans. We will then see even strictly human actions in a different light – such as the cruel exchange system of slavery which forever changed landscape ecologies on both sides of the Atlantic (Protevi 2006; The Long Now Foundation 2019; Yusoff 2018). But we can go back even further in history and inquire about the origins of matter. Karen Barad's 'agential realism', one could say, conceives of matter as frozen action, i.e. 'not snapshots of preexisting *things* frozen in time – caught in the act as it were – but rather condensations of multiple material *practices* across space and time' (Barad 2007: 360 emphasis added). We tend to fall, Barad observes, for 'the illusion of the self-evidentiary nature of "the given"'. But humans, non-human lifeforms and things are not 'given'; they arise from actions.

If we, for instance, look at energy production with planet-human relations and their metabolic quality in mind, we find that this relation is characterized by imitation when we resort to established strategies and, say, build new coal power plants. Other options are exnovation (abolishing the outdated), innovation (introduction of the new) or renovation (upgrading the existing). Depending on the combination of such relations we can assess a civilization's level of development. According to the Soviet astronomer Nikolai Kardashev (1964), a civilization, whether on this or another planet, can progress through successful innovation. The scale that Kardashev devised (and that bears his name) 'classif[ies] technologically developed civilizations in three types: I. technological level close to the level presently [i.e. in 1964] attained on earth, II. a civilization capable of harnessing the energy radiated by its own star and III. a civilization in possession of energy on the scale of its own galaxy' (Kardashev 1964: 219). In other words, the distinction concerns the degree to which a civilization makes use of energy sources beyond its own planet (Gray 2020).

Second, how far is planetary thinking *recentring*, dispossessing humans of their self-awarded special status among the inhabitants of Earth? Chakrabarty's take on this is somewhat paradoxical: with the advent of the planetary, he holds, 'humans have become a question for themselves, but they are not sure what this question really entails' (Chakrabarty 2019: 31). *Homo sapiens* is neither the culmination nor the endpoint of evolution. We are a species that vacillates between omnipotence and a fear of losing control, between driving and being driven, a species that has produced a new geological era called the Anthropocene and no longer believes that the Earth system can be restored to the Holocene or even the Pleistocene – or moved forward, thanks to human ingenuity and vigour. Humanity's status as a 'super-agent' controlling 'nature' is up for negotiation. Consequently, radical strands of post- or transhumanism want to get rid of humanity for good (Aydin 2017; Bauer 2010; Winner 2004), whereas more moderate variants still view humans as 'companions' (Gane 2006; Haraway 2003). We are inclined to champion the latter solution and thus avoid going straight from an anthropocentrism to a planetocentrism in which humans are but one of an ecosystem's many and equally important components – as posited by more extreme forms of an object-oriented ontology that regards 'objects – whether real, fictional, natural, artificial, human or non-human' as 'mutually autonomous' (Harman 2018: 12). While gradually extending the definition of consciousness and of the possibility of intentional action is possible (one thinks of artificial intelligence, but also of animals), there are clear limits to this when it comes to inanimate matter (Alaimo 2016).

Intellectually more ambitious than a levelling of all differences between the human and the non-human are attempts to detect and integrate the 'more than human' (Bellacasa 2017). There appear to be two ways of approaching such an extended anthropology: first, an 'embodied anthropology' (Fuchs 2021) offers the chance of conceiving of bodiliness, aliveness and embodied freedom as the foundation for an autonomous human existence in which (other than in transhumanism) technologies remain mere instruments. And secondly, a 'geotropic astronautics' (Blumenberg 1987: 675) enables us to turn the camera and point it back to Earth, to make the planet our protagonist – if only because it takes too long to reach other potentially habitable planets.

A realist anthropology will not privilege humans due to certain features they display, and yet it will hold on to them as an essential variable (Bajohr 2019). A complete levelling would be possible only if we considered ourselves responsible for assigning an equal or unequal status to ourselves and others. Against the demand of overcoming the human, the German philosopher Otfried Höffe makes the case for a concept he calls '*oikopoiesis*': humans, Höffe holds, shape (*poiesis*) their natural and social environment in such a way that it becomes for them a home (*oikos*) (Höffe 2020: 11). Such a recentring anthropology accounts for its in- or exclusions of certain knowledges and ways of being (Dolphijn and Tuin 2012; Garske 2014), including the integration of 'companion species' (Haraway 2003) – for instance, by granting legal rights to rivers, mountains or forests, and as with joint-stock companies or homeowner associations, choosing human spokespeople to represent their interests (O'Donnell and Talbot-Jones 2018). In such models, human institutions take control and responsibility, which is necessary since the efficacy of non-humans does not amount to an intentional agency. After all, even the 'parliament of things' would have to be convened by humans (Latour 2009; Leinen and Bummel 2018).

Now for the third feature of planet-human relations: *transversality*. The dichotomy of nature and culture, and of science and the humanities, was constitutive of modern knowledge landscapes and immensely productive; but in planetary thinking it is being wound up. Juxtapositions between inside and outside, actor and structure, local and global are becoming largely irrelevant from this perspective. Instead, the in-between is what matters, the interconnections of such categories, interdependency and co-constitution (all of which played a subordinate role in modern science). Planetary vocabulary includes such neologisms as 'natureculture' (Haraway 2003), 'spacetimematter' and 'ethico-onto-epistemology' (Barad 2007), which signal the inseparability of its components. Barad's concept of 'intra-action' goes beyond interaction in that it 'recognizes that distinct agencies do not precede, but rather emerge through, their intra-action' (Barad 2007: 33). For example, a lichen intra-acts with the earth, air and water that surround it in two ways: the lichen influences the elements by means of its photosynthesis and the processing of carbon dioxide; while in turn the elements predetermine the nutrients and moisture available to it. Only through this intra-action do all these entities really come into being in the first place. Old cosmologies and animistic belief systems around the world have been aware of these interconnections for millennia.

Many things have effects on reality: quasi-objects (like a ball) can tie collectives together (like a team); hybrids constitute a mixing of hitherto separate ontological spheres (Latour 1993) and boundary objects can connect knowledges and practices by way of their malleable adaptability and yet coherent identity (Star and Griesemer 1989). Of particular relevance are hyperobjects, i.e. real objects that spatially and temporally expand beyond human cognitive capacities, so that we can only observe and depict them by way of technological measurements. They elude human perception, can only be captured in their effects and thus open up a space of 'interobjectivity' (Morton 2013). Examples of hyperobjects are black holes, the biosphere, the solar system, all plutonium existing on Earth or plastic bags. The spiritual tie between humans and such hyperobjects may seem odd, but it is a corrective to viewing planet-human relations solely in terms of the sensually perceptible.

As pointed out in feminist contributions to the debate (Irni 2013; TallBear 2017; Willey 2016), we can learn a lot in this regard from knowledge systems that never conceived of the world as a formation of nature and culture, or practice and theory, to begin with (Cusicanqui 2012; Sundberg 2014; Todd 2016). Indigenous knowledge, in all its breadth and diversity, is not (only) an archive of practices of thought, faith and action; it is relevant 'situated knowledge' which remains closely tied to the practices of those who produce it (Nakashima et al. 2012; Whyte 2013; Wildcat 2009). An example of a transversal 'thing' would be the root bridges of Meghalaya in Northeast India which are usually formed by guiding the aerial roots of rubber trees across creeks and which can last for many decades, despite severe monsoon rains – a living planet-human relation that bursts the limits of the nature-culture dichotomy (Watson 2019).

In his books *The Life of Plants* and *Metamorphoses*, the philosopher Emanuele Coccia (2019; 2021) turns the anthropocentric view upside down by revealing that plants – due to their production of oxygen, without which both us animals and Earth's all-important atmosphere would not exist – are the real sources of life on Earth. The proud illusion of humanity's special place in the cosmos, substantiated by the humanities, thus dissolves in a continuous stream of all life forms, in which humans are but ephemeral minor components. It is plants, argues Coccia, that engage the planet in 'an endless cosmic contemplation' (Coccia 2019: 87):

> If the Earth physically rotates around the Sun, it is plants and thanks to them that this connection produces life and matter [...]. Plants are the metaphysical transfiguration of the rotation of the planet around the Sun, the step that transforms a purely mechanical phenomenon into a metaphysical event.

2 Interdependencies

To add another level of abstraction to this: planet-human relations do not exist separately from one another but are involved in all kinds of interdependencies. Since the beginning of the 'Great Acceleration' – the simultaneous rise of various socio-economic indicators (e.g. world population, transport, telecommunication) and Earth-system measurands of human activity (e.g. domesticated land, ocean acidification, loss of tropical forests) – such interdependencies have exponentially gained in importance (Steffen et al. 2015). The science network 'Future Earth', for instance, names as the greatest threat for the well-being of future generations the 'potential integration and interaction' of the risks that the planet is currently facing (Fuchs 2020: 15):

> A survey by Future Earth of more than 200 scientists has revealed five global risks that have the potential to impact and amplify one another, in ways that might cascade to create global systemic crisis: failure of climate change mitigation and adaptation; extreme weather events; major biodiversity loss and ecosystem collapse; food crises; and water crises. These are issues that already consume huge amounts of press and academic attention […]. But the emphasis of the survey results is that it is the interplay between these five risks that is most concerning.

That interplay of planetary elements can result in unexpected, hard-to-predict phenomena. Indications may be found in research on the Earth system's tipping points (Lenton et al. 2008), the collapse of large ecosystems (Cooper, Willcock and Dearing 2020), planetary boundaries (Rockström et al. 2009) and on the trajectories of the Earth system (Steffen et al. 2018). Von Humboldt realized that 'everything is interaction' (Wulf 2016: 59); and the sociologist Georg Simmel had something similar in mind when he talked about the process of 'socialization'.

Two schools of thought in particular have made contributions towards a better understanding of planet-human relations: relationalism and complexity research. According to relationalism, one can only productively interpret the properties of something in relation to something else; it follows that entities are for the biggest part relational in nature. Such assumptions are typical of some hermeneutic and qualitative approaches – from systems theory (Capra and Luisi 2014) to theoretical ecology (Ulanowicz 2009) – which focus on networks of complex relations (mostly metanetworks and assemblages of assemblages). The relationalist hypothesis can aid us in clarifying what interdependencies in planet-human relations are (relational ontology), how we can recognize such relations (relational epistemology) and how we should shape them (relational ethics) – as well as how to combine all three in terms of an 'ethico-onto-epistemology' (Walsh, Böhme and Wamsler 2021).

Such a relationalist approach could be applied, for example, to the bacterium *Chroococcidiopsis*. In the barren and extremely dry landscape of the Atacama Desert, strains of the bacterium extract water from gypsum rocks, thus transforming them into anhydrites. By means of X-ray diffraction, researchers were able to track which rocks the microorganisms had already colonized. As the authors of the study note, these findings 'may also offer potential strategies for water storage technologies in extreme environments, including extraterrestrial habitats' (Huang et al. 2020). Robert Kokoska of the US Army Research Office foresees concrete applications in materials synthesis and power generation (Bell 2020), while other research on *Chroococcidiopsis* has found that they tend to survive even when exposed to space and Mars-like conditions (Billi et al. 2019). The spectrum of interdependencies that planetary thinking identifies, interprets and ethically evaluates thus ranges from tiny bacteria to entire planets.

This interpretative work needs to be complemented by complexity research which aims to empirically grasp, formalize and generalize interdependencies in planet–human relations. Four concepts of complexity theory are of particular importance in this regard (Cilliers 2001; Thurner, Hanel and Klimek 2018; Woermann and Cilliers 2012). First, there is *emergence*, which means that the properties of interdependencies in planet–human relations 'are very different, and often unexpected, from properties of their individual components', i.e. the planet and humanity (Domenico and Sayama 2019: 3). Examples of such properties include the concepts of non-linearity, indirect effects and non-reducibility (all known from chaos theory), the significance of different scales, as well as a phase transition to new states, as it occurs with respect to the tipping points of the Earth system. Secondly, there is *non-equilibrium*: Planet Earth and its societies 'may have stable states at which they can stay the same even if perturbed' but more common are 'unstable states at which [they] can be disrupted by a small perturbation' or entirely 'unpredictable' and 'chaotic' behaviours (Domenico and Sayama 2019: 8–9). Thirdly, interdependencies engender a considerable degree of *self-organization* with no or only little control – as in swarms, order arising from disorder, or self-similarity. Finally, there is the possibility of *adaptation* through mechanisms like learning, sharing information transfer, psychological or social development, or variation and selection (Domenico and Sayama 2019: 10–13). In other words, when 'functioning' interdependencies are disrupted, as was the case multiple times in the history of life on Earth, they might be capable of regenerating or modifying, so as to survive.

3 Normativities

Since planetary thinking 'disclos[es] vast processes of unhuman dimensions', it 'cannot be grasped by recourse to any ideal form' (Chakrabarty 2019: 25).

Decisions about how to keep living (well) and cope with the loss of life – on this or another planet – depend on insights about how the universe as a whole functions, 'reflections on the moral status of life in the universe, be it earthbound or extraterrestrial' (Losch 2019: 264). And since planetary thinking not only alters the descriptive premises but may shake the prescriptive foundations as well, normative principles will have to be negotiated anew: the planetary must be factored in, but without obliterating the human. As mentioned before, ultimately it is up to humans to decide how they want to fashion their relationship with the planet (Bellacasa 2017; Giraud 2019). Normative approaches extend from a 'Planet First' anti-anthropocentrism (Hayward et al. 2019; Lynch and Norris 2016) to 'Interaction First' demands for humans to take responsibility for their relations with the planet (Dryzek and Pickering 2019; Whitmee et al. 2015) to 'Humans First' calls for placing the planet entirely at the disposal of humans (Machan 2004; Servigne and Stevens 2020) to, finally, a 'Technology First' posthumanism (Lovelock 2020). One could also say that the spectrum ranges from 'submit to the earth' to 'subdue the earth' (Genesis 1:28). However, following on the heels of the anthropocentric fixation of Western modernity, a radical posthumanist turn would seem to us like a form of escapism, a flight from responsibility.

Two vital concepts that could open up the discursive space for a planetary ethics are habitability and hospitality. That Earth is habitable is not just due to its distance from the sun. Other requirements for enabling and sustaining life that Earth meets include the presence of enough energy to sustain the metabolism and reproduction of living beings, of water and of carbon to build complex molecules (Cockell et al. 2016). An ethics of habitability expounds how humans should approach their relationship with (the habitability of) Earth (Denoual 2020) as well as other potentially habitable planets which they might want to colonize in the future, when Earth's sustainability has worn thin because of their own behaviour. Indications that this will happen abound, and the 'Earth Overshoot Day' project brings this message home to us every year: in 2025, the point where all resources that Earth is capable of producing within a given year were exhausted was reached on July 24 – with the Global South somewhat compensating for the wastage of the North.

The second factor, hospitality, has a long philosophical tradition. This is what Kant wrote in 1795 on the subject in *Perpetual Peace* (Kant 2003: 16):

> [T]he *right to visit*, to associate, belongs to all men by virtue of their common ownership of the earth's surface; for since the earth is a globe, they cannot scatter

themselves infinitely, but must, finally, tolerate living in close proximity, because originally no one had a greater right to any region of the earth than anyone else.

Today, hospitality is considered something like a human right (Cavallar 2002). People who have to leave their homes due to natural disasters, for instance, are not visitors. They cannot return; they will stay; and they thus rely on an 'absolute hospitality', which consists in 'open[ing] up [our] home' and 'giv[ing] place to them […] without asking of them […] reciprocity' (Derrida and Dufourmantelle 2000: 25). Once we combine (planetocentric) habitability and (anthropocentric) hospitality, the 'guest' *Homo sapiens* and the 'host' Earth are no longer across from one another but symmetrically joined (Dikeç, Clark and Barnett 2009).

It follows that in the context of the planetary, ethical questions must be addressed beyond the usual temporal and spatial horizons, and without falling for primitive fallacies, e.g. with respect to population policies (Coole 2018; Gesang 2013). A planetary ethics factors in interplanetary space and deep time, transforming cosmopolitanism into a chronopolitism. Needed are approaches that combine habitability and hospitality; for instance, by developing convivialism (Adloff and Clarke 2014) into a 'cosmovivialism' (Cadena 2015: 285-6):

> [C]osmovivir may be a proposal for a partially connected commons achieved without canceling out the uncommonalities among worlds because the latter are the condition of possibility of the former: a commons across worlds whose interest in common is uncommon to each other.

4 Planetary health

Except for, fortunately rare, cases of disaster, astronauts 'return to Earth safely'. Had a space flight taken off in 2019, returned a few years later, and remained without connection to Earth in the interim, that mission's astronauts would have found themselves in a radically different world – like those that left during the Cold War and returned after the fall of the Berlin Wall in November 1989. Covid-19 is neither past nor exceptional. A pandemic of this magnitude, which shook the Global North way more than the rest of the world, was expectable and had been anticipated in national emergency management guides, but (apart from some East Asian states) no actual precautions had been taken. The epidemiological evidence indicates that similar pandemics are probable in the near future, perhaps with even more severe impacts. From a planetary perspective, then, 2020 was quite a normal year; incidentally, it was also the year in which 'global human-made mass' for the first time exceeded 'all living biomass' (Elhacham et al. 2020). The Covid-19 pandemic is a classic example of a metaphysical hyperobject (Morton 2013) which, for all its very substantial consequences, eludes customary human experience and perfectly illustrates the workings of planet-human relations. It provides us with the opportunity to summarize key elements of planetary thinking and – apropos of 'Down to Earth' – draw conclusions for the *universitas* and for political action.

In the autumn of 2020, the government of Denmark decreed for 17 million mink, farmed for the production of fur coats and collars, to be mass-slaughtered (or, as the technical euphemism goes, 'culled'). At this point, animal-borne SARS-CoV-2 variants had already been reported in Spain, the Netherlands and the United States and in Denmark, 200 people had fallen ill due to the mutation. The region of Nordjylland was locked down and PCR mass testing was ordered. Although disease progression was rather mild, the possibility that mutated virus variants could undermine the vaccination efforts has been worrying authorities ever since, and animal rights activists used the incident to once again call for mink breeding to be prohibited. This typical case of zoonosis (or rather: amphixenosis, with the pathogen being passed from humans to animals and then back) makes it painfully obvious that as animals, humans are biologically prone to interspecies transmission. Nearly two-thirds of all human diseases are transmitted by organisms, among them plague, tuberculosis, swine flu, rabies, anthrax, Lyme disease, avian influenza, taeniasis (tapeworms) and Ebola. According to the most convincing hypothesis to date regarding the emergence of SARS-CoV-2, horseshoe bats are the likely natural reservoir of the virus, with an intermediary host or direct transmission to humans still under investigation. The wet market of Wuhan in Central China has been identified as the place where that outbreak most likely occurred: a marketplace in which living and dead wild animals are penned up in the tightest of spaces, get butchered, sold and cooked, so that human contact with their body fluids is all but inevitable.

We will not pursue the Danish case any further here – in which the legality of the mass killing of mink was disputed and an exhumation of carcasses became necessary when their decomposition released phosphorus and nitrogen in the soil, threatening to contaminate drinking water and a bathing lake. Also, we will not pursue the question of which negative environmental effects – besides the well-known *positive* effect of a short-term reduction of CO_2 emissions from aviation – the pandemic has resulted in (think single-use face masks, etc.). From a planetary standpoint, the precarious relation of human, animal and environmental health requires a triangulation of methods that go beyond the anthropocentric 'global health' approach. It is not by chance that *The Lancet*, arguably the world's most prestigious medical journal, has called for a transition from 'public' or 'global health' to 'planetary health' (Horton et al. 2014; Whitmee et al. 2015). The WHO's 'One Health' concept is another instance of a dehierarchization within the human-animal-environment triangle. Traditional interpretations of the interdependencies between the three have come under some criticism as of late.

The human-animal relationship is marked by notable closeness to, even idolization of pets but also by ruthless practices of butchering for meat behind the walls of slaughterhouses. For the longest time, practices like industrial animal farming and animal testing in the context of drug and cosmetics production were carried out without much opposition, and their victims were seen as unavoidable collateral damage on the path to growth and progress. The protests by animal rights activists and recently also by human-animal studies scholars have so far been unable to stop this massive-scale cruelty (Radhakrishna and Sengupta 2020). The German author of children's fiction, Cornelia Funke, asserts that in Indigenous fables and fairy tales, animals and plants are quite naturally talked to, sometimes even worshipped, whereas in our Western literary tradition they degenerated into 'dumb creatures' hopelessly inferior to humans (Spreckelsen 2021). But more and more it dawns on us that animals, too, have consciousness, a will and emotions, and that they should be treated as fellow inhabitants of planet Earth, i.e. with respect. Compassion for animals and the fact that meat production strongly contributes to climate change are two of the main drivers behind the current proliferation of vegetarianism and veganism (Willett et al. 2019).

Now for the second axis in our triangle, that connecting animals and the environment, or animate and inanimate nature: The exploitation and destruction of resources, industrial production and an ever-denser service network have dramatically diminished animals' habitats and caused a mass extinction of species. The extent of the accompanying destruction of nature can be appreciated thanks to the documentary work by photographers like Sebastião Salgado (2013) or Edward Burtynsky (Burtynsky, Baichwal and Pencier 2018). While in Africa, Ebola epidemics used to be symbolically interpreted as an aggressive act of nature against humans, we now know that viruses tend to go back to grave human interference in ecosystems (Quammen 2013; 2020; Rulli et al. 2017; Vidal 2020). Viruses are highly dangerous, but the main offenders are humans, who increasingly invade nature reserves for the sake of raw material extraction, put the local animal world in a flurry, and thus increase the risk of zoonosis. If one ascribed agency to animals (as many exponents of human-animal studies do), this could be interpreted in terms of self-defence.

Only if we adequately acknowledge this suffering can we avoid remaining one-sidedly focused on the third axis: environment–human. This focus still dominates most global health programmes, whose aim it is to battle 'scourges of mankind' like malaria, smallpox and TB, reduce mortality rates and protect humans from hazards springing from the environment or animals. Or to prevent the increased incidence of tropical diseases in the Global North (which is partly due to global warming) as well as of non-communicable diseases caused by an unhealthy diet, lack of physical activity, smoking, excessive consumption of alcohol and environmental pollution. In this view, the culprit is usually some manifestation of nature, not the anthropogenic disruption of natural cycles. By contrast, in planetary thinking the three poles are treated as equal and relational. The SARS-CoV-2 pandemic, rather than being a divine punishment, signals the dysfunctionality within the relations between humanity, animals and the environment. Non-human animals and inanimate nature display agency to a degree that makes it impossible to any longer treat them as objects.

This also implies a major change in the parameters of politics and education. Intergovernmental agreements for securing the widespread availability of vaccines, medical emergency services, health education and preventive healthcare, as well as financial resources for medical research and treatment will remain the business as usual of global health institutions. Two of the UN's seventeen sustainable development goals are 'good health and well-being' (# 3) and 'clean water and sanitation' (# 6), and the achievement of a 'universal health coverage' one of the main targets of the WHO (World Health Organization 2023).

To commit to planetary health means to regard animate and inanimate nature as fellow agents. In this holistic perspective, different interests must be balanced in radically new ways. In the past, demands by environmentalists, for instance, concerning toad migration or the preservation of forest ecosystems, were ridiculed by the mainstream of public opinion. But such biotopes and a stewardship for endangered species are essential in today's world. Sometimes, these efforts come into conflict with other 'green' endeavours: in Eastern Germany, environmentalists successfully opposed the clearance of a forest owned by the Tesla company because it was a habitat of snakes and sand lizards (Chazan 2020).

5 Instead of a resume: flourishing as a comprehensive praxis – the case of coral reefs

An important environmental concern is the preservation and restoration of coral reefs. Coral reefs are the biggest animal-made structures in the ocean and, somewhat like rainforests, they exercise considerable ecological influence on their environment. Corals are, contrary to what the layperson might think at first glance, not plants but animals, more precisely cnidarians, i.e. sessile, colony-forming organisms that occur in hard form as stony corals and in soft form. Corals are polyps whose structure resembles a cylinder or a phallus. At the top, a polyp has tentacles with nematocysts, which it needs to capture food. A gullet leads through a large cavity to the foot plates that anchor the polyp in the reef. Corals obtain their food by filtering out microplankton, nutrients and trace elements from the current-rich seawater and by endosymbionts, embedded in the polyp cells, which are responsible for the intensive colouring of the coral's living tissue. Corals are found exclusively below the surface in marine environments, in temperature zones favourable to them. Their outer shape sometimes resembles widely branching plants with gracile or stronger branches, at other times they look like flat plates, patches or bunches of flowers, some of them resemble mushrooms or cactuses. The appearance of a large garden is enhanced by the sometimes strong colours.

Some coral reefs are enthusiastically called 'cathedrals of the deep sea' because of their vertical struts and diagonal connections. The famous Great Barrier Reef, the world's largest contiguous collection of over 2900 individual coral reefs off the north-east coast of Australia has a length of over 2,300 kilometres and covers an area of approximately 344,400 square kilometres. It was 'discovered' by British seafarers in the eighteenth century, it is visible to the naked eye from space and is regarded as a 'marine wonder of the world' and declared a 'natural world heritage site'. This hints to the symbolic significance of corals and opens up a debate about the threat posed by acidification and warming of the sea, in particular, but also the introduction of microplastics and excrement as well as through excessive fishing and diving tourism. The visible manifestation of this degradation is a breathtaking coral bleaching and even the full destruction of the reefs.

From the point of view of 'human flourishing', corals are not only a delightful visual enrichment, they have also been used for centuries for jewellery. Corals also have a useful effect: their substances are used for folk medicine and as a building material. Furthermore, they are also attributed magical properties, which leads to a less anthropocentric view of them. Beyond the symbolic happiness of humans, the intrinsic value of corals as more-than-human beings has to be acknowledged; in particular, their agency in the oceans and coastal areas plays a long underestimated role. What appeared for a long time to be just a magical imagination must be redefined as a permanent real interaction with the human neighbourhood which can be proven by scientific evidence.

As coral reefs are under massive threat of extinction, thousands of advocates, stewards and volunteers have come together to dedicate themselves to lay research and active conservation. The Global Coral Reef Alliance is an example for a non-profit foundation founded in 1990 and financed by donations. It has set itself the task of researching coral reefs, having them declared as nature reserves and rebuilding damaged or extinct reefs with the help of Biorock technology.

Another way of approaching planetary flourishing (or any other subject, for that matter) is by art. Corals are very prominent in photography and even sculpture, see for instance, *Crochet Coral Reef*, a political work-in-progress by the Australian twin-sisters Margaret and Christine Wertheim, a science writer and a feminist poet based in Los Angeles. Their 'large-scale coralline landscapes', to which women from all over the world contribute, are created through 'techniques of crochet to mimic in yarn the curling crenelated forms of actual reef organisms' (Wertheim 2009). In different manifestations, the installation was shown, among many other places, at the Smithsonian National Museum of Natural History and at the 2019 Venice Biennale, furthermore at the Helsinki Biennial, the Andy Warhol Museum, the Hayward Gallery and at the Museum Frieder Burda, Baden-Baden.

The Wertheim's long-term project combines first-rate artful handicraft, experimental science, ecological reflection, math and political engagement. The crochet work reproduces the hyperbolic shapes and colours of living reefs and a non-Euclidean geometry found all over the cosmos – 'a beautiful impassioned response to dual calamities devastating marine life: climate change and plastic trash' (Wertheim 2009; Wertheim and Wertheim 2022). Ocean warming and acidification, eutrophication, pollution and overfishing all contribute to the endangerment of coral reefs, a complex marine ecosystem that provides a safe habitat for a biocenosis of plants (mainly algae) and animals (e.g. fish, invertebrates, worms, sponges and the corals themselves) (Leinfelder 2018).

The observation of the state of coral reefs, chiefly by diving and taking pictures, amounts to a transdisciplinary science, which also includes laypeople as 'stewards' of marine lifeforms, their environment, and the history of their habitats. Such transdisciplinarity is an important key to planetary thinking; it uses insights from various fields and knowledge cultures to identify and examine complex problems, and it enlists non-scientists to participate in these efforts (Padmanabhan 2018; Sass 2019).

13

Reclaiming Care – *Homo Curans* as Vision for Human Flourishing and Sustainability Transformation

Ariel Macaspac Hernandez

1 Introduction – care in the core of the polycrisis and sustainability transformation pathways

The polycrisis – spanning climate change, biodiversity loss, poverty, inequality, geopolitical conflicts and threats to human well-being – demands innovative approaches to understanding and addressing these interlinked challenges. *Pathway thinking*, supported by scenarios, is one approach that envisions sustainable futures where human and planetary well-being are interconnected (Rosenbloom 2017; Vuuren et al. 2015).

Pathway thinking offers action-oriented roadmaps for transitions to sustainable outcomes (IPCC 2022; Soergel et al. 2024). For instance, decarbonized urban transport systems require shifting investments towards public transit, active mobility infrastructure like charging stations, while enacting policies that reduce car dependency. Such initiatives not only cut emissions but also enhance air quality, public health and urban liveability.

Sustainable Development Pathways offer transformative shifts between development trajectories to achieve multiple goals including climate mitigation (Hernández 2021b; 2021a; Soergel et al. 2024). While sustainability focuses on long-term goals, transformation concerns the processes required to realize these goals, underlining their interdependence (Hernández 2022a; Rees 1995). Amid deepening global crises, such pathways are critical for addressing interconnected challenges.

The author acknowledges the support of THE NEW INSTITUTE through a research fellowship as well as of the Transfer for Transformation (T4T) project at the German Institute for Global and Area Studies with the support of the Leibniz-Gemeinschaft through its Leibniz-Wettbewerb grant, T95/2021e.

The concept of polycrisis further underscores critiques of Western anthropocentrism and instrumentalism, which perpetuate binaries such as 'Western vs Non-Western' science or culture. This framing often depicts Western approaches as modernist and non-Western as traditional (Godrej 2016). Moreover, sustainability pathways frequently overlook human security, raising concerns over securitization. Emergency policies may infringe on human rights, and non-state actors might resort to disruptive actions (Wæver 2011). By integrating human security frameworks, the complexities of achieving deep transformation become evident (Zhou et al. 2020).

This chapter calls for rethinking sustainability beyond positivist and deterministic paradigms that prioritize quantifiable metrics or standardized SDG indicators (Riahi et al. 2017). Although these approaches hold value for global governance, they often fail to account for local contexts and narratives, challenging the reliance on objectivity and empirical evidence as sole knowledge criteria (Freistein 2022; Morgan et al. 2022). Furthermore, this chapter advocates for a transformative reorientation towards human flourishing, underpinned by the concept of care (Cebral-Loureda, Tamés-Muñoz and Hernández-Baqueiro 2022; McMahon 2023). By placing care at the forefront, this approach expands sustainability efforts to include qualitative insights and local narratives, challenging reductionist metrics and standardized indicators.

As a response to the polycrisis, care offers a holistic approach that transcends simplistic threat-elimination strategies. By mobilizing care resources, initiatives and institutions can foster interconnected solutions to complex challenges. Building on prior research (Hernández 2014b; 2017; 2021a; 2022b), this chapter places care at the heart of sustainability, proposing a shifting from survival-based frameworks to those focused on thriving for all life forms.

Section 2 introduces *Homo curans* – the 'caring human' – as a vision of human nature centred on care, providing an analytical framework for sustainability transformation. Section 3 connects this concept to the *Deep Institutional Innovation for Sustainability and Human Development (DIIS)* framework, reimagining social institutions – including politics, economics, technology, religion, gender and education – through the lens of care. This approach seeks to transform these systems to prioritize holistic well-being.

Section 4 explores how care-centred policies and innovations can foster a resilient and equitable world. By leveraging 'key stories of care' and a transformation wheel, it demonstrates the practical applications of care across diverse domains. This reimagined pathway redefines sustainability as more than survival, shifting the focus to thriving and creating conditions in which all forms of life can flourish.

The conclusion reaffirms the centrality of care in advancing human flourishing and sustainability transformations. It underscores the importance of dialogue on the pivotal role of care in shaping coherent strategies for transformative change, emphasizing the imperative to overcome structural dominance and power asymmetries. This highlights the necessity of establishing ethical boundaries and value systems that underpin holistic development.

By placing care at the core of sustainability efforts, the chapter offers a compelling alternative care-centred pathway that prioritizes well-being and interconnectedness. This approach fosters resilience and adaptability in the face of global challenges, envisioning a just and sustainable future for all.

2 Reclaiming care for human flourishing and sustainability transformation

Homo curans refers to the very humanness as essentially caring, which in itself has various dimensions of self-care, care for community, ecological care, planetary care and transcendental care. *Homo curans* presents not only a compelling vision of a possible future characterized by sustainability and equity but also actionable pathways to achieve it. By situating the concept of care in its broader context, the figure facilitates the reclaiming of care as a key value while interrogating the implicit and explicit assumptions that often disconnect its dimensions. The objective is to deepen understanding of care and demonstrate its integration into a sustainability perspective, offering an integrated view of human flourishing.

Care is a multilayered concept, intersecting several fields of knowledge, each with its specific interpretations and implications for human flourishing, societal well-being and sustainability. To grasp the full significance of care, it is necessary to explore its varied meanings across academic, professional, traditional, Indigenous and literary domains. This inter- and transdisciplinary perspective enriches understanding and highlights care's pivotal role in building a sustainable and equitable future.

Over recent decades, socio-political mobilization of care has gained momentum as a pathway to achieving human flourishing and sustainability (Daly 2021; Held 2006; Tronto 1993). Increasingly, many funding agencies, international development programmes, and national and local governments embed care, happiness and the impact of relationships/relationality into their project criteria and as prerequisites for international cooperation (Bruton and Sorin 2022).

For instance, Bhutan's Gross National Happiness (GNH) framework and its GNH index have globally influenced government policies, encouraging the integration of well-being metrics into national accounting systems (Thinley and Hartz-Karp 2019). Similarly, the Inner Development Goals (IDG) initiative outlines transformative skills for sustainable development, placing care – for self, others and the planet – alongside co-creation, at the core of its framework (IDG 2023).

2.1 Discourses and entry points in sustainability transformation

In recent years, scholars have increasingly linked care, happiness, well-being, human flourishing and sustainability transformation (Beraldo and Bruni 2020; Fine 2024). Notable works have focused on care (Ehrenfeld 2019; Singer, Ricard and Karius 2019) and related concepts like social innovation (Backhaus et al. 2017), well-being (O'Mahony 2022), happiness (Musikanski et al. 2021) and societal development (Lintsen et al. 2018). For instance, John Ehrenfeld (2019) defines flourishing as a qualitative system property rooted in care for the immaterial or transcendental (spiritual) aspects of existence. He underscores the importance of care as integral to sustainable and equitable development.

Despite its growing prominence in sustainability policy, care remains theoretically and methodologically underexplored. While psychology has acknowledged its role in happiness, disciplines of economics, political science and sociology continue to grapple with conceptualizing and incorporating care into key areas, including health, entrepreneurship, work, migration, inequality, power and climate change (Ley 2023; Lynch 2022; Moriggi et al. 2020). Overcoming these difficulties is critical to recognizing care as a socio-political driver for sustainability.

A significant gap in sustainability transformation research lies in the incorporating care as a key pathway to sustainable futures. Longstanding debates in economics and sociology have juxtaposed concepts like *Homo economicus, Homo sociologicus* and *Homo sustinens*, offering avenues beyond purely economic understanding of human nature (OECD 2020; Tronto 2017). Transformation processes towards sustainability are non-linear and power-driven, necessitating inclusive and affirmative action to integrate care into theories and frameworks of change.

The terms employed in sustainability transformation research are often normative and poorly defined, leading to ambiguity. Concepts such as life satisfaction, well-being, decent living and contentment are frequently used interchangeably, despite their etymologically distinct meanings rooted in diverse research traditions. These differences are further complicated by 'loss in translation' when terms acquire varying meanings in local contexts and languages. As Howarth (2015) and Ruggeri et al. (2020) emphasize, critical engagement with these conceptual differences is essential for achieving rigor and coherence in sustainability studies and fostering meaningful interdisciplinarity.

Paradoxically, the ambiguity of 'care' is not a weakness but a strength, enabling disciplinary, cultural and epistemic diversity. This multiplicity offers valuable entry points for the understanding of care as a central element in human flourishing and sustainability transformation.

2.2 Academic and professional definitions of care

Care is a multifaceted term studied across disciplines, each offering distinct perspectives and contributions. It encompasses individual well-being as framed by positive psychology, systemic equity highlighted by feminist theories, and innovation and sustainability within the fields of science, technology, engineering and mathematics (STEM). Despite differing emphases, all these views converge in recognizing care as a fundamental basis for human flourishing and advancement of society.

Positive psychology defines care as prosocial behaviour characterized by empathy, compassion and supportive relationships, focusing on measurable outcomes like reduced suffering and enhanced well-being (Emmons 2020; Seligman 2011). However, this approach often prioritizes individualism, overlooking socio-cultural contexts and the burdens of caregiving, such as stress and emotional exhaustion (Held 2004; Lyubomirsky 2008).

Feminist theories, particularly the ethics of care, address care's gendered nature and societal undervaluation (Gilligan 2008; Tronto 2013). Unlike utilitarian and Kantian ethics, these perspectives prioritize relationships and interdependence (Noddings 2013). They advocate for a caregiving oriented society that combats systemic inequalities, challenges hypermasculinity and fosters equity and sustainability (Tronto 2017).

Religious traditions present care as a moral imperative tied to ethical living. Christian agape love, Buddhist karuna (compassion), and Islamic zakat (charity) connect care to spiritual development and concern for others

(Esposito 2011; McGrath 2001). Philosophical perspectives complement this view, incorporating existentialist authenticity in relationships (Sartre 2018), the utilitarian assessment of care's societal benefits (Mill 1863), and the Confucian ideals of *ren* (benevolence) and familial duty (Confucius 1979).

In STEM disciplines, care manifests through ethical innovation, human-centred design and sustainability practices. Scenarios and system modelling offer methods to address global challenges, proposing technological pathways aligned with human and environmental well-being (IPCC 2022; Marx et al. 2024). These practical approaches extend care into actionable strategies for equity and sustainability.

Each of these perspectives underlines distinct aspects of care: personal flourishing in positive psychology; relational and structural dimensions in feminist and socio-cultural theories; moral grounding in religious and philosophical traditions; and problem-solving in STEM. Together, they reveal the multilayered significance of care at personal, social and global levels, illustrating its centrality in fostering a sustainable and equitable future.

2.3 Indigenous wisdom and holistic perspectives on care

Indigenous wisdom traditions provide a holistic and relational approach that complements modernist academic and professional conceptions of care. Indigeneity, as Herget (2007) and Pratt (2002) argue, is often perceived in opposition to modernity or modernism. Although the SDGs claim universality through multilateral negotiation, they remain grounded in European notions of individuality, growth and separation between humans and nature, as van Norren (2022) suggests. Conversely, Indigenous perspectives are frequently framed as anti-development, as evidenced by the labelling of Mexican Indigenous resistance to wind energy projects (Godoy 2021).

Reclaiming or decolonizing the terms 'Indigenous' and 'traditional' is necessary for reframing their contributions to human flourishing and sustainability. Historical marginalization, limited access to modern infrastructure and the preservation of traditions and rituals have often led to the mischaracterization of Indigenous communities as 'backward' (Smith 2012). The UN Declaration on the Rights of Indigenous People defines Indigenous peoples as those native to a geographical region, with distinct traditions, languages and social systems deeply tied to their ancestral lands. However, this term remains contested, as legal definitions often clash with socio-cultural understandings (Bello-Bravo 2019).

Indigenous wisdom emphasizes relationality and interconnectedness, framing care as a symbiotic practice that integrates community, environment and spirituality (Storm 2021). For example, African philosophy, particularly ubuntu, embodies the principle 'I am because we are', emphasizing communal relationships, shared humanity, mutual care and environmental stewardship. Ubuntu regards environmental destruction as a violation of the interconnected web of life, underscoring sustainability and guardianship of nature for future generations (Chigangaidze 2021; van Norren 2022).

Similarly, Japan's Indigenous spiritual tradition of Shintoism prioritizes care for nature, including non-living entities, through principles such as *Magokoro* (true heart) and *Mottainai* (regret over waste). Reverence for *kami* – spirits and deities inhabiting natural objects and phenomena such as mountains, rivers, trees, and animals – instils respect for the environment. Rituals such as *Misogi* (ritual purification) and *Harae* (ritual purification to expel evil spirits), along with shrine worship, highlight the importance of harmony with nature (Toshio, Dobbins and Gay 1981).

Other Indigenous knowledge systems, such as those of Aboriginal Australians and the Igorot people of the Philippines (Adonis 2011), provide valuable insights on care, human flourishing and sustainability, offering complementary perspectives to modernist approaches.

Indigenous values are also reflected in literature and poetry, which articulate the relational and emotional dimensions of care. For instance, the principles of community and relational bonds central to Indigenous care resonate in Toni Morrison's novel *Beloved* (1987), while Mary Oliver's (1986) poem *Wild Geese* celebrates the human-nature relationship at the heart of Indigenous values.

2.4 Care in conservation, natural resources management and resilience

The concept of care intersects with discourses on conservation, natural resource management and resilience, highlighting the interdependence between humans and natural systems (Lockwood et al. 2010; Phalan 2018; Rockström et al. 2009). Ecological care becomes necessary in supporting sustainability and social change, particularly as human practices increasingly threaten global ecosystems (Cote and Nightingale 2012).

In this context, care involves not only minimizing harm to the 'natural world' – including climate, seas, rainforests, mountains and deserts – but also enhancing the adaptive capacity of natural systems. This includes supporting critical ecological processes, such as bird migration and thermohaline (ocean) circulation (Vogel 2016).

As a resilience strategy, care focuses on sustaining essential ecosystem services in the face of recurring disturbances, such as diseases and hurricanes (Garmestani and Benson 2013; Hernández 2024). This proactive approach ensures ecosystems remain functional and biodiverse under changing conditions. Conservation-driven care also acknowledges that motivations are not purely altruistic, as human survival and well-being are intrinsically tied to healthy ecosystems.

Strategies such as 'sharing, caring and sparing', outlined by Rhys Green et al. (2005), demonstrate how organizations, governance structures, incentives, power relations and societal norms influence conservation outcomes. Caring as a conservation practice refers to harnessing interspecies relationships to safeguard the well-being of both the biota and the natural environment (Garmestani and Benson 2013). This approach emphasizes maintaining the health of ecosystems not only for current but also for future generations.

This perspective underscores the ethical and practical dimensions of conservation, recognizing that human well-being is inseparable from ecosystem health. By building caring relationships within ecological systems, conservation practices can enhance resilience, enabling ecosystems to self-restore and adapt to perturbations.

Integrating care into conservation necessitates proactive stewardship and long-term thinking. This approach balances immediate human needs with the imperative to protect and sustain ecosystems, ensuring their health and the services they provide for future generations.

2.5 Interim conclusion – reclaim care for human flourishing and sustainability transformation?

Care must be reclaimed from its narrow and fragmented interpretations and reimagined as a robust driver of ethical, ecological and relational change. This section contextualizes care through six interconnected approaches, bridging theoretical insights with practical applications to foster meaningful impact.

(1) **Expand Knowledge Beyond Conventional Scientific Criteria**

Broaden the scope of knowledge by incorporating lived experiences and contextual complexities accessible through qualitative insights. Foster collaboration across diverse knowledge communities, breaking disciplinary silos and democratizing access to knowledge. Science must serve an equitable society, avoiding political capture and technocratic dominance.

(2) **Expose and Counter the Manipulation of Care Under Authoritarian Regimes**

Critically examine how authoritarian regimes weaponize care to mask inequalities and consolidate control. Develop strategies to identify and counter such manipulations while assessing whether flourishing is possible within unequal societies. Establish relational infrastructures and feedback mechanisms to ensure care is accessible, equitable and contextually relevant.

(3) **Integrate Non-Western Knowledge Systems**

Incorporate a diverse range of stakeholders to co-produce inclusive knowledge frameworks that challenge technocratic and Western-centric paradigms. Embed Indigenous and traditional wisdom into the design of care practices to enhance resilience, equity and sustainability across scales and communities.

(4) **Address Disconnects in Care Mobilization**

Bridge gaps in self-care, community care, ecological care and planetary care. Investigate the causes of pseudo-care and how it reinforces injustices, particularly from the perspective of marginalized populations. Clearly define actionable research priorities to develop holistic and integrated solutions.

(5) **Resist Care-Resistive Power Configurations**

Deconstruct power relations that constrain equitable care. Prioritize consensual knowledge creation and interdisciplinary collaboration to empower marginalized voices and challenge dominant structures. Ensure the equitable distribution of knowledge and elevate underrepresented perspectives in care-related discourses.

(6) **Apply Marginalized Wisdom to Inclusive Solutions**

Leverage the insights of marginalized communities to develop representative and equitable care practices. Revisit traditional concepts of care, ensuring autonomy – particularly in public health and disability contexts – is preserved, and preventing the emergence of new injustices.

When operationalized, these approaches position care as a transformative framework that reimagines social institutions to prioritize sustainability, equity and collective well-being. This action-oriented model situates care as an agent of systemic change and meaningful social transformation.

3 *Homo curans*: care as a vision for reimaging social institutions

Human flourishing, a cornerstone of sustainable development, is profoundly interconnected with the concept of care. The notion of *Homo curans* – the caring human – provides a normative vision that emphasizes care – for oneself, others, society, the environment and transcendental dimensions – as foundational to human identity and societal organization. This framework offers an alternative pathway to transform institutions and address the polycrisis.

In this framework, care reimagines institutions such as politics, economics, technology, religion, gender and education. It transitions these systems from perpetuating un-care to becoming equitable, empathetic and sustainable. This transformation challenges entrenched power dynamics, aligning with Audre Lorde's (2018) metaphorical warning that the 'master's tools will never dismantle the master's house'. Instead, care-centred frameworks lay the foundation for collective care, dismantling inequitable systems and building structures that foster human flourishing and sustainability.

3.1 New ways of seeing

The concept of *Homo curans* challenges dominant worldviews by introducing care as an existential element of human identity and behaviour. Philosophers like Ehrenfeld (2019) argue that care is intrinsic to human existence, with apathy and hatred viewed as distortions of this essence. However, to fully appreciate the transformative potential of care, it must be contextualized within broader perspectives.

Self-care, often associated with individualization in Foucault's concept of the 'care of the self', can also be understood as part of collective practices. Indigenous traditions, for instance, root self-care within relational and ecological interconnectedness, emphasizing a shared sense of responsibility. Collective care, as exemplified by ubuntu's ethos of 'I am because we are', highlights care as a practice that integrates human and environmental well-being (Chigangaidze 2021; van Norren 2022).

Similarly, planetary care extends this relational understanding by embedding care within environmental stewardship. It calls for ethical guardrails to balance human needs with ecological sustainability, fostering empathy, shared responsibility and holistic well-being. By situating care at the centre of human identity and societal practices, *Homo curans* offers a pathway for addressing the interconnected challenges of human and ecological flourishing.

3.2 Transforming institutions through ethical boundaries

Transforming institutions necessitates a fundamental re-evaluation of ethical frameworks. Dominance-based values such as hierarchy, coercion and exploitation must be replaced by partnership values rooted in equity, collaboration and mutual respect. The ethics of care challenges classical moral theories that prioritize autonomy and rational self-interest, proposing relational ethics that recognize individuals as empathic, interdependent beings connected within broader ecosystems. This paradigm shift calls for a global sense of responsibility, where caring for strangers, future generations and the biosphere embodies a transition from individual gain to collective flourishing.

The 'partnership model' articulated by Eisler and Fry (2019) illustrates the transformative potential of embedding care into institutional values. This model confronts systemic violence and promotes inclusivity by fostering values of equity, cooperation and the public good. Sustainability transformation, therefore, demands a shift in the foundational values that drive major social institutions – from dominance-based principles of hierarchy, inequality, coercion and private gain to partnership-based principles of equity, collaboration and shared benefit.

Social systems grounded in domination are characterized by rigid top-down hierarchies, including the ranking of one group of people over another. These systems perpetuate cultural acceptance of abuse and violence, reinforce beliefs that dominance and inequality are inevitable or even moral, and utilize fear and force to preserve structural violence. By contrast, partnership values challenge these dynamics, offering pathways for more inclusive, equitable and sustainable institutions.

3.3 Reimagining narratives and institutions through *Homo curans*

The narratives of modernity, which emphasize separation and techno-optimism, have contributed to systems of un-care by fostering isolation and hyperindividualism. The concept of *Homo curans* critiques these assumptions and proposes alternative narratives. Revaluing non-monetary contributions, such as caregiving and relationship-building, elevates these activities as essential societal pillars rather than peripheral endeavours. Inclusive growth paradigms that prioritize well-being over GDP redefine economic success, while decolonizing knowledge integrates Indigenous wisdoms that challenge Western dichotomies between humans and the environment. For example, Shintoism's reverence for nature underscores the interconnectedness of care practices and sustainability. By emphasizing interconnection, *Homo curans* promotes systems that prioritize equity, sustainability and shared purpose.

Placing care at the heart of social institutions requires systemic innovation, necessitating a critical examination of the narratives underpinning modernity. These narratives often perpetuate an ethos of separation, isolated individualization and a lack of empathy, which form the foundation of un-care in current social institutions. Examples include conceptions of humanity as separate from nature, techno-optimism and the uncritical acceptance of technological 'creative destruction' (Nolan and Croson 1995). Additionally, social media and IT-mediated interactions contribute to disconnection, materialism is upheld as the measure of human success and hypermasculine competitiveness and aggression dominate societal paradigms.

To counter these dynamics, care must be embedded into transformative narratives that reframe societal goals and guide systemic change. Such care-based narratives could include the following:

- **Politics** must foster participatory governance, amplifying marginalized voices and community resilience.
- **Economics** must acknowledge and adequately remunerate care work, ensuring fair and equitable economic structures.
- **Education** must integrate relational and ecological care into curricula, fostering empathy, critical thinking and a sense of interconnectedness.
- **Climate policy** must prioritize the rights and needs of vulnerable communities, embedding care into both mitigation and adaptation strategies.

Care-centred institutions not only address structural inequities but also enable human and planetary flourishing.

Reimagining social institutions through the lens of *Homo curans* shifts the focus from mere survival to thriving. Rooted in collective, empathetic care, this framework challenges individualistic paradigms and integrates Indigenous and feminist philosophies. By embedding care into institutional narratives and practices, humanity can work towards a sustainable, equitable future.

The following section explores how care-centred narratives of modernity can be operationalized to shape social institutions.

4 The *Homo curans* pathway – care in the narratives and dimensions of transformation

Achieving sustainability and human flourishing requires positioning care as a central driver of systemic change. This transformative approach incorporates 'key stories of care', moving beyond conventional modernity and dominant sustainability narratives to establish a comprehensive framework for shaping and evaluating policies. By embedding care into transformative narratives, we can chart a pathway towards a more sustainable and equitable future.

Figure 13.1 illustrates six interconnected systems that shape this transformative journey. Each system exerts both direct and indirect influences on the process of change:

(1) Economy and Technology: embedding care into work, technological innovation, finance and banking, and economic paradigm and growth.
(2) Nature: recognizing the interconnectedness of ecosystems, redefining ownership structures and integrating care practices that safeguard environmental health and biodiversity.
(3) Resource Provision: ensuring the equitable and sustainable distribution of essential resources, emphasizing care for vulnerable communities and future generations.
(4) Material Well-Being: promoting access to basic needs, including housing, healthcare and education, while reducing inequality and poverty.
(5) Transcendental Well-Being: addressing spiritual and existential dimensions of human flourishing, emphasizing interconnectedness and meaning-making.
(6) Society and Relationship Infrastructures: strengthening community ties and fostering relational care through participatory governance, social cohesion, cultural inclusivity and empathy-driven public policy.

The following sections outline new care-centred narratives for these dimensions.

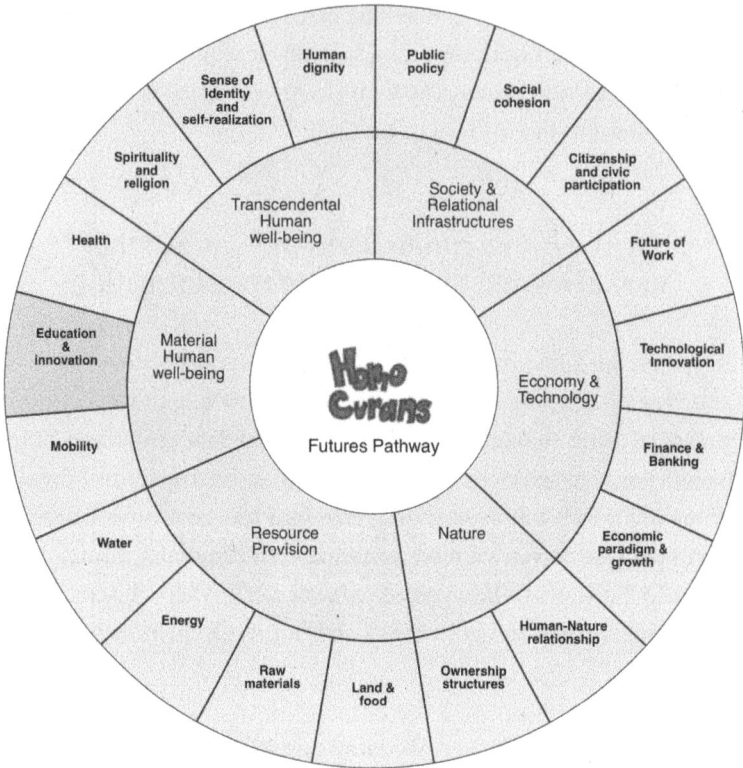

Figure 13.1 The *Homo Curans* pathway – transformation wheel. (Source: the authors.)

4.1 Economy and technology

This dimension explores how economic systems, technology and work can foster sustainability through care-centred narratives:

- **Future of Work**: shift the focus from viewing work solely as income generation to emphasizing interpersonal relationships, job satisfaction and well-being within workplaces (Ley 2023). Envisioning the workplace as a space of encounter positions care at its centre, prioritizing human interaction and collaboration. Employee contributions to collective care – for both colleagues and customers – should complement material productivity in performance appraisals, reinforcing care as a core workplace value.

- **Technological Innovations**: emphasize user-centred design that prioritizes societal and environmental needs over profit (Ramvi et al. 2023). Innovations should actively involve users throughout the design and development process to ensure accessibility and responsiveness to their needs. Technology must serve as a tool for enhancing care, equity and sustainability.
- **Finance and Banking**: advocate for fair risk-sharing and equitable financial practices that prioritize societal value creation over profit maximization (Dieye 2020). For example, instead of rigid fixed-interest rates, financial institutions could adopt models such as *Mudarabah* and *Musharakah*, where profits are shared equitably between parties based on successful investments. Financial institutions should profit only from successful investments, ensuring equitable financial relations and promoting fairness in the financial sector.
- **Economic Paradigms and Growth**: recognize care work, including childcare and elder care, as integral to economic systems (del Castillo 2024; Glenn 2021). Advocate for fair wages, improved working conditions and increased public investment in care services to create a care-focused socio-economic model. This model measures economic growth by how economic activities create value for human well-being. This model values both paid and unpaid contributions. Beyond market-based exchanges, it integrates reciprocity, barter, cooperatives and solidarity. In contrast to a 'war economy', which produces destructive outputs like weapons, a care-centred economy fosters sustainable growth by creating value for society and the planet.

4.2 Nature

Human–nature relationships and ownership structures are pivotal in fostering care and sustainability:

- **Human–Nature Relationship**: recognize nature as sacred and inseparable from humanity, advocating for stewardship and planetary well-being (Falkner and Buzan 2019). This perspective contrasts sharply with the anthropocentric view of nature as a resource for efficient exploitation. The concept of planetary well-being complements care ethics by promoting ethical transformation that balances human needs with those of nature.

This harmonious integration aims to sustain both humanity and the planet, promoting a shared responsibility for ecological preservation (Gough 2017).
- **Ownership Structures**: reimagine the natural world and its outputs as commons, recognizing public goods as unsuitable for privatization (Miller et al. 2020). Care-based property regimes should be established through collaborative partnerships among governments, the private sector and societal actors. These frameworks recognize the value of diverse formal and informal resource management practices, fostering collective responsibility for shared resources. This approach promotes sustainable use, equitable distribution, and a long-term ecological stewardship (Lawson 2019).

4.3 Resource provision

Effective stewardship of water, energy, land and materials is essential for the well-being of all life forms, ecosystems and the planet. Caring for these resources involves conservation, ethical management, and equitable access:

- **Water:** ensure ecosystems and biodiversity are protected while guaranteeing clean water as a universal right for all life forms (Hall, Van Koppen and Van Houweling 2014). Recognize water's cultural and spiritual significance by integrating community engagement into conservation efforts (Godden and Ison 2019).
- **Energy:** balance human needs, ecological health and planetary sustainability by optimizing production, distribution and consumption of energy to minimize harm and enable planetary recuperation (Graczyk 2021). Ensure reliable, affordable and clean energy to support social justice and environmental integrity (Hernández and Prakoso 2021; Rao, Min, and Mastrucci 2019). Emphasize holistic strategies that include technological innovation, supportive policies and active community participation.
- **Land and Food:** promote long-term soil health, biodiversity, and food security by addressing trade-offs between human food needs and ecosystem well-being (Mbow and Rosenzweig 2019). Reduce food waste, diversify diets and strengthen local food systems. Encourage conservation practices such as organic farming, terracing and precision agriculture to enhance resilience to climate change (Duru, Therond, and Fares 2015).
- **Raw Materials:** advocate for the ethical sourcing of raw materials such as steel, aluminium, wood and plastic. Emphasize responsible life-cycle management, including sustainable extraction practices that respect human rights and support affected communities (Marx et al. 2024; Thorstensen

et al. 2024). Foster interconnected care for human and ecological well-being, ensuring raw materials contribute to sustainable and equitable development.

4.4 Material well-being

Material well-being addresses essential needs that enable human thriving through inclusive, equitable, and sustainable systems:

- **Mobility:** promote inclusive, accessible and environmentally sustainable transportation systems that enhance community connections and improve access to essential resources and services like healthcare, education, and employment (Russel 2014; Wang and Sun 2019). Prioritize smart urban planning to provide affordable, reliable, accessible and inclusive transportation options tailored to the needs of people with disabilities, the elderly and economically disadvantaged groups, fostering community integration (Meunier 2020).
- **Education and Innovation:** co-design education curricula and innovation projects with communities to align R&D with local identities and needs (Brand and Jax 2007). Embrace a broader and more inclusive approach to education by involving target groups as co-creators of curricula and activities (Rassool, Edwards and Bloch 2006). Tailor programmes to community profiles, incorporating native languages as mediums of instruction to foster inclusivity and accessibility (Tsebe 2021). Measure the success of innovation and R&D by their tangible impact on communities and the planet, rendering innovations that lack meaningful community benefits obsolete.
- **Health:** advocate for a holistic, equitable healthcare system that prioritizes human dignity, solidarity and equity (del Castillo 2024; Velasquez 2014). Such a system treats patients as whole individuals, addressing their physical, emotional, social and spiritual needs. It ensures access to quality healthcare for all, regardless of socio-economic status, ethnicity, or location (Fine 2024). By addressing social determinants of health – such as poverty, education, housing and environmental factors – the system tackles root causes and advances health equity.

4.5 Transcendental (immaterial) well-being

This dimension emphasizes the non-material conditions necessary for individuals and communities to thrive:

- **Spirituality and Religion**: provide meaning, hope and resilience while critiquing consumerism, exploitation of others and the environment. Mobilize non-material virtues such as courage to navigate challenges and question injustices (Florea and Caudill 2014). Religion, as the institutionalized aspect of spirituality, plays a pivotal role in sustainability transformations by influencing the worldviews and cosmologies of adherents (Koehrsen and Huber 2021). Religious institutions, with their extensive resources and economic influence (Stork and Öhlmann 2021; Tirosh-Samuelson 2020) can also drive social and environmental activism by leveraging their social connectivity and ability to mobilize collective action.
- **Human Dignity:** uphold human rights and shape societal policies that ensure dignity and agency for all communities. Centring human dignity in policy discourses fosters inclusivity and equity, creating the conditions for people to live with respect and purpose (Kleinig and Evans 2013).
- **Identity and Self-Realization**: promote interconnectedness, belonging and personal growth, fostering mutual care within communities and across generations (Crocetti et al. 2023; Hernández 2014a). Facilitate self-realization by helping individuals discover their *ikigai* – a sense of life purpose that brings meaning to their existence (Mogi 2022). Encourage mutual support within communities, where individuals not only pursue their personal growth but also aid others in their journeys (Noddings 2013). This collective approach ensures that individuals and communities thrive together, fostering shared purpose and mutual support.

4.6 Society and relationship infrastructures

Social structures and networks serve as the foundation for care-driven transformations by fostering inclusivity, cohesion and participatory governance:

- **Public Policy:** collaborate with non-state actors to integrate local needs, traditions and identities into policy-making. Framing power as identity-based, rooted in value systems and measuring its effectiveness by translating diverse values into relevant, quantifiable, and comparable terms (Penetrante 2012). Strengthen specialized coordination between state and government institutions to address the needs of diverse groups, while granting local governments significant autonomy in policy development and implementation to ensure tailored and context-sensitive solutions.

- **Social cohesion:** mobilizing a sense of belonging among community members by balancing adaptability with preserving shared identity and values (Manca 2014). Policy-making must carefully maintain this balance, recognizing that social cohesion is dynamic. Avoid stagnation in social systems that resist necessary change, while mitigating rapid or extreme shifts that risk disenfranchising individuals and communities. This approach fosters a resilient and inclusive social fabric.
- **Citizenship and civic participation:** empower civil society to actively shape policy-making, ensuring inclusivity and responsiveness in governance (Spain 2009). Citizenship and civic participation should be seen as assets of persuasion in political decision-making (Kingdon 1984). Leverage the convening power of civil society groups to identify pressing issues and develop evidence-based solutions. Facilitate fluid mobility between civil society leaders and political leaders, recognizing civil society as a recruitment pool for future policymakers. This ensures that civil society perspectives are well-represented in governance, fostering a more inclusive and responsive governance process.

5 Conclusion – reimagining sustainability through the lens of care

This chapter presents a transformative vision and framework for sustainability, rooted in the centrality of care as a multidimensional and integral component of human flourishing. By redefining human flourishing through the lens of *Homo curans* – the caring human – this analysis introduces a novel approach to addressing the polycrisis. Care, as conceptualized in this framework, encompasses self-care, care for others, societal care, environmental stewardship and attention to transcendental aspects of life, collectively forming a holistic pathway towards sustainable and equitable futures.

The proposed approach highlights the potential of care to reshape socio-economic, political and cultural institutions. Moving away from dominance-based paradigms, it fosters values of equity, partnership and cooperation. Embedding care into all facets of society – from economic systems and technological innovations to public policy, education and religion – offers a practical pathway for addressing the root causes of the polycrisis, including climate change, biodiversity loss, social inequalities and geopolitical instability.

This reimagining of institutional structures prioritizes long-term thinking, proactive stewardship and inclusive decision-making.

Critically, this chapter challenges the over-reliance on Western-centric paradigms and quantifiable indicators in sustainability discourse. By incorporating Indigenous and traditional wisdoms and emphasizing qualitative insights, it promotes culturally relevant and context-sensitive approaches to care. This inclusive perspective enriches the sustainability agenda, ensuring that it addresses local realities while fostering global solidarity.

Placing care at the heart of sustainability transformations reshapes both the process and goals of transformation, offering an alternative to reductionist models that prioritize growth and survival over holistic well-being. The *Homo curans* framework not only advocates for ethical boundaries and interconnected value systems but also equips societies to confront power asymmetries and structural dominance, fostering a global society where care, empathy and interconnectedness drive human flourishing and environmental sustainability.

This chapter calls for sustained debate and research to further explore the transformative potential of care in fostering systemic change. Mobilizing care as a resource and guiding principle offers a compelling alternative for addressing the intersecting challenges of the polycrisis. By anchoring sustainability in the ethos of care, actions are needed to mobilize care resources, initiatives and institutions to address the root causes of various crises. Ultimately, this approach advances a vision of human flourishing that is resilient, equitable and aligned with the interconnected realities of a globalized world.

14

Reimagining a World Fit for our Humanity

Ian Hughes

1 Escaping the Neolithic

Philosopher Alain Badiou challenges us to imagine that throughout the long history of humanity, we have been living in societies that prevent human flourishing. Badiou calls the social structures we have been living in, at least since the advent of the state and probably much longer, 'the Neolithic' (Badiou 2022). At the top of this hierarchical structure, power and wealth are concentrated in the hands of a small elite. At the centre, a sizeable middle class command sufficient agency and resources to live reasonably comfortable lives and play the defining role in maintaining the system's structure by acquiescing to the greater power of the elite. At the bottom of the hierarchy, a majority of people live in powerlessness and poverty, living lives that are focused primarily on maintaining the lifestyles of those better off than them.

This structure has persisted for millennia although the identities of the elites have changed throughout history, from chieftains to god-kings, pharaohs, sultans, emperors, party chairpersons and presidents. The challenge for contemporary politics in the age of the contemporary polycrisis, Badiou asserts, is for humanity to finally escape the unjust hierarchical structure of the Neolithic.

The Enlightenment and Modernity, of course, promised just such an escape. The ideas of the Enlightenment – of individual liberty, progress and representative government – laid the foundations for democracy and ignited the hope of breaking free from the long age of tyranny of authoritarian structures. The Scientific Revolution and the Industrial Revolution that followed have given us a profoundly different view of our place in the cosmos and have enabled a level of prosperity for many which was previously unimaginable. But modernity

---

I would like to sincerely thank Bob Grumiau, Andrej Zwitter and Karma Ura for their deep insights which helped shape the arguments in this chapter.

has also been imbued with pathologies. Gross inequalities, violence, and mass exploitation, including slavery and colonialism, have been the essential foundations of Western progress. Money and material possessions have become the arbiter of human worth and the horizon by which human lives should be lived. The desacralization of nature was a further foundational belief of modernity. To be civilized was to accept that the Earth was inert and machine-like, just as Newton told us, there for us to master and exploit.

Modernity's failure to escape the Neolithic can be understood as the result of a selective co-option by hierarchical power structures of the positives of modernity – science and reason, tolerance of dissent, human rights and equality, and accountable democratic government – alongside the retention of modernity's pathologies – violence, materialism, greed and a hypermasculinity which denigrates love and care. This selective appropriation has ensured that gender continues to be a defining feature of social hierarchies, and of the Neolithic as a whole. 'Inferior' women and 'feminine' men who do not display the aggressive hypermasculinity necessary to oversee and maintain a violent and cruel order are unsuited for positions of power within such a social structure.

Sociologist Ulrich Beck has described the conflict between the positives and the pathologies of modernity as generating what he calls an ongoing metamorphosis of contemporary society (Beck 2016) – an attempted fundamental reformation of the foundational social institutions of politics, economics, technology, religion, gender and education, from within. This metamorphosis is not a planned, consciously intended systemic reconfiguration. It is instead occurring through myriad ongoing conflicts, progressions and regressions across all social institutions. It might end in either progress or catastrophe. The list of crises currently facing humanity is long, of course, including climate change, species extinction, ecological destruction, technological disruption, socially destabilizing levels of inequality, steeply rising geopolitical tensions and war. Within this polycrisis (Lawrence et al. 2024), Beck places hope in what he calls the possibility of 'emancipatory catastrophe' – the possibility that an impending catastrophe could make changes possible that seem impossible today.

Badiou, however, warns us to be cautious in our optimism that such an escape can be realized. He recalls how previous attempts at such an escape, among which he counts early Christianity and communism, have resulted, so far, in failure. With regard to early Christianity, Badiou sees Saint Paul's interpretation of Jesus as both God and man as signalling the divine within every human person, and the foundation for reshaping human relations on the basis of a radical equality. Instead, Christianity was co-opted by the Roman, and subsequent, states to justify radical inequality and violent conquest. Communism too promised radical social equality, but it contained the seeds of its own failure even more starkly, as its promise of justice was premised on violent revolution and unending class war.

How then to approach Badiou's challenge to escape the Neolithic, a challenge to fundamentally reimagine the foundations of society? Such a reimagining has now become an existential necessity for our species, given the contemporary polycrisis, particularly the threats of climate change and nuclear war.

A growing body of opinion is viewing the threats we face as deeply interlinked and a signal of the urgent necessity for deep systemic change across the whole of society (Hughes et al. 2021; Kanger and Schot 2019). Existing narratives and proposed solutions, however, tend to focus primarily on climate, and emphasize innovations in technology, economics, and (to a lesser extent) politics. This narrow scope limits recognition of the diversity of crises we face and our ability to develop appropriate strategies in response.

In this chapter, I will discuss one narrative of transformation that is central to addressing the current polycrisis, namely the transition to global peace. The escalating occurrences of war, particularly in Ukraine and the Middle East, and the rising confrontation between the United States and China, are undermining the global cooperation needed to collaboratively address global challenges. Reducing war and violence are central to Badiou's challenge to escape the Neolithic, made all the more difficult as violence is built into the Neolithic's structural design.

To help us think about Badiou's challenge, colleagues and I have been developing a 'thought model' which attempts to grapple with the daunting challenge of reimagining entirely different social structures – in effect reimaging new forms of civilization. Our thought model, Deep Institutional Innovation for Sustainability and Human Development (DIIS), aims to help reframe the dominant narrative of transformation from techno-social transitions to deep global cultural transformation (Hughes et al. 2021; 2024). Based on critiques from social science, our DIIS model begins by identifying abject failures in key social institutions, including economics, democracy, technology, religion, gender and education, as the source of our current malaise. It argues that achieving greater sustainability and human flourishing requires more than just technological advancements, as the current dominant narrative would have it; it necessitates a profound and global re-evaluation and reinvention of all of these foundational social institutions. The DIIS thought model aims to deeply challenge conventional assumptions in current sustainability narratives, and is intended to act as a generative framework, inviting further contributions towards its aim of developing a holistic conceptualization of deep global transformation.

As our model has been described previously (Hughes et al. 2024), a brief outline here will suffice. At the centre of the model is the 'Reimagining of Social Institutions' that we believe constitutes the core of transformation. As Badiou argues, the social structures of all of our social institutions need to be fundamentally critiqued and reimagined. This deep reimagining is informed by four elements of our thought model – 'ethical directionality'; 'ways of seeing'; 'dynamics of transformation'; and 'narratives of transformation'. These four sides of the DIIS framing reflect the underlying premises of the model: that the transformation to sustainability will extend beyond the narrative of techno-social transition; that the transformation will be inherently normative; that considerations of the dynamics of transformation must include a diversity of types of change, including changes in culture, practices and values; and that any model of transformation must be deeply transdisciplinary and inclusive of currently marginalized worldviews.

To take each of these briefly in turn, the model argues that an emphasis on 'ethical directionality' is essential to ensure that the transformational journey towards sustainability and human flourishing is guided by a moral compass that respects human rights, equality and the dignity of all life forms. Crucially, a reimagining of democracy must form a core element of this. As history and contemporary events illustrate, conditions of crisis and uncertainty often result in the rise of destructive leaders on a wave of mass support, and the potential alteration of societies in a direction that is severely detrimental to both sustainability and the public good. One need only consider Hitler's Germany, Stalin's Russia or Mao's China to see tragic examples of this. An ethical grounding in core democratic values of non-violence, participation, transparency, inclusion, human rights and solidarity is crucial if we are to stem the current reemergence of such destructive authoritarian leadership, and the ongoing dismantling of established democracies, and avoid the repeat of the catastrophes of the twentieth century.

With regard to 'Ways of Seeing', the DIIS framing considers change in the current dominant paradigm of human subjectivity to be key to global transformation. This move beyond the subjectivity of modernity, or '*Homo economicus*', will involve retaining the positives of modernity, such as rationality and science, human rights and democracy, but it will also involve challenging the negatives of modernity, such as dominance hierarchies, humanity's separation from nature, structural violence, and gender and racial inequalities. It will also require giving effective voice to stakeholders, disciplines, and perspectives critical of existing paradigms that are currently excluded or marginalized in public discourses and decision-making, such as feminist and Indigenous perspectives.

With regard to 'dynamics of transformation', the model asserts that it is necessary to formulate an understanding of the pathologies within actually existing political, economic, technological and cultural systems, since currently existing systems are both the root cause of the contemporary polycrisis and the systems upon which interventions must be made to bring about urgently needed changes. The DIIS model sees large-scale global transformation, to be achieved within rapid timeframes, as essential for avoiding mounting crises and preventing systemic rupture of existing political, economic and social systems (Hughes et al. 2024).

The final part of the DIIS framing is 'Narratives of Transformation'. As stated, the current dominant global discourse on sustainability is based on a narrative that places technological change in the form of renewable energy technologies and the phasing out of fossil fuels at the centre of the transformation. Such a narrative, we contend, is grossly insufficient. Even if it were possible to bring about such technological substitution, in the absence of accompanying changes in the structures, values and practices of all our social institutions, we would still remain within the deeply unjust, hierarchical structures of the Neolithic. The DIIS thought model therefore urges us to develop alternative narratives of what the transformation to sustainability must entail, including, for example, narratives which aim to restore nature as sacred (Miner, Dowson, and Devenish 2012), post-growth and post-capitalist models of economics (Hughes et al. 2021), and critiques of structures of power based on hypermasculine constructions of gender (Cohen and Karim 2022).

This chapter will focus on using our DIIS framing to explore one particular interpretation of 'Narratives of Transformation', namely, that the transformation to greater sustainability and human flourishing is not simply a narrative of changes in techno-social systems, but a wholesale dismantling of the violence system that lies at the heart of the Neolithic.

2 Dismantling the violence system as a precondition for sustainability and human flourishing

"The practice of violence changes the world, but the most probable change is a more violent world."

(Arendt 2002)

Violence is intrinsic to the Neolithic and manifests at multiple levels, including war, structural violence, systemic exploitation and interpersonal violence. Decommissioning this violence system is an essential part of the transformation to greater sustainability and human flourishing. Such a decommissioning would allow the reallocation of the vast resources currently focused on war, both to reduce violence at the other levels in the system and to enable the global cooperation which is a prerequisite for addressing the polycrisis.

This section draws on recent analysis by war historian Richard Overy, who has summarized what research across multiple disciplines can tell us about the causes and motives for war (Overy 2024). Overy addresses questions of human biological, psychological and cultural propensities for collective violence, as well as the main motives that have animated warfare throughout history.

2.1 The war system

The likelihood that humans have adapted biologically to engage in collective violence, when necessary to ensure survival, seems to be the first building block in explaining war. It is plausible that during our evolution, coalitional violence would have increased the likelihood that particular groups survived under conditions where survival was a struggle. During our long prehistory, humans have experienced repeated periods of glaciation and other severe environmental challenges where populations were reduced significantly. Violence in this context may have had adaptive value by increasing access to resources such as food and territory, as well as fending off threats from humans and other species. During human evolution, traits for violence, when necessary, as well as traits for sociality and cooperation, were therefore likely selected for at both individual and group levels. In this view, violence and cooperation are not opposites, but two elements of an evolutionary package humans developed over hundreds of thousands of years. As a consequence, human nature is both potentially aggressive and destructive, and potentially cooperative and empathic.

Overy dismisses, however, two dangerous ideas, mistakenly derived from Darwin's theory of natural selection. These are that we are genetically programmed for violence and war, and that war is a natural biological phenomenon, a means of evolutionary competition to sort the strong from the weak. These were Hitler's views, for example, and they underpinned his programmes of extermination of the old and infirm, the Holocaust, and his wars to the death between 'races'. There is no scientific evidence for either idea.

The human biological imperative to fight when necessary has been reinforced by an evolved psychology which enables such violence (Cashdan and Downes 2012). Our psychological makeup includes a disposition to accept collective violence, particularly by males, alongside a capacity for dehumanizing others, particularly those we deem to be a threat. As is the case with our biology, our psychology has developed in response to selection pressures related to both collaborative living and competition for survival and, therefore, allows us to be either cooperative or confrontational. An important insight from evolutionary psychology is that our behaviour is a conditional response to the environment. Particular environments can stimulate violent responses – including othering, dehumanizing and mobilizing for collective violence. Other environments can stimulate sociality and cooperativeness.

In our species' more recent history, culture rather than natural selection has come to be the key influence on our collective behaviour (Mead 2015). Tragically, the evidence strongly suggests that the co-evolution of culture with our biological and cultural dispositions towards violence, created early in our species' history the social structures and practices characteristic of the Neolithic. The archaeological and anthropological evidence points to the emergence everywhere of hierarchical societies headed by warrior elites where violence was embedded in societal design, with men acting overwhelmingly as the perpetrators of violence. Badiou's Neolithic social structure, it seems, although not universal, emerged tens of thousands of years ago, even before the establishment of the first large-scale states (Wengrow and Graeber 2021). Social psychologists point to factors such as our social susceptibility to accept authority, to identify with and favour in-groups, and to blame victims for our acts of aggression towards them, as possible factors in the emergence and subsequent consolidation of the Neolithic up until the present day (Brewer 1999).

Once consolidated, various motives for war ensured the continuity of violence that has been required to maintain the Neolithic's stratified hierarchical structure. The major motives that have animated warfare throughout the ages have included defence of beliefs, the quest for resources, and pursuit of individual power and glory. All of these, as Overy notes, are built into the human condition.

Defence of religious faith, supernatural beliefs or political ideologies, represents a motive for war that has persisted for thousands of years. Few ancient cultures did not have a god or gods of war and warrior cults where violence was divinely sanctioned. Sacrificial offerings of people seized during warfare were a common feature of societies from early China to central America. Christianity and Islam, which Overy describes as the world's two most bellicose religions, have long histories of divinely sanctioned violence. Such religiously inspired warfare continues today, not least in the conflict within Islam between Sunni and Shia which has spanned the Islamic world (Nasr 2007) as well as the conflict between Israel and Palestine. Secular religions too act as motives for war. Historian Emilio Gentile has characterized as 'political religions' ideologies such as communism, fascism and National Socialism, which aim for social reconstruction, distinguish between true believers and heretics, and demand rigid conformity from adherents under threat of violent retribution (Gentile 2006).

The quest for resources in the form of land, material resources and people, is a second enduring motive for war. Oil and mineral resources have been at the centre of recent wars, for example, from Angola to Sierra Leone, the Congo, and the two Gulf Wars. Throughout the history of warfare, human beings have also been one of the 'resources' that have been fought for. It is estimated that around 100 million men, women and children were enslaved in conquests during the age of Roman imperialism (Overy 2024:170), while almost 10 million African slaves were forcibly taken to the New World during the trans-Atlantic slave trade.

Of particular salience for the emergence of the early Neolithic structure, of hierarchical societies ruled by warrior elites, is what Overy (2024) calls hubristic warfare – war led by men in pursuit of individual power and glory. This Overy describes as the most dangerous and unpredictable form of war from the classical world to the modern day. Alexander the Great, Napoleon and Hitler are just three examples he cites of hubristic individuals who led violent conquests for empire, fuelled by unbridled narcissism and an insatiable appetite for glory. In 'A Theory on the Origin of the State', anthropologist Robert Carneiro (1970) also points to hubristic leadership as a possible defining element in the emergence of the Neolithic, claiming that the origin of the state lay in the extension of power of one tribal community over others, usually through men adept in warfare. Everything else that followed from chiefdom to kingdom to empire was a change in degree rather than kind.

In his description of hubristic narcissistic leaders, the psychoanalyst Otto Kernberg (2003) writes that such a leader 'experiences and expresses an innate grandiosity, needs to be loved, admired, feared and submitted to at the same time, cannot accept submission from others except when it is accompanied by intense idealising loyalty and abandonment of all independent judgement, and experiences any manifestation contrary to his wishes as a sadistic, wilful, grave attack upon himself'. The society that such hubristic leaders create, Kernberg continues, is one of 'totally subservient, idealising subjects, with totally corrupted ruthless antisocial characters whose pretence of loving and submitting to the leader permits their parasitic enjoyment of his power'. Kernberg's description is apt for hubristic narcissists and the societies they have created across history, from Napoleon to Hitler, to Vladimir Putin (Wood 2024) and Donald Trump (Lee 2019a). Such dangerous personalities are the focus of my book *Disordered Minds* which argues that such individuals, although a minority, have had a catalytic effect in fomenting war and violence throughout history (Hughes 2018).

2.2 Structural violence in the Neolithic

The war system that Overy describes is part of a wider global violence system which also includes structural violence. Structural violence refers to the avoidable deprivations that society imposes on groups of people that constrain them from meeting their basic needs, severely impacting on their duration and quality of life. The harm is structural because it is the product of the social institutions and practices of the Neolithic; it is violent because it causes death and injury (Lee 2019b). Structural violence is by far the most lethal form of violence, as well as the most potent cause of other forms of violence. Between ten and twenty million deaths per year can be attributed to structural violence (Høivik 1977). This is more than ten times the number due to warfare, homicide and suicide combined.

Examples of structural violence include deaths from poverty, lack of access to clean water, undernutrition and lack of healthcare. It manifests starkly in the form of gender disparities and racial disparities. With regard to the former, in low-income countries, almost four million excess deaths occur each year as a result of gender inequality. Structural violence also manifests as modern-day slavery, which impacts predominantly on the poorest and most vulnerable. Incredibly, there are around twice as many slaves today as were forced from Africa during the entire four hundred years of the trans-Atlantic slave trade, almost all of them desperately poor. Slavery has not ended, it has simply changed form (Haugen and Boutros 2015).

Like war, structural violence is not an aberration but a consequence of the foundational principles of the Neolithic. Theologian Walter Wink describes what he calls the 'Myth of Redemptive Violence' as the foundational myth upon which the entire system is founded (Wink 2010). According to this myth, human beings are incapable of peaceful co-existence; order must continually be imposed from above; unquestioning obedience is the highest virtue and order the highest value; peace can only be achieved through violence; security can only be achieved through strength; and social institutions exist to legitimize power and privilege and maintain the Neolithic. This 'orientation toward evil', Wink writes, is one into which virtually all modern children, especially boys, are socialized in the process of maturation. A violent hypermasculinity, and a denigration of empathy and care, characterize the entire system. This system continues to act, as it has throughout history, in the way it was designed to do, namely, to enable violent hubristic leaders to thrive, at the expense of the flourishing of the majority of humanity.

3 Leveraging the forces of reason, beauty and morality

If nature were not beautiful, it would not be worth knowing, and if nature were not worth knowing, life would not be worth living.

(Poincaré 2003)

Overy's analysis points to the fact that human genetics, psychology and culture are not fixed in favour of either violent confrontation or peaceful, cooperative collaboration. Instead, both propensities coexist, allowing for the possibility of global peace. Let us turn now to consider how such a peace might be imagined.

3.1 Arendt and Kant

In her biography of Hannah Arendt, 'We Are Free to Change the World', Lyndsey Stonebridge highlights the core principles that Arendt used to navigate the descent into violence that she experienced during the rise of Nazism (Stonebridge 2024). The lessons she urges upon us are threefold: to think deeply, to name truth courageously and to love the world that is being lost hard enough to bring it back into being.

It was at the trial of Adolf Eichmann in Jerusalem, one of the leading architects of the Holocaust, that Arendt grasped what she believed to be the true origins of totalitarianism, and the real meaning of the term 'crimes against humanity'. The origins of totalitarianism, Arendt argued, lay in Eichmann's process of totalizing thinking. Totalizing thinking works as follows: it strips an individual of the plurality of their identities, feelings and relationalities; it replaces their true unique identity with a false unidimensional identity as a hated, threatening other; it groups that inhumanly reduced individual with all other individuals whose identities and humanity have similarly been impoverished; and it passionately desires to expel and exterminate such groups whose presence, and very existence, are seen as intolerable. Such totalizing thinking is a crime against humanity in the twin sense that it strips targeted individuals of their humanity, and it aims to deprive humanity as a whole – 'the people' – of our species' true diversity (Stonebridge, 2024).

Struck by the falsity and danger of such thoughtlessness, Arendt spent her life thinking about thinking. Could thinking, she asked, save us from complicity in bureaucratically regulated evil, like the administrative extermination of six million Jews? For Arendt, the inability to think enables ordinary people to commit evil deeds on a gigantic scale. Thought must therefore enable not simply knowledge, but the ability to tell right from wrong, especially in times of crisis.

Arendt's view on the power of thinking to resist evil was based on her reading of Kant whose transcendent philosophy deeply influenced her. According to Kant, thinking – human reason – is intrinsically imbued with morality. To think clearly is to think morally.

Kant rejected both the rationalist view that knowledge reflects an absolute description of the world independent of the observer, as Newtonian science believed, and the empiricist view that knowledge is derived from the senses alone. For Kant, all knowledge comes both through the senses which provide sense data, and reason which translates that sense data into knowledge. Neither sense experience nor reason alone is able to provide knowledge.

Kant's doctrine of transcendental idealism further states that because all knowledge is obtained through our human cognitive and perceptual apparatus, we can have knowledge only of what he calls 'appearances' and not of 'things as they are in themselves' (Kant 2007). The former is the world of nature, or phenomenon; the latter is the unknowable noumena, the world 'as it is' independent of human perception and cognition. That such a transcendent world exists, but that we can know nothing of it, is the foundation of Kant's transcendental philosophy.

Kant further asserts that human reason contains within it a priori concepts which are not in the world as it is 'in itself' but are rather the grounds upon which human knowledge is based. These fundamental concepts, which Kant called 'categories', include space, time and causality. To aspire to know the world 'as it is in itself', free from the categories by which we come to know it, such as cause, time and space, is to aspire to what Kant calls the 'unconditioned'. We cannot, Kant says, gain such 'unconditioned' knowledge. But at the same time, he argues, our cognition is such that we are compelled to do so. The scope of the powers of human reason is such that we are driven to ask unanswerable questions.

Kant's transcendental philosophy is not simply a philosophy of knowledge, however. It is a philosophy which also locates morality and beauty as intrinsic to the human condition. Kant argued that, like all other objects in nature, human beings exist both as an 'empirical self' within the realm of nature, and as 'transcendental self', outside it. Every person is both part of nature and part of a transcendental world. While knowledge of what I am independent of my cognitive and perceptual apparatus can never be mine, Kant insists, human beings can get intimations of such transcendental knowledge. These intimations, he argues, are to be found in moral life and aesthetic experience. Morality and beauty point towards the noumena.

With regard to morality, what Kant called moral imperatives infuse the way human beings know the world, just as much as the categories of space, time and causality are functions of our cognition (Kant 2002). These fundamental moral principles are intrinsic to our knowledge making of ourselves and the world and can be formulated in a number of different ways – as the imperative to treat others as we would be treated ourselves; to treat myself and others as an end and never as a means only; and that we should act to bring about a world as it ought to be. Morality, Kant argues, based on these imperatives, arises from the transcendental self, unknowable in itself but manifesting through our compulsion to act in accord with their command.

Kant's philosophy also holds that in experiencing beauty, as with morality, we gain intimations of nature's, including our own, transcendent origins. As Roger Scruton, in describing Kant's position, writes, 'a person who cannot see the beauty of nature, including the beauty of other human beings, lacks the sense of the "transcendental" from which all true morality springs' (Scruton 2001).

For Kant, because morality is built into the way we perceive and understand the world, an unjust system like the Neolithic offends our innate sense of reason. In the unjust Neolithic, the gulf between 'ought' and 'is' is an abyss, an abyss that is built into the foundational design of the Neolithic.

Most people, as Susan Neiman writes, are inclined to deny this. Others either try to convince themselves that what 'is' is reasonable (by denying injustice), or by saying that 'ought' is impossible (by denying hope) (Neiman 2015). Morality, in stark contrast, refuses to either deny or excuse, but instead confronts the injustices of the world and refuses to let them stand. Drawing on Kant, we argue that it is upon our innate and universal morality that our escape from the Neolithic must be based.

4 Dynamics of transformation – moral agency and dismantling the Neolithic

To recap briefly, this chapter has argued that in order to address the multiple challenges of the contemporary polycrisis, we must escape the unjust hierarchical social structures of what Badiou calls the Neolithic, in which power and wealth are concentrated in the hands of the few, while the majority live in conditions of oppression and exploitation. A central part of Badiou's challenge is to dismantle the violence system which maintains the power asymmetry and inequalities of contemporary societies. These power asymmetries and gross injustices are the root cause of war and violence, while the global investment in technologies of war diverts resources away from addressing the challenges of the polycrisis, including climate change and global poverty.

Richard Overy's analysis of the causes of war strongly suggests that human beings are not violent as a result of our biology or psychology, but that human nature is both potentially aggressive and destructive, and potentially cooperative and empathic. In determining whether we act aggressively or cooperatively, culture plays a critical role. Our escape from the Neolithic, with its culture which encourages violence and inequality, will therefore require a transition to cultures of peaceful coexistence, empathy and cooperation which bring about more egalitarian social structures (Eisler and Fry 2019).

How then can we envisage such a transition? In this final section, I will argue that such a transformation in values and culture will encompass not only a change in social structures but a corresponding change in consciousness. Following Kant, the transformation to sustainability and greater human flourishing will be based on leveraging the innate and universal nature of reason, beauty and morality.

4.1 Meshworks

Anthropologist Arturo Escobar has written extensively about how we might think about social structure in the 'post-Neolithic' (Escobar 2018a). For Escobar, the transformation to a post-Neolithic society means that the centrality of Neolithic dominance structures and values has been weakened, so that the range of possible alternatives that are considered valid and credible is significantly enlarged. Transformation involves a weakening and a displacement of currently dominant paradigms, across all of society's social institutions.

The post-Neolithic also demands, of course, a different logic of social organization, not one organized on the principles of order, centralization and hierarchy building. It requires instead forms of social organization based on partnership values of democratic and egalitarian social practices; cultural rejection of abuse and violence; and beliefs about human nature that support equality, compassion, caring and cooperation (Eisler and Fry, 2019).

Escobar (2014) points to nature as providing a useful metaphor for such possible post-Neolithic social organization. Networks, rather than hierarchies, constitute the basic architecture of nature's complex systems, he points out. Self-organization, complexity and emergence are the principles upon which life itself is based. Escobar uses the metaphor of meshworks to describe these principles applied to social organization. Meshworks are based on decentralized decision-making, self-organization, heterogeneity and diversity. They are non-hierarchical, they develop through their encounter with their environments, and have multiple, rather than one single overt goal.

For Escobar, meshworks represent an altogether new framework of social interaction, a profoundly relational model in which negotiated views of reality may be built, that foster the circulation of ideas that are not so subject to centralized controls, and that enable the emergence of sub-cultures that are aware of the need to re-invent social and political orders. (Escobar, 2014). Through the operation of meshworks, Escobar asserts, other worldviews, and other voices, can shape our social realities (Escobar 2009).

4.2 States of mind

In the aftermath of the rise of Hitler and the barbarism of the Second World War, a generation of child psychoanalysts sought to understand the cataclysm in terms of human psychic development. One of these psychoanalysts, Melanie Klein, formulated a theory of human psychic development that is not linear and progressive, moving over time to ever higher stages of maturity. Instead, Klein viewed the human mind as developing capacities for various 'states of mind' which we can revert to at any age depending on circumstances (Klein 1996).

In early infancy, the child's cognitive state of development is such that it can only experience the world in black and white. The infant can therefore only see the world as either all good or all bad – either a blissful paradise or a nightmarish, paranoid hell. This state of mind is characterized by aggression, fear, disregard for the rights of others, an idealization of the good and a catastrophizing of the bad, and a relationality based on self-preservation and paranoia. Klein termed this phase of psychic development the paranoid-schizoid position.

For Klein, the child's capacity to develop an alternative state of mind, which ameliorates the extremes of the paranoid-schizoid position, depends crucially on the parents' loving care. By repeatedly holding the child's extreme emotions, caregivers enable the infant to gradually come to experience the world not starkly in terms of black and white, but as it is in reality – a world of both suffering and joy, of both empathy and violent exploitation, of both love and cruel indifference. Over time, through repeated experiences of having the chaos of their destructive emotions 'contained', the baby begins to internalize their carer's capacity within its own mind. In this way, the infant develops the capacity for a second state of mind, which Klein called the 'depressive' state, in which it is able to see the world more clearly as it is, to mourn the fact that reality is not as perfect as it would wish, and to use that realization as the basis for creative engagement with the world. Klein's model of the human mind is one in which we can move into either paranoid-schizoid or depressive mindsets at any time, depending on the circumstances, or culture, we find ourselves in.

4.3 Moral agency in transformation

Escobar's metaphor of meshworks and Klein's view of human nature as oscillating between paranoid-aggressive and depressive-creative states of mind give us a language for imaging our escape from the Neolithic. If we consider the structure of the Neolithic as being constructed and maintained by immoral agents, that of the Meshwork is constructed and maintained by moral agents. The transition from Neolithic to Meshwork is brought about by the example and actions of moral agents.

In this view, the Neolithic can be thought of as the infantile paranoid-schizoid state of mind projected and institutionalized in the real world – a hierarchical structure characterized by top-down control through violence and coercion, constructed and led almost exclusively by males socialized into a hypercompetitive, narcissistic, status-driven masculinity, and imbued with values of materialism, coloniality, inequality and injustice.

Living within such a social order is experienced by moral agents, as Kant described, as offending our innate sense of reason, beauty and morality. While many, as stated earlier, either try to convince themselves that what 'is' is reasonable (by denying injustice), or by saying that a more just social order is impossible (by denying hope), moral leaders refuse to either deny or excuse, but instead confront the injustices of the world and refuse to let them stand.

While violent, hubristic, narcissistic leaders rise to power by inciting fear and vilifying others, moral leaders do the opposite. In their ability to peacefully contain intense emotion, even in times of extreme crisis, such leaders allow us to contain our own intense emotions and, following Arendt, remind us that to think clearly is to think morally.

Moral leaders show us that morality is possible within contexts of violence and injustice. And through the support they receive from many, they reflect the fact that such morality is already widespread among us. In doing so, they give us hope that fundamental change is possible. As they engage in the twin processes of changing minds and changing institutions, they often become protesters, sometimes practical politicians, as well as persuaders and teachers of the vision for equality they are pursuing.

Moral leaders insist that we are capable of creating a world better fit for our humanity and draw to our attention the fact that such a society already exists and is simply marginalized. Their role in leading the way to sustainability and greater human flourishing, by acting on the innate and universal nature of reason, beauty and morality, and enjoining us to do likewise, is central to the dynamics of transformation.

Flourishing in the International Sphere – From Sovereignty to Solidarity: Challenges to 'Modern' International Law

Hans-Joachim Heintze

1 Introduction

The Sustainable Development Goal (SDG) No. 16 demands: 'Promote peaceful and inclusive societies for sustainable development, provide access to justice for all and build effective, accountable and inclusive institutions at all levels' (UN Department of Economic and Social Affairs, n.d.). The objectives included in this goal can only be achieved with and through the law. Since these are not only requirements for individual states, but are addressed to the international community, international law must support these tasks in a special way. It is therefore necessary to examine the extent to which this legal system is able to cope with the challenges associated with it. At first glance, the state's claim to sovereignty, which dominates international law, seems to stand in the way of the SDGs. A deeper analysis, however, reveals developments that can enable international law to become an instrument for overcoming global challenges.

This requirement of international law has its origins in the concepts of natural law and the philosophy of the Enlightenment, because these demanded tolerance and humanity in the dealings of people and states with each other. The application of these principles necessarily leads to solidarity between individuals and states. Solidarity is understood as:

a form of unity that mediates between an individual and community and entails positive duties; different forms of solidarity differ a great deal in how they motivate and manifest these relations.

(Scholz 2011: 1022)

In practical terms, this means that all people and states must treat each other in a way that includes mutual respect and accountability, so that global justice can ultimately emerge. Even the humanist thinkers therefore saw the state, which sets the law, as having a responsibility to serve humanity and solidarity. This historical body of ideas has also found its way into the codification of modern international law, for example, by incorporating collective human rights such as the right of peoples to self-determination and the right to development into international treaties. These rights are no longer classic individual rights, but 'solidarity rights' due to their collective dimension (Ipsen 2024: 394). The rights of solidarity make clear the departure from the 'traditional' sovereignty-oriented international law in some areas, e.g. human rights. However, it must be questioned to what extent this applies to the entirety of the international legal system. In particular, the consent character of international law must be taken into account.

2 International law – a different legal order

National law is enacted by the legislature and regulates the relations between its subjects – natural and legal persons – in a special way. In contrast to the often highly effective moral norms, the legal norms are characterized by the enforceability that exists within the framework of the state's monopoly on power.

In order for the law to fulfil its regulatory function, its rules must be precise and applied for a long period to enable stable and predictable relationships. Those subject to the law must be able to understand and apply the norms. They trust that they are universally applicable, and changes to the law must be made through a formal procedure. These procedures are complicated and usually lengthy, as it is always necessary to ensure that new regulations are in line with the Constitution. International law, like any law, is characterized by its binding nature. However, its establishment and enforcement are fundamentally different from national law, because international law does not recognize a legislator. This is due to the fact that all states enjoy sovereign equality. Because of their sovereignty, they cannot be compelled to behave in a certain way, nor can they be brought to justice without their consent to legal proceedings. Consequently, international law can only be created through agreements between states. States must accede to international treaties in a formal procedure and this usually requires ratification by national parliaments. In this way, international obligations are transformed into national law (Dupuy 2011).

It is easy to imagine that the agreement of legal norms between the very differently constituted states is a lengthy process. After all, national interests have to be taken into account. Even in the member states of communities of values – such as the EU – the agreement of treaties often proves to be difficult and lengthy.

The Law of the Sea of Nations is a good example of the problems of agreeing on binding international standards. The United Nations Convention on the Law of the Sea (UNCLOS), also called the Law of the Sea Convention or the Law of the Sea Treaty, is an international treaty that establishes a legal framework for all marine and maritime activities. As of July 2024, 169 states and the European Union are parties. While the EU is a member of the treaty, the United States has not yet been able to decide whether to become a member, which shows that states with the same system do not automatically have to belong to the same treaties.

The convention resulted from the third United Nations Conference on the Law of the Sea (UNCLOS III), which took place between 1973 and 1982. UNCLOS was the largest and longest codification conference of the UN because it covered all aspects of the use and protection of the sea. This comprehensive approach resulted in a modern treaty that also anchored numerous new concepts – such as that of solidarity. In view of the complex drafting process and the need for acceptance of stable and predictable standards, it is understandable that the standards should apply over a longer period of time. Against this background, it is understandable that international treaties are generally quite static. Nevertheless, they can be supplemented or replaced. However, there are high hurdles for this.

For example, the UN Charter may be amended and supplemented, as Article 108 of the UN Charter allows. However, any change of the UN Charter requires ratification by two-thirds of UN member states, i.e. currently at least 129 states. Among those states must also be the ratifications of the five permanent members of the Security Council. As practice shows, the five permanent members can rarely bring themselves to vote together and therefore the UN Charter remains in its traditional form, although it needs to be changed. For example, the UN Trusteeship Council would have to be abolished, because there are no longer any trusteeship territories, and therefore there are no tasks left for the organ of the Trusteeship Council. Nevertheless, this UN body continues to exist as part of the UN Charter because it is difficult to get 129 states to ratify the treaty amendment, which eliminates the whole trusteeship concept that made sense at the end of traditional colonialism. This example demonstrates the inertia of international law. In addition to the procedural problems of amending treaties, the insistence on what has been agreed upon – i.e. certain statics – is primarily due to the states' invocation of their sovereignty.

3 Justice – a challenge to static international law

The law must not only create stability and predictability – i.e. legal certainty – but it must also be fair. Justice is a social value, and the turn to solidarity is an expression of the idea of justice: 'with a strong ethical underpinning' (Wet 2006: 612). This idea is not new. As early as 350 years ago, Emer de Vattel spoke in his monograph 'La codification du droit international' of the moral duty of states to show solidarity as a justified expectation, and ICJ judge Alejandro Álvarez stated in the middle of the last century that states 'had to behave in a more cooperative manner, based on solidarity' (Zobel 2006). This requirement is reflected in the UN Charter, whose concept of peacekeeping was codified in Chapter VII:

> In the UN law on the maintenance of international peace and security and in international humanitarian law ... solidarity operates as an instrument to achieve common objectives through the imposition of common obligations.
>
> (Wellens 2010: 13)

The entire system is thus based on the collective action of the international community against the lawbreaker. This is expressed in Article 49 of the Charter (Novak and Reinisch 2012: 1386): 'The central concern of Article 49 is the duty of the members mutually to assist each other'.

Together with Article 50, it regulates the modalities of solidarity, burden sharing, and equitable distribution of costs for the Security Council's sanctions against the state that violates international law. This provision is also applied in practice, as Klein demonstrates:

> A perfect application of the principle of solidarity has been the temporary denial to Macedonia of the right to benefit from any Art. 50 assistance because it had not lived up to its duty of solidarity by non-compliance with SC sanctions.
>
> (Wellens 2010, note 7: 29)

It is no wonder that the principle of solidarity has found its way into numerous UN General Assembly Resolutions after the end of the East-West confrontation. An example of this is Resolution 59/193 (2005), which describes solidarity as a 'fundamental value'. Wolfrum concluded from this in 2010: 'The principle of solidarity reflects the transformation of international law into a value-based international legal order' (Wolfrum 2006: 1087). Inevitably, the concept of solidarity had to play a fundamental role in the upcoming UN Charter reform attempts at the turn of the millennium, in addition to the security aspects.

The UN High-Level Panel on Threats, Challenges and Change 'A More Secure World: Our Shared Responsibility' argues clearly:

> Today, more than ever before, threats are interrelated and a threat to one is a threat to all. The mutual vulnerability of weak and strong has never been clearer.
> (UN-Doc. A/59/565, para. 17)

Slaughter supports an obligation that the Community of States holds to intervene in cases of grave violations of laws, viewing this as an embodiment of the evolving values within international law.[1] She posits that contemporary interpretations of sovereignty must align with principles of justice and require a departure from the conventional, static understanding of international law.

Indeed, the focus on human rights and the right of peoples to self-determination brings a dynamic element to the international legal order. New actors – peoples and human rights defenders – are involved in the development of the law and challenge the rigid sense of sovereignty. The most recent example is the codification project 'Business and Human Rights',[2] which is clearly based on the principle of solidarity. International rules are intended to prevent inhumane methods of exploitation of employees by transnational companies in the countries of the Global South.

The states and their rigid claim to sovereignty also come under pressure when peoples assert their claim to the right to self-determination. That is the right of peoples to demand the creation of their own statehood and domestic orders elected by the people. This has led to the formation of hundreds of newly independent states in the last century and is a success story of the UN (Fisch 2009: 45ff).

However, the right of peoples to self-determination is not only historically significant in terms of the international community. Rather, very topical questions arise again and again, for example with regard to the permanent sovereignty of peoples over their natural resources (Schrijver 2011). The legal status of Western

Sahara clearly shows that the international community has different positions on the enforcement of the right to self-determination. As early as 1974, the UN GA applied in Resolution 3292 the right to self-determination to the people of Western Sahara. The ICJ concluded in its Advisory Opinion on Western Sahara:

> that the decolonization process envisaged by the General Assembly is one which will respect the right of the population of Western Sahara to determine their future political status by their own freely expressed will.[3]

Since 1963, the UN Special Committee on Decolonization considered Western Sahara to be a non-self-governing territory and called on states to do everything possible for the well-being of the inhabitants. This included supporting self-government and giving 'due account of the political aspirations of the people' (United Nations Department of Political and Peacebuilding Affairs, 2022).

Against this background, most states were neutral with regard to the status of this area of Western Sahara, so a relatively uniform position dominated. With the assumption of Trump's presidency, the United States departed from this by unilaterally recognizing Moroccan sovereignty over Western Sahara ('United States Recognizes Morocco's Sovereignty Over Western Sahara' 2021). The US decision also referred to the fact that a negotiated solution would be sought within the framework of the UN.

The UN Security Council acknowledges in its Resolution 2703 (2023) 'that the status quo is not acceptable, and ... progress in negotiations is essential in order to improve the quality of life of the people in Western Sahara'. From this assessment, the resolution in §4 leads over to the negotiations, with 'a just, lasting, and mutually acceptable political solution, which will provide for the self-determination of the people of Western Sahara'. This phrasing differs from the wording of Article 1 of the *UN Covenant on Civil and Political Rights* on the right to self-determination, which grants all peoples a 'free' choice about their political status. The Security Council limits the 'free choice' insofar as it restricts it to the successful conclusion of negotiations between the participants in the 'informal consultations: Morocco, the Frente Polisario, Algeria, Mauritania as well as France, the Russian Federation, the United Kingdom and United States'.

This formulation is a precondition for the implementation of the right of self-determination of the people of Western Sahara. It has become necessary for two reasons. On the one hand, the UN wanted to determine the Saharan people's will to self-determination through a referendum. However, it failed because immigration from neighbouring countries made it impossible to determine who was entitled to vote in the referendum. On the other hand, the armed conflict between the Polisario, Morocco, and other states in the region, which began immediately after the withdrawal of the Spanish colonial power, showed that the enforcement of the right to self-determination required 'international control' (Saxer 2010: 763ff).

However, the need for a negotiated solution to clarify the political status of the territory does not mean that other states can seize Western Sahara's natural resources. For example, the attempt of the EU states to fish off the resource-rich coast of Western Sahara on the basis of a treaty with Morocco failed due to a ruling by the European Court of Justice (ECJ).

The ECJ ruled that the Polisario, as the representative of the people of Western Sahara, can claim their rights before this court and that Morocco cannot dispose of the natural riches of this area. The European Court of Justice also attaches fundamental importance to the right of self-determination. It decided that the application of an agreement concluded between the EU and Morocco to the territory of Western Sahara is incompatible with the right of self-determination of peoples (European Court of Justice 2018: para. 63).[4] Ultimately, the Polisario, as the UN and internationally recognized representation, must give its consent to such uses of resources.

The assertion of the Palestinian people's right to self-determination has been on the agenda of international politics for decades, yet it has been repeatedly neglected. This has led to accusations in the UN that the United States applies double standards in international politics and, for example, in the war against Ukraine, only intensively defends Western interests, but not those of the Palestinians, who are fighting for their independence (Gowan 2024).

It was only the cruel war in defence waged by the Israeli armed forces against the terrorist organization Hamas that put the open Palestinian question back on the agenda. There is now an increasing tendency in Western states to recognize the right to the state of Palestine (Milanovic 2024). The prerequisite, of course, is the creation of a recognized representative body of the Palestinians and the overcoming of their division across different countries.

The Western Sahara and Palestine are two internationally recognized open questions of the enforcement of the right of self-determination of these peoples. Both examples show that the international community is forced to find solutions, otherwise international peace is endangered. In this respect, the legitimate demand for self-determination of peoples brings a dynamic element to international law, which ultimately finds its basis in the postulate of justice.

4 Humanity – a codified obligation under international law

Modern international law prohibits states from using military force in interstate relations. Of course, every legal system is also confronted with violations of the law, which leads to armed conflicts such as Russia's aggression against Ukraine.

Nevertheless, these conflicts do not take place in a legal vacuum. Rather, international humanitarian law – formerly known as the law of war – applies. This legal system is characterized by the fundamental tension between humanity and military necessity. This means that the combatants are also bound by the commandments of humanity, which is why not everything that is militarily possible is allowed in armed conflict. Rather, there are rules for the means and methods of warfare that oblige the parties to respect humanity. These rules are codified in Protocol Additional to the Geneva Conventions of 12 August 1949, and Relating to the Protection of Victims of International Armed Conflicts (Protocol I).[5] For example, the use of weapons that cause unnecessary and superfluous suffering to soldiers is prohibited.[6]

Most of the rules of international law refer to the need to protect civilians. If they are nevertheless affected by the hostilities, then it is collateral damage that must be proportionate to the military necessities of the war. Part of this protection is that the Geneva Conventions have created regulations that entitle civilians and hors de combat (persons who do not take part in combat) to humanitarian assistance from the occupying power or the international community. This is where the link between humanity and solidarity can be found. Regulations are detailed and include an obligation to carry out assistance measures under Article 70 of the Code of Civil Procedure. This obligation applies to states that are in a position to provide assistance and to those states through whose territory the aid supplies reach the destination area. The party exercising control where the help is needed must also agree. Since this is an obligation under Article 70 of the AP I, consent may not be arbitrarily refused (Bothe and Ernst 2023).

This may seem surprising at first, because, in the end, humanitarian aid is provided to the population of the territory controlled by the enemy, which can be seen as an advantage for the enemy. However, Article 70 of the AP I clearly states that such assistance cannot be regarded as interference in the conflict or as an unfriendly act. Rather, it is important to ensure that aid is impartial and humanitarian. Therefore, the parties to the conflict are allowed to impose conditions on the way in which aid can reach the civilian population, and the aid supplies may also be controlled. If the party to the conflict exercising sovereignty is unable to provide the civilian population with sufficient food and medicine, the party has to allow humanitarian aid.

Wars are not the only challenge facing the international community that must be tackled in a spirit of humanity. Since even parties to an armed conflict are obliged to provide a minimum of humanity, the question arises as to the extent to which states are also obliged under international peace law to request foreign assistance in the event of natural or technical disasters if they themselves cannot adequately care for the victims (McDermott and Natoli 2018: 197ff).

Some authors see the legal framework for such an obligation in human rights, because these form an objective system of values that establish the obligations of cooperation and protection of other states.[7] Nevertheless, it is controversial whether there is a human right to humanitarian aid (Cedervall Lauta 2016). Indeed, the legal framework for this is 'largely frayed and inconsistent' (Spieker 2013: 68). However, in 1991 the UN General Assembly adopted general guidelines for strengthening emergency humanitarian aid, which ascribes the primary role to the affected state, but calls on foreign countries to help if the victims are undersupplied.[8]

Because of the obvious loophole, the International Law Commission of the UN General Assembly was given the task of dealing with the topic 'Protection of Persons in the Event of Disasters' in 2007. The Special Rapporteur on the subject, Eduardo Valencia-Ospina, presented his first report in 2008 and stated that it would be guided by the regime of international humanitarian law:

> in conflict situations there exist a large body of law dealing with assistance that may not only inspire rules on the protection in the event of disasters, but may even be applied by analogy to the extent that a rule is relevant to disaster situations other than armed conflict.[9]

The Special Rapporteur initially assumed that there was already a right *erga omnes* of victims to be assisted and a corresponding universal obligation *erga omnes* for states to provide assistance. This was rejected by the UN Secretariat:

> Notwithstanding assertions of the existence of a generalized "right to humanitarian assistance", such position, to the extent that it imposes a "duty" (as opposed to a "right") on the international community to provide assistance is not yet definitively maintained as a matter of positive law at the global level.[10]

In the end, the Special Rapporteur found a solution to this problem. The presented draft codification allows the conclusion, that there is a tendency toward the humanization of international law. The EU rightly welcomed the document, saying it represented a successful compromise between the principle of sovereignty and the need for international cooperation to protect people in humanitarian emergencies.[11]

The approach copies the provisions of international humanitarian law, according to which a state may not arbitrarily refuse international aid if it cannot adequately help the victims of a disaster with its own resources. This obliges the state concerned to justify itself when it rejects offers of assistance, in fact, a restriction of its national sovereignty in the interests of humanity and solidarity.

5 Solidarity – a necessity in the face of global challenges

The UN claims the status of a world organization to ensure global peace and well-being. However, it was not a supranational world government that was created. Rather, the UN is based on the voluntary cooperation of sovereign states in the interest of world peace. The protection of the sovereignty of the member states is expressed in Article 2 (7) of the UN Charter, according to which the UN may not interfere in the internal affairs of states. Nevertheless, the organization, which is made up of such differently constituted and developed states, could not do without a reference to solidarity among the member states.

This is a constant demand on the member states and is reflected in the competencies of the UN Security Council. In accordance with the Charter, the Security Council may impose coercive measures against a state that violates international law and order to induce the infringing state to behave in accordance with the law. These coercive measures, such as economic sanctions, can result in considerable economic disadvantages for individual states. For richer countries, it is certainly possible to compensate for the losses to the national economy. Poorer countries, on the other hand, cannot bear the brunt of such restrictions on their economic opportunities. Therefore, solidarity is used according to Articles 49 and 50 of the UN Charter for actions to benefit particular states. For any resulting economic difficulties, Article 50 provides 'to consult the Security Council with regard to a solution of those problems'.

This provision was first applied systematically and comprehensively during the Kuwait War in 1990: 'Immediately after sanctions had been imposed numerous states … requested assistance under Art. 50' (Novak and Reinisch 2024, n. 8, para 13: 1392). Since then, UN sanctions have become a widely used instrument of peacebuilding,[12] and the UN's associated solidarity obligations are occupying numerous bodies in order to make sanctions more effective on the one hand,

and to compensate states for negative consequences. Nevertheless, this duty of solidarity enshrined in the Charter has not led to any significant constitutional discussion in the Council's deliberations. Moreover, demands in accordance with Article 50 have not been raised in recent years (UN Department of Political and Peacebuilding Affairs 2022): 'None of the Council-mandated sanctions committees received formal requests for assistance under Article 50 of the Charter'. While practice with regard to the Security Council and its application of Articles 49/50 of the UN Charter is limited, interesting developments have occurred with regard to the codification of international law.

6 Common heritage principle and the law of the sea as a pacesetter

It shows the growing influence of the concepts of solidarity and humanity on the work of the world organization that the principle of the Common Heritage of Mankind (CHM) was anchored in the codification of international law in the 1980s. The principle has found its way into sovereignty-oriented international law. An important breakthrough is the Law of the Sea Convention of 1982.[13] As Wolfrum (2011) argues:

> ... the common heritage principle introduces a revolutionary new positive element into the law of the sea by indication that the control and management of the seabed is vested in mankind as a whole. Mankind in turn, is represented as far as the deep seabed is concerned by the International Seabed Authority, which is the organization through which the States Parties organize and control deep sea activities. Thus, State Parties are meant to act as a kind of trustees on behalf of mankind as a whole.

It is important to note that the holder of this principle is not the state – as is usually the case with other principles of international law – but mankind. This represents a clear shift towards solidarity, because 'mankind' must use this common heritage in solidarity so that all people can benefit from it.

This principle developed in connection with the use of territories that are not under the jurisdiction and control of any state. They are *res communes* and no state can make sovereign claims. Examples include deep seabed, high seas, Antarctica, the atmosphere, the Moon and outer space. Therefore, there is a need for international agreements that regulate the use of such areas. However, there is no general legal regime that covers all these sovereign areas. Rather, special regulations apply in each case.

The development of the CHM principle regarding the use of the deep seabed has a particularly clear link to solidarity:

> The principle of CHM was, however, transformed into a concrete regime for the Area, under Part XI of UNCLOS. It was a radical shift from the open access regime of the high seas.
>
> (Oral 2023: 33f)

In this respect, it was a radical departure from the open access regime of the High Seas, which had been codified in the Convention on the High Seas in 1958.[14] The convention spoke of the freedom of the seas, but that meant first-come, first-served. Although claims to sovereignty were rejected, unlimited exploitation was permitted.

UNCLOS put an end to this state of *laissez-faire regime* with reference to the CHM with its Part XI by combining the exploitation and sharing of economic benefits. To this end, the Convention defined in Article 1 (1) the 'Area' as 'the seabed and ocean floor and subsoil thereof, beyond the limits of national jurisdiction'. At the heart of this is the obligation that no state should be allowed to claim sovereign rights over the territory and its resources. There is no private ownership, because the resources of the area belong to humanity as a whole. In order to enforce this claim and at the same time enable the exploitation of resources in the interest of humanity, the International Seabed Authority (ISA) was created, which manages the area and issues licenses for the extraction of mineral resources. At the same time, the ISA checks that the area is used exclusively for peaceful purposes. The ISA is made up of all member states of UNCLOS, but acts in accordance with the interests of all humanity. Thus, it is an instrumentalization of the CHM. The member states of the ISA organize and manage the area, approve the activities of companies, and create their own company for deep-sea mining. At the same time, it organizes a distribution of the economic results of deep-sea exploitation (Oral 2023: n. 42): 'The equitable sharing of the benefits from the exploitation of the natural resources of the Area is core of the CHM regime under Part XI'.

The idea of solidarity is expressed in particular in the mechanism of deep-sea mining. A company that wants to carry out this type of mining is assigned two areas by the ISA, which it must first explore with a view to possible exploitation. The results must then be reported to the ISA. Once the research results have been submitted, the ISA assigns a territory to the company. The other area explored can be exploited by the ISA's mining company itself. In this way, the ISA avoids costly research and thus enables the exploitation of deep-sea mineral resources in the interest and benefit of all humanity. The rich states and companies pass on their knowledge to the international community, a genuine application of the concept of the common heritage of humanity. At the same time, however, the revolutionary regulation found with the Convention on the Law of the Sea of 1982 also shows its limits. At that time, a regulation had been laid down that the states and companies capable of mining in this region were also obliged to pass on their technologies to other interested parties, but it became clear that the highly developed industrial countries were not prepared to accept this obligation and refused to accede to the Convention. The consequence was a change in the originally envisaged obligation to transfer on technology (Ipsen 2024: 891).

This shows that the international law of solidarity repeatedly comes up against the limits not only of states' claims to sovereignty but also of capitalist economic interests. Nevertheless, the principle of the common heritage of humanity is gradually conquering traditional international law and creating practical consequences such as the payment mechanism, which 'means to ensure that all humankind, including future generations, enjoys the benefits of its shared stewardship of the Area' (Thiele, Damian and Singh 2021). This idea is followed by the 'The Mining Code', which is the draft exploitation regulations (International Seabed Authority 2022b). Of course, the revolutionary idea of commonality does not automatically prevail. For this reason, numerous conferences are currently taking place on the further legal design of the concept (International Seabed Authority 2022a).

It is also interesting to note that the provisions of the Convention on the Law of the Sea on the status of the seabed also include the conservation of marine life as part of the CHM.

The principle of solidarity is also mentioned in other international agreements. Reference should be made to the Convention to Combat Desertification in Those Countries Experiencing Serious Drought and/or Desertification[15] and the United Nations Framework Convention on Climate Change of 9 May 1992.[16]

This treaty is the basis for the annual World Climate Conferences (COP). On the first day of the 28th session, member states agreed on a practical step to enforce the principle of solidarity by creating a voluntary fund to address climate-related loss and damage. Five states, as well as the European Union, immediately provided a combined total of about $414 million. It was also agreed that by 2030 the fund would be funded with at least $100 billion per year. The invocation of the principle of solidarity has thus led to practical consequences (Geinitz 2024).

7 Concluding remarks

The challenges of the present make it clear that they cannot be solved exclusively by states that invoke their sovereignty. This is the approach of classical international law. This classical understanding of sovereignty was already called into question in 1945 with the UN Charter, because the Charter is based on the principle of justice, which means that all states, whether small or large, enjoy sovereign equality. But it is not only states that bear rights and obligations in the sense of the UN Charter. Rather, the document also mentions the self-determination and equality of peoples for the first time and mentions respect for human rights as a task of the member states (Mahoney and Mahoney 1993: 59). Justice is thus understood in the UN in a broader sense than just the equality of rights between states. Peoples and human beings also have a right to justice, and in this respect, the UN Charter also includes these groups.

Through the participation of peoples and individuals, international law took on a dynamic component and overcame its static moment. In fact, the idea of humanity was anchored above all in international humanitarian law, which is applied in armed conflicts. The prohibition of inflicting unnecessary suffering on combatants is due to the concept of humanity. This erects a barrier that not everything that would be possible is allowed, even in war. Humanity is also clearly expressed in the treatment of civilians in conflict. Thus, the commitment of humanitarian aid to civilians in occupied territories reflects a deeply humanitarian idea.

The turn to justice and humanity, which international law had been implementing since 1945, also meant a departure from an absolute understanding of sovereignty. This development took place gradually and was also dependent on political factors. Therefore, the process had also been interrupted by phases of standstill and confrontation. Nevertheless, this development has continued because there is a need for cooperation in tackling global challenges. The present now poses special challenges for the cooperation of the international community.

What is meant here is above all the urgently needed protection of the environment, global warming and the end of natural resources. These challenges can only be tackled by the international community as a whole, as they take no account of national borders and affect all states to a greater or lesser extent. Consequently, invoking state sovereignty is not expedient. Cooperation is needed in an organized and codified form. In practice, there is talk of an international law of solidarity. This has found a clear legal expression in the concept of the common heritage of mankind.

This concept is currently facing its practical test. In UNCLOS, it was legally agreed that the deep-sea floor is a common heritage of humanity, that no state or other entity may unilaterally appropriate. Rather, the decision on its exploitation is the responsibility of an international body. Now comes the practical test, because the first application from a company for deep-sea mining has been submitted (Rosenberg 2023). The new international law of solidarity is thus demanded. It is a challenge that is in line with SDG 16. It is about peaceful cooperation that includes all states and societies. Marine mining should produce sustainable development and be fair through the direct and indirect participation of all. To make this possible, an accountable and inclusive institution was created with the Seabed Authority. Whether this organization is also 'effective' in the sense of SDG 16 remains to be seen. Of course, this is a political question. For the time being, it can be seen as a success that the international community has at least accepted the need to turn to an international law of solidarity. The gradual implementation of the right of solidarity could contribute to flourishing in the international sphere.

Notes

1. Slaughter, A.-M. (2005) 'Security, Solidarity, and Sovereignty: The Grand Themes of UN-Reform', *The American Journal of International Law* 99 (3): 624.
2. Working Group on Business and Human Rights, OHCHR, OHCHR website, https://www.ohchr.org/en/special-procedures/wg-business (accessed 18 April 2024).
3. Western Sahara Advisory Opinion, https://icj-cij.org/sites/default/files/case-related/61/9467.pdf (accessed 11 October 2025).
4. European Court, C-266/16 of 27.2.2018, para. 63.
5. ('Protocol Additional to the Geneva Conventions of 12 August 1949, and Relating to the Protection of Victims of International Armed Conflicts (Protocol I)' 1977).
6. Art. 35 (2) AP I.
7. UN General Secretary regarding the New International Humanitarian Order, UN-Doc. A/61/234, para. 5, 7 August 2016.
8. UN-Doc, A/Res. 46/182, 11 March 2013.
9. UN-Doc. A/CN.4/598, 5 May 2008.
10. UN-Doc. A/CN.4/590, 11 December 2007, para. 251.
11. UN-Doc. A/71/10, 17 March 2016, para. 22.
12. According to the Resolution 'Implementation of the provisions of the Charter of the United Nations related to assistance to third states affected by the application of sanctions'. UN-Doc. 54/107, 9 December 1999.

13 United Nations Convention on the Law of the Sea, 10 December 1982, 450 UNTS 11.
14 Convention on the High Seas, 29 April 1958, 450 UNTS, Art. 2.
15 United Nations Convention to Combat Desertification in those Countries Experiencing Serious Drought and/or Desertification, Particularly in Africa, '1954 UNTS', 3.
16 United Nations Framework Convention on Climate Change, '1771 UNTS', 107.

16

Human Flourishing: An Integrated Systems Approach to Development Post 2030

Andrej Zwitter, Carole Bloch, George Ellis, Richard Hecht, Ariel Hernandez, Wakanyi Hoffman, Dean Rickles, Victoria Sukhomlinova, Karma Ura

1 Introduction

Introducing this perspective article in the year 2025, a mere five years remain to meet the Sustainable Development Goals (SDGs) – a set of targets, indicators and objectives established for the global community. As per the UN's SDG-Midterm review (UNStats 2023), the global achievement of these indicators stands at less than 15 per cent. Predictive analyses of UNStats cast a shadow on the ultimate efficacy of the SDGs. This might suggest that their impact may be confined largely to elevating awareness around the critical need for enhanced sustainability practices across all societal sectors. Many authors argue that we should redouble the efforts and extend the duration of the SDGs (Fuso Nerini et al. 2024), and that it remains crucial to move the impact beyond discursive to more profound policy and institutional impact (Biermann et al. 2022). Simultaneously, there is a rising call – predominantly from regions outside the Northern and Western developed world – for the integration of marginalized perspectives, neglected epistemologies such as Indigenous knowledge systems, and a shift towards empowering local levels through a transition from top-down to bottom-up approaches (besides capacity building efforts) in global governance, as Yap and Watene (2019) argue. Their argument further underscores the need for an integrated approach to development that transcends traditional metrics and methodologies, highlighting the complexity and dynamism of global systems. The shortcomings of the SDGs, particularly in addressing non-material

This chapter was previously published as an open access perspective article in the journal *Earth System Governance* (Zwitter et al. 2025).

aspects of human flourishing and adapting to a rapidly changing global context, highlight the imperative for innovative development paradigms.

In short: *We argue that the current SDGs framework is important but insufficient to lead to human flourishing. Post-2030 human development needs to adopt non-material aspects of flourishing and adapt accordingly its approach to governance.* The purpose of this perspective article is, therefore, to propose a new conceptualization of human development as human flourishing as an integrated systems approach utilizing multi-level governance to inform the starting debate of a post-2030 (post-SDGs) development framework. This integrated systems approach encompasses both material and non-material aspects of human flourishing. It emphasizes the necessity of incorporating local and regional specific aspects of human flourishing, such as emotional well-being, cultural richness and spiritual fulfilment to capture the full spectrum of societal health and human experience. The novelty of our approach lies specifically in its application to the renegotiation process of the SDGs post 2030. This approach challenges the current development narrative, advocating for a paradigmatic shift towards harmonizing material conditions of development with the richness of human experience encompassing non-material conditions of human flourishing (Greek: *eudaimonia*).

2 Theoretical background and method

We propose an integrated systems approach (Fiksel 2006) that addresses development challenges by recognizing the interconnectedness of economic, psycho-social, cultural and environmental factors (i.e. material and non-material aspects of the human condition), ensuring positive influences across these areas. By expanding the discourse in this way, we address the knowledge-action gap hindering the feasibility of sustainability policy implementations, as noted by Brutschin et al. (2021) and Jewell and Cherp (2020). This proposal not only enriches the scientific debate on Earth system governance but also advances the thinking on governance through global goals, offering a more integrated and resilient approach to addressing global challenges.

Current sustainability research often relies heavily on quantitative metrics like GDP to gauge progress (Stiglitz, Sen and Fitoussi 2010), which tend to focus on material aspects of the human condition. This narrow focus has overlooked the immaterial dimensions of human flourishing, such as mental health, community resilience, cultural and spiritual fulfilment, and a sense of purpose. These aspects are crucial for a more complete understanding of human well-being and sustainability. We call for moving beyond a perspective rooted in 'mechanistic thinking', which emphasizes the availability of material resources and technological innovation as keys to achieving sustainability (Peng et al. 2021; Trutnevyte et al. 2019). In addition, integrating non-material aspects of human flourishing into the global development agenda allows researchers to explore the suitability of qualitative indicators for themes such as: mental health prevalence, availability of mental health services, and programmes focusing on emotional intelligence, (inter-)cultural education and social skills within school curricula. This approach emphasizes creating a comprehensive framework that acknowledges the complex, interconnected factors contributing to human well-being in sustainable development. The integrated systems approach is foundational to the Integrated Flourishing Measures (IFMs) which we propose in this perspective article, and which aim to assess and promote progress by considering psycho-social resilience, cultural richness and non-material fulfilment alongside non-traditional economic indicators such as volunteering services. We introduce a significantly different governance structure than the SDGs, proposing new IFMs based on bottom-up cascading targets and indicators towards global level principles. These measures integrate human flourishing with the flourishing of non-human species, rooted in the Virtue Ethics' concept of *eudaimonia* (flourishing) (Hursthouse 2002). This prioritizes a process-oriented view of sustainable development as human flourishing with localized, achievable endpoints.

The results presented in this perspective article are based on multi-stakeholder consultations and workshop-based focus groups, organized through the research programme on 'Conceptions of Human Flourishing', at THE NEW INSTITUTE, Hamburg, from September 2023 to June 2024. The programme focused on conceptualizing non-material aspects of human well-being, aiming to reflect a diversity of epistemic perspectives that more accurately mirrors the vast array of cultural, religious, spiritual, emotional and social realities worldwide than are currently captured by many tools of global governance. The insights were drawn from regular working group meetings, invited scholars and experts, and four workshops subdivided into focus groups with a wide array of external participants.[1] Cross-disciplinary, transdisciplinary and intercultural results were ensured by inviting guests and workshop participants from academia, Indigenous communities, civil society and policy-making from many countries (Table 16.1).

The results were subsequently reviewed and served as a basis for this perspective article. It revealed a logically consistent set of interrelated problems encountered in centralized developmental approaches such as the SDGs. The

Table 16.1 Regions and Countries

Region	Countries/Entities
Africa	Kenya, Nigeria, South Africa, Zimbabwe
Americas	Apache Nation (USA), Navajo Nation (USA), Brazil, Mexico, USA
Asia	Bhutan, China, India, Japan, Philippines, Hong Kong
Europe	Austria, France, Germany, Ireland, Netherlands, Russia, Spain, Switzerland, Ukraine, United Kingdom
Oceania	Australia

overarching finding informs also the structure of the subsequent argument based on three principal points of departure:

(1) Incorporating **Non-Material Conditions of Flourishing** (Section 3) as part of lived local experiences requires
(2) **New Indicators and Measurement** (Section 4) that can accommodate local specificities, non-material conditions and help dismantle structural inequalities. Such an understanding of the importance of local diversity in the expression of non-material conditions necessitates
(3) **Multi-level Governance** (Section 5) that replaces global goals with a set of global principles and develops goals and indicators bottom-up.

This structure promotes a dynamic, participatory model of development based on an integrated systems approach of human flourishing advocating a balance between material and non-material aspects.

3 Non-material conditions of flourishing

In considering non-material conditions of human flourishing, mental health, spiritual health and community health are inextricably linked, particularly when considering the impact of cultural disorientation on psycho-social well-being. Cultural disorientation, which can arise from rapid social changes, migration or the loss of cultural identity, significantly affects mental health not only on an individual but also on a community level. Addressing these issues requires an understanding of how cultural context shapes psychological well-being and the integration of culturally sensitive mental health as well as spiritual practices into larger policy approaches to human flourishing.

Many debates in our expert focus groups delved into the critical distinction between material and non-material conditions for human flourishing, revealing an underrepresented area in current development paradigms. This debate highlighted that while material conditions such as economic stability and access to resources are essential, they represent only one aspect of what it means to live a fulfilled and meaningful life. For example, Layard (2006) in his landmark book illustrates psychological and sociological findings pointing at non-material conditions of happiness that are collectively attainable. Non-material conditions – encompassing religious, spiritual, emotional, imaginative, intellectual, linguistic and cultural well-being – play an equally vital role in fostering a flourishing society (Zwitter and Dome 2023). This link between human flourishing and non-material aspects of the human condition has been recognized in positive psychology (Seligman 2011) and empirically verified. Similar research in the same domain inextricably connects human character traits (see also Virtue Ethics [Chapter 5]) and *eudaimonia* (flourishing) to each other, as was originally argued by Seligman and Peterson (2004) and which led to a virtue-based approach to positive psychology. This dual perspective challenges the existing development focus that predominantly targets material achievements, suggesting a more nuanced approach that equally values inner growth, happiness and community bonds (Dalai Lama and Cuttler 1999).

Acknowledging this dichotomy between material and non-material aspects of development and the absence of the latter in the SDGs, new initiatives like the 'Inner Development Goals' (2024) were proposed to compensate for this omission and to provide a more inclusive framework for future policies to foster human flourishing. Such a framework would not only aim at improving physical and economic standards of living but also at enriching the spectrum of human experience through fostering emotional resilience, religious and spiritual anchoring, emotional contentment, meaningful imaginative and intellectual opportunities, and cultural expression and appreciation. Similar proposals such as the work of Dowson, Devenish and Miner (2012) followed suit, emphasizing the need to include spiritual components to the aspects of human flourishing. Similarly, the Harvard initiative on Global Flourishing Goals initiated by Karthikeya et al. (2024) as well as the latest book of Mountbatten-O'Malley (2024), one of the co-authors of the Global Flourishing Goals, indicates that we are transitioning from human development grounded in a materialist worldview to one of human flourishing that integrates material and non-material aspects into the conception of *eudaimonia*.

While this might sound abstract and outside the realm of policy-making the following suggestions further elaborated in the subsequent sections seem pertinent. One way to achieve this integration of non-material aspects into the future post-2030 SDGs, policies must encourage practices and initiatives that nurture these non-material dimensions, such as promoting mental health, preserving cultural heritages and spiritual practices, and supporting community engagement and solidarity. By integrating these broader aspects of human well-being into the development agenda, policies can aspire to create societies where individuals not only survive materially but also thrive non-materially, ensuring a balanced approach to development that truly caters to the integrated systems nature of the human condition. This policy recommendation urges for a paradigm shift towards embracing the full spectrum of human needs and aspirations, paving the way for a development strategy that harmonizes material success with the richness of human experience. This richness of human experience, as will be discussed below, is fundamentally linked to local actualization and practices, often found in Indigenous communities and regional cultural expressions.

4 Towards integrated flourishing measures

True human flourishing encompasses far more than economic growth and the fulfilment material aspects; it includes the non-material dimensions of life such as emotional well-being, cultural and linguistic richness, and psychological as well as religious, spiritual and emotional fulfilment. This was acknowledged two decades ago by intergovernmental institutions (OECD 2001). However, the continuing reliance on quantitative metrics, such as Gross Domestic Product, specifically in the Human Development Index, for example, for assessing progress towards human development is flawed at best and counter-productive at worst (Stiglitz, Sen and Fitoussi 2010). Many of the most important aspects of human flourishing are difficult to measure quantitatively; happiness, friendship, cultural expression, spiritual fulfilment and so on, require not only the adaptation of non-quantitative metrics, they require a rethinking of the aspects of human flourishing that include non-material aspects and acknowledgement that these aspects can differ fundamentally in their expression on the local level. Therefore, limiting human development to measurable material conditions seems to miss the purpose of human development, that is: *enabling people, beyond human security, to achieve a fulfilled life*. In this respect, Bhutan's Gross National Happiness (Kim, Richardson and Tenzin 2023) serves as a radiant example moving away from traditional material indicators and re-centring the focus of attention from wealth to well-being (i.e. an understanding of *eudaimonia* as the goal of any development). These non-material aspects, often qualitative and aspirational in nature, escape traditional centralized measurement frameworks being both non-material and expressing themselves locally differently. But they are crucial for a comprehensive view of progress and well-being.

Consequently, there is a pressing need to expand the development paradigm to include broader, more inclusive indicators that capture the full spectrum of human experience and societal health (IISD – SDG Knowledge Hub 2021) and to localize these measurements. To address this measurement problem and to transcend the limitations of GDP-focused metrics we recommend developing new indicators, so-called **Integrated Flourishing Measures** or **IFMs**, that can more accurately reflect the non-material aspects of development. These indicators should aim to capture qualities such as community cohesion, cultural and linguistic vitality, environmental stewardship, and individual and collective well-being (Jimenez 2008). To address the need for indicators capturing the non-material aspects of development, participants in the renegotiation of the SDGs for a post-2030 framework could advocate for the development of new localized indicators such as:

- **Community Cohesion Index:** strength of social bonds and the sense of belonging within communities.
- **Cultural Vitality Index:** diversity and vitality of cultural expressions and the accessibility of cultural experiences to all segments of society.
- **Language Vitality Index:** uses of multilingualism and language, both in oral and written forms.
- **Environmental Stewardship Indicator:** public engagement with environmental conservation efforts and sustainable lifestyle practices.
- **Psycho-social Well-being and Happiness Indexes:** life satisfaction, emotional well-being, and fulfilment beyond economic status, including mindfulness, religious and spiritual well-being.

Furthermore, integrating ecological integrity and climate adaptivity into these localized indicators can help create policies that promote a renewed understanding of the human condition as being intrinsically and symbiotically linked with the natural environment. Such an understanding is common in Indigenous knowledge systems. The localization of indicators to include non-Western and Indigenous epistemic frameworks might yield further insight on how an integrated systems approach views human-nature interaction and interdependency. By recognizing the complexity and interdependence of these factors, our policy recommendations aim to guide the formulation of a post-2030 development agenda to truly reflect the aspirations, needs and challenges of meaningful human flourishing.

The same quantitative market-based and other material indicators of development also have a tendency to mask structural inequalities. A nuanced understanding of the structural inequities that remain unaddressed within the SDGs framework emphasizes the need for a more inclusive approach that better accounts for regional differences, vulnerabilities and the diverse starting points of various communities. It encapsulated insights from Amartya Sen's Capability Approach (Sen 2001) and the Multi-Dimensional Poverty Index developed by Alkire and Foster (Alkire 2002). This perspective highlights that the SDGs often focus on income inequalities through measurable economic indicators, which may not fully capture the breadth of disparities affecting communities worldwide. This oversight can lead to policies that, while well-intentioned, may not address the root causes of inequality or the multifaceted nature of disadvantage and marginalization (Atkinson 2015; Stewart 2005).

To effectively integrate non-material dimensions and IFMs into development strategies, we suggest several key approaches. Implementing comprehensive mental health services and programmes can be aimed at enhancing emotional well-being in communities, while at the same time ensuring these initiatives respect local cultural contexts. Support initiatives that protect, celebrate and facilitate multilingualism, language diversity and cultural practices, fostering appreciation for cultural heritage thereby become crucial cornerstones of such a policy. We consider the enhancement of community bonds and social cohesion through the promotion of community centres, events and participatory activities as essential components of a policy that includes non-material conditions of flourishing. Furthermore, to recognize and support diverse spiritual practices and spaces as integral to community and individual well-being, ensuring such support originates from local initiatives rather than state imposition is of crucial value to non-material well-being conditions. Additionally, we suggest promoting artistic and cultural expressions as vital means for individual creativity and community identity. These strategies collectively contribute to an integrated systems approach to development that values both material and non-material aspects of human flourishing.

In light of these discussions, there is a compelling case for policy recommendations on how to integrate the IFMs into the post-2030 development agenda. Such policies should prioritize a deeper engagement with the unique strengths and challenges of different regions and communities, acknowledging that a one-size-fits-all approach to development is inadequate. Specifically, development policies must consider the multifaceted dimensions of inequality, including access to and quality of education, healthcare, cultural and linguistic resources, and political representation. To achieve this, there must be a concerted effort to incorporate the voices and experiences of marginalized groups into the decision-making process, ensuring that development strategies are co-created with those most affected by disparities. Table 16.2 highlights the need for development policies to be tailored to regional and community-specific contexts, incorporating both quantitative and qualitative assessments of locally specific material and non-material needs and practices. It emphasizes expanding well-being indicators beyond material indicators, focusing on disparities in education, health and inclusion, while also involving marginalized voices in decision-making. Localization of these indicators has the added benefit of more accurately reflect community-specific aspects of flourishing. Lastly, this calls for the creation of comprehensive inequality indicators that reflect various dimensions, validated through collaboration with academic and global organizations across diverse communities.

Table 16.2 Policy Recommendations Regarding Implementing Integrated Flourishing Measures

	Policy	Measurement Strategy
Regional and Community Specificities	Tailor development policies to the unique environmental, cultural, spiritual, linguistic and economic contexts of different regions and communities	Conduct regular, comprehensive community assessments to identify local practices, needs and strengths, using both quantitative data and qualitative insights from community engagement forums
Beyond Material Metrics	Incorporate broader indicators of well-being and access to opportunities, beyond traditional material metrics	Utilize surveys and census data to assess disparities in conducive environments for lifelong learning, education, health and living standards; revisit the existing implicit and explicit assumptions made about material and non-material aspects of human flourishing in the collection and interpretation of data; e.g. implement participatory approaches to measure social and cultural inclusion

Marginalized Voices in Decision-Making	Ensure that development strategies and policies are co-created with the participation of marginalized and vulnerable groups	Establish and track metrics milestones and pathways to impact for community participation in governance, using both attendance records and qualitative evaluations of the inclusivity and effectiveness of participatory processes
Comprehensive Indicators of Inequality	Advocate for and support the development of comprehensive indicators that reflect the complex realities of inequality, capturing economic, social, cultural, linguistic, educational and political dimensions	Collaborate with academic institutions, NGOs and international organizations to research and validate new measures of inequality; pilot these measures in diverse communities to ensure their relevance and accuracy

5 Multi-level governance

We have already demonstrated the need to localize indicators to better reflect cultural expressions of human flourishing and to redress some of the embedded structural inequalities and exclusion of local epistemologies. Our integrated systems approach represents a shift in the understanding of human flourishing of being intrinsically dependent on non-material conditions that can often only be found in local cultural and spiritual expressions. This raises some core concerns when it comes to global and centralized governance exemplified by the seventeen SDGs with its 169 targets and 247 indicators. One of the main critiques raised during the workshops and particularly from Indigenous community leaders about current development approaches was the lack of recognition for diverse local life experiences and goals ultimately expressed in indicators. The concept of 'localizing flourishing' emphasizes the significance of empowering communities to define their own sustainable futures and derives directly from the shift to include also non-material conditions of human flourishing. This conversation highlights a critical gap in the current SDG framework, which often adopts a top-down approach, potentially overlooking the nuanced needs, aspirations and capabilities of local contexts. Sustainable development cannot be uniformly prescribed; rather, it must be fluid and adaptable, reflecting the rich variety of global communities. This perspective advocates for a development model that is rooted in local realities and leverages Indigenous knowledges and practices, thereby ensuring that solutions are culturally sensitive, contextually relevant, and sustainable and accepted amongst affected communities (Ki-moon 2015).

In response to this critique, we recommend that policies for human flourishing post-2030 actively promote the decentralization of goals, targets and indicators, encouraging the customization of goals to fit local conditions and priorities (see also Mansuri and Rao 2013). This policy recommendation suggests establishing structural mechanisms for local communities to regularly participate in the decision-making processes through feedback loops, ensuring their voices are heard and their knowledge is integrated into sustainable development strategies, as the UNESCO convention on cultural diversity implies (Beukelaer, Pyykkönen and Singh 2014). Furthermore, it calls for the development of flexible multi-level governance frameworks that can accommodate local innovations in sustainability science, which in turn increase efficiency and effectiveness in responding to crises and recognizes the value of grassroots initiatives and community-led solutions. By fostering a bottom-up approach to development, where local actors are seen as equal and essential partners in the pursuit of sustainability, we can create a more inclusive, equitable and effective path towards achieving global development goals (Zürn 2010). This approach not only respects the autonomy and wisdom of local communities as valid knowledge, but also harnesses their potential to contribute unique solutions to global challenges.

The policy framework for a post-2030 development agenda of human flourishing emphasizes a multi-level governance system, similar to the Multi-Stakeholder Partnership model proposed by Glass, Newig and Ruf (2023), which takes into consideration the complementary roles of different stakeholders on different governance levels. This approach aids in integrating global principles with meso-level goals and specific regional and local targets and indicators. The global principles which derive from the discussion above can be summarized as:

(1) **Human Development as Flourishing**: prioritizing a comprehensive approach to human well-being that balances material conditions with non-material aspects like emotional, cultural and spiritual health.
(2) **Ecological Integrity and Climate Adaptivity**: ensuring that human development aligns with the preservation of ecological systems and adapts to the challenges posed by climate change.
(3) **Cultural and Epistemological Diversity**: valuing the diverse cultural, religious and Indigenous knowledge systems, incorporating them into global governance to foster sustainable and contextually relevant solutions.

(4) **Equity and Social Justice**: promoting fairness and dismantling structural inequalities, ensuring marginalized voices are included in decision-making processes, and providing equal access to resources for all communities.
(5) **Interconnectedness and Global Solidarity**: emphasizing the interconnected nature of human and planetary systems, fostering empathy, cooperation and collective action to address global challenges collaboratively.

These global principles are contextualized through medium-level strategies addressing regional nuances and challenges. At the local level, our framework promotes the development of specific goals and indicators informed by community needs, resources, as well as cultural and religious contexts. This dynamic and participatory model values local innovations and solutions, ensuring a cohesive yet adaptable structure that accommodates global community diversity (see Figure 16.1).

The objective of our integrated systems approach is to achieve inclusive, equitable and effective human flourishing by localizing the development and

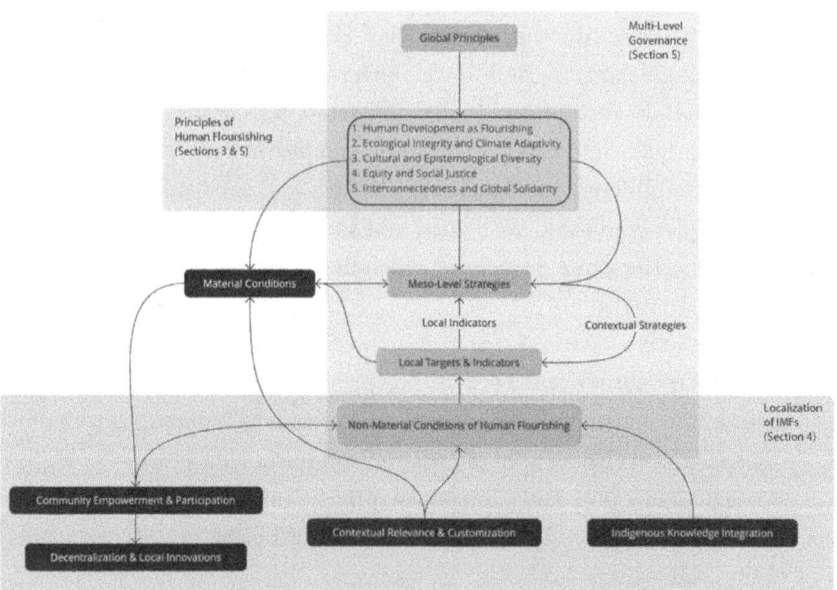

Figure 16.1 Multi-level governance for human flourishing combining global principles, meso-level strategies and local-level target and indicator definition and implementation.

implementation of targets and indicators in order to empower communities and to foster a bottom-up approach that accommodates cultural and psycho-social as well as spiritual specificities of local communities. This further requires the following procedural norms:

(1) **Empowerment and Participation**, ensuring local communities actively define their sustainable futures.
(2) **Contextual Relevance**, customizing indicators and targets to fit local conditions and cultural necessities and specificities.
(3) **Decentralization**, promoting local innovations and solutions.
(4) **Integration of Indigenous knowledge**, leveraging grassroots initiatives for sustainable solutions.
(5) **Multi-Level Governance**, aligning global aspirations (principles) with meso-level strategies and local capabilities and needs.

Strategies to achieve this include developing localized indicators, enhancing participatory decision-making, building local capacity, creating flexible frameworks, and fostering partnerships among governments, NGOs and community organizations. Implementation steps necessitate comprehensive quantitative and qualitative assessments of local capabilities and risks, community engagement and an adaptive perspective on goals not as endpoints but as a process.

6 Conclusion: an integrated systems approach

An integrated systems approach to human well-being and sustainability is imperative for the renegotiation of the future of the SDGs post 2030. This integrated systems approach acknowledges the interconnectedness of *non-material and material aspects of human flourishing* (see Section 3), societal welfare and environmental integrity, advocating for strategies that recognize the interdependence and nurture all aspects of life systems as also advocated by Rockström et al. (2023) for safe and just Earth System Boundaries. It challenges traditional models that isolate human, societal and environmental needs, promoting synergies that enhance both human and planetary health without causing harm to any dimension (Raworth 2017). Furthermore, it advocates for the importance of non-material conditions of human flourishing, as proposed by the pioneer of positive psychology Martin Seligman (2011), deriving not only from a general idea of positive psychology but from its embeddedness in local realities of cultural, social and spiritual practices.

Paradigms, which overly rely on material and economic indicators fail to capture the full spectrum of well-being, including emotional, psycho-social, spiritual and environmental dimensions. An integrated systems approach must, therefore, transcend traditional metrics by incorporating indicators that reflect relationships, community resilience and ecosystem health. This aligns with modern understandings of prosperity, which values harmony between human activities and the natural world. This integrated systems approach to human flourishing based on an understanding of the human condition consisting of material *and* non-material aspects requires new indicators and measurement approaches which we have proposed as Integrated Flourishing Measures – IFMs (see Section 4).

Moreover, this perspective article demonstrated (in Section 5) how adopting such an integrated systems approach logically requires acknowledging regional differences. It advocates for frameworks that enable the translation of global principles into localized actions, allowing communities the flexibility to adapt these overarching values to their specific contexts and priorities. By focusing on global principles rather than prescriptive global goals, the development agenda can become more adaptable, inclusive, and responsive to the dynamic and complex nature of global and local challenges and the local and regional diversity of cultural, religious and collective expressions of non-material conditions of human flourishing. This approach not only respects the autonomy and wisdom of local communities, but it also harnesses their potential to contribute innovative solutions to global issues, fostering a more connected and harmonious integrated system of planet and people.

Sustainability involves not only environmental conservation and material welfare but a fundamental rethinking of humanity's symbiotic role in the ecosystem of the planet and a re-imagination of the human condition as consisting of both material and non-material aspects. A healthy planet is foundational to human flourishing, and linguistic and cultural diversity as well as social cohesion are crucial for resilient communities. Post-2030 principles and goals should advocate for preserving climate adaptive, ecological integrity and nurturing cultural and social conditions that bond human agency with an integrated understanding of larger planetary processes in order to comprehend what sustainable societies truly need to become. Specifically, the promotion of psycho-social and moral components together with aspects of mindfulness and spiritual aspects of positive psychology cultivate empathy and solidarity and are essential for a just, sustainable and flourishing society on local and global levels (Ura 2017) and in harmony with planetary systems.

Connecting to forthcoming discussions on the future of the SDGs beyond 2030, our proposal underscores a critical perspective on the current framework of SDGs while acknowledging the important achievements that they represent. We advocate for the need for a paradigmatic shift towards emphasizing principles of living together over discrete quantifiable targets. This recognizes the need for a development agenda that transcends the limitations of narrowly defined goals, advocating for an integrated systems approach that considers the interconnectedness of global challenges and the diverse needs of local communities worldwide. Our perspective article highlights the importance of reimagining human development as human flourishing in terms of shared values and principles that acknowledge local diversity and cultural richness. Future development efforts hinge on our ability to foster a sense of solidarity with each other and the planetary ecosystem and an understanding of our inevitable interdependence to this ecosystem through a recognition of diverse cultural practices that have recognized this interdependence beyond its material dimensions. The SDGs represent a foundational milestone in international collaboration towards common development goals. It is up to us to shape the post-2030 SDGs to make them fit for human flourishing beyond mere material needs.

Note

1   For more details on the project, workshops and events see: https://thenew.institute/en/programs/conceptions-of-human-flourishing (accessed 11 October 2025).

Conclusion

Andrej Zwitter

1 Reaffirming the necessity of a meta-scientific approach

This volume began with a simple but profound question: what does it mean to flourish as a human being in the twenty-first century? The answers, as the chapters demonstrate, cannot be found within the confines of disciplinary silos or the narrow epistemic bounds of empiricist materialism. Instead, flourishing must be rethought as a multidimensional and relational phenomenon that spans material, social, psychological, cultural and spiritual domains. Such rethinking necessitates a meta-scientific approach, i.e. an approach that steps back from any single methodology or worldview to explore the assumptions, limits and interdependencies of different ways of knowing.

As George Ellis reminds us in Chapter 1, the dominant scientific worldview, shaped by reductionist physicalism, has neglected core aspects of the human condition: meaning, agency and narrative. The rise of scientism, a dogmatic view of the twentieth century's scientific findings and methods as dogmatic truths and which seeks to explain all phenomena through impersonal laws and quantifiable data, has left us with powerful technologies but impoverished understandings of value, purpose and the good life. The 'middle view' Ellis proposes, that is neither naive narrative nor pure mechanism, offers a framework where emergent complexity and symbolic agency are acknowledged as real and causally efficacious. Markus Gabriel (Chapter 2) expands this notion by arguing that meaning itself is ontologically indispensable. We are not observers of a self-contained universe but participants in overlapping fields of sense, where reality is co-constituted by mind and world. This view reframes flourishing not as mere survival or utility maximization, but as spiritual and existential alignment within a meaningful cosmos.

Dean Rickles (Chapter 3) adds to this foundation by demonstrating how worldviews shape ontological security. A society that denies agency, teleology and interconnectedness risks ontological disorientation and collective alienation. In contrast, a meta-scientific perspective restores the possibility of seeing ourselves not as accidents of matter but as moral agents embedded in a living system. Together, these contributions outline the groundwork for a renewed science – not anti-scientific, but one that includes science within a broader ecology of knowledge. Such a science (or *meta-science*) is purpose-driven, open to ontological pluralism and responsive to the normative dimensions of human life.

In short, if we are to build frameworks for human flourishing that address the full complexity of the human condition, we must transcend scientism and develop a meta-scientific sensibility – one that allows us to see not only how the world works, but also what the world means and what it demands of us.

2 Embracing epistemic and ontological pluralism

Human flourishing, as this volume illustrates, is not a culturally neutral or epistemically monolithic concept. It is lived and interpreted through diverse ontologies, worldviews and Lebenswelten. A key insight emerging from Parts Two and Three is that different academic disciplines and cultural traditions offer distinct but equally valid perspectives on what it means to flourish. This calls for an epistemic humility and a willingness to learn across paradigms.

Theo Bouman (Chapter 4) presents insights from positive psychology that complement, rather than compete with, philosophical and spiritual approaches; it emphasizes the psychological role of meaning and a process-oriented view of what it means to flourish. Andrej Zwitter (Chapter 5) revives the Neo-Platonic tradition, suggesting that intellectual and spiritual virtues are not historical curiosities but vital components of well-being. This perspective extends the set of cardinal virtues emphasized in positive psychology to the realm of spiritual virtues. These spiritual virtues enable the individual to reflect upon themselves as part of a larger cosmos and as being on a teleological trajectory that aims towards transcendent harmony in an ultimate oneness. Harald Atmanspacher (Chapter 6) and Richard Hecht (Chapter 7) further widen the scope, offering frameworks in which intuition, transcendence and utopian hope form essential dimensions of the good life. Atmanspacher takes a deep dive into the question of what it means to flourish from a Spinozist perspective underlining the importance of intuition to comprehend the well-being of individuals, societies and cultures in their interdependency with nature. Hecht demonstrates how the concept of hope in particular helps to constructively overcome adversities and traumas and aids in the process of flourishing. This fundamental role of hope in face of crises can be seen as an important addition to positive psychology, which as Bouman (Chapter 4) argues can be criticized for not sufficiently taking into account processes that are outside the influence of the individual.

In Part Three, Wakanyi Hoffman (Chapter 8) draws on Indigenous cosmologies that centre relationality, ecological stewardship and the sacred. Through narratives and oral traditions, she illustrates how wisdom keepers through the ages and across different cultures have recognized the reciprocal relationships among diverse life communities, including humans, animals, plants and other beings – summarized in Ubuntu's deeply philosophical core tenet 'I am because we are, and since we are, therefore I am'. Victoria Sukhomlinova (Chapter 9) offers a Confucian ethics of relational agency grounded in education and ritual through an exploration of the *Daxue* (The Great Learning). She treats the *Daxue* not merely as a philosophical text but as a living artefact that illuminates the Confucian language of flourishing by illustrating how the tradition of engaging with this text shapes the reader itself. Maggie O'Neill (Chapter 10) articulates a feminist and radically democratic ethics of flourishing that attends to biographical narrative, structural violence and epistemic justice. In her chapter, she stresses the vital importance of working with, through and in gender studies, and paying attention, through active listening, to the voices of disenfranchised and suppressed communities. Her insights illustrate (and foreshadow Hernandez's *Homo Curans*) that true flourishing can only proceed if we give vulnerable individuals and communities a central place in our thinking and practice of virtue ethics and concepts of care.

Together, these chapters argue for the legitimacy of ontologies that are often marginalized in mainstream policy and science. The point is not to relativize truth but to expand the circle of valid knowledge traditions. Each ontology is internally coherent and capable of generating meaning and guidance for its adherents. To ignore this diversity in favour of a monocultural or technocratic model of flourishing is both ethically and epistemologically impoverishing. Thus, a pluralistic approach to flourishing does not flatten difference but honours it as essential. The challenge for global frameworks is to create space for these differences while cultivating shared commitments to human dignity, ecological balance and the search for meaning.

3 Towards a reimagined SDG framework post 2030

The United Nations Sustainable Development Goals (SDGs) were a major milestone in the history of global governance, expanding the development agenda beyond economic growth to include environmental sustainability, gender equality and social justice. However, as many contributors to this volume have noted, the current framework remains largely tethered to material indicators and quantifiable metrics. This reflects not only a political bias towards technocratic governance, but also a deeper epistemological limitation rooted in materialist empiricism.

The critique articulated across this book does not reject the SDGs but calls for their evolution. Human flourishing, as reimagined here, must be grounded in a broader set of indicators that include non-material dimensions such as spiritual fulfilment, cultural vitality, emotional resilience and narrative coherence. Chapter 16, co-authored by the participants in the Programme on Human Flourishing at The New Institute, lays out a systems-based framework for post-2030 development that integrates these dimensions. It proposes a multilevel governance model that balances universal ethical commitments with cultural specificity and local agency.

Karma Ura (Chapter 11) draws on Bhutan's Gross National Happiness (GNH) index as a concrete alternative to GDP. He demonstrates with more than two decades of Bhutanese experience with GNH that embedding well-being in ethical and spiritual principles of Buddhism is not only possible but can yield stable and measurable policy results. Frederic Hanusch and Claus Leggewie (Chapter 12) contribute to this reimagining through their planetary thinking framework, which situates human flourishing within Earth systems and intergenerational responsibilities. By looking at flourishing as a metabolic and thus interdependent process between all living creatures and their environment, they illustrate that putting humans in the centre of creation is not only logically faulty but leads to inherently unbalanced results. Ariel Hernandez (Chapter 13) offers the concept of *Homo Curans*, the caring human, as an alternative to the *Homo Economicus* and as a new framework of sustainability thinking that integrates Indigenous knowledge and feminist ethics. The argument is simple and powerful – the inherent fundamental socio-economic assumption needs to shift from efficiency and gain optimization to care. Ian Hughes (Chapter 14) reinforces this call by introducing a model of Deep Institutional Innovation for Sustainability and Human Flourishing (DIIS), arguing that the polycrisis demands a systemic cultural transformation across key social institutions – from politics and economics to gender and education. Finally, Hans-Joachim Heintze (Chapter 15) extends this vision of a normative refocusing of our shared principles into international law. He calls for a post-sovereign framework of international law based on the emergent legal principle of global solidarity. This illustrates that the process of recentering human priorities has not only just begun but is already reaching one of the domains slowest to adapt, international norm setting.

These contributions converge on the need for a new global agenda that is not only sustainable but also meaningful. The future of the SDGs lies in their ability to incorporate epistemic diversity, foster care-based ethics and embrace a pluralistic yet integrated conception of the good life. A reimagined SDG framework post-2030 must be attuned to both measurable outcomes and unmeasurable values – to both the tangible and the transcendent in human flourishing.

4 The non-material foundations of human flourishing

Amidst the wide-ranging cultural and disciplinary perspectives represented in this volume, a shared insight becomes unmistakably clear: the non-material dimensions of human flourishing are indispensable. Meaning, hope, intuition, spirituality and reverence recur not as decorative or subjective supplements, but as foundational elements of a life well lived. These non-material aspects resist easy quantification, yet they shape our deepest experiences, motivations and values. Rickles (Chapter 3) explicitly critiques the dominant scientific worldview for stripping away purpose, meaning, empathy, joy and love, emphasizing that these are not reducible to measurable or repeatable phenomena. He argues for a restorative worldview in which humans are 'cosmic custodians', co-participants in shaping reality rather than passive matter. He sees meaning as foundational to ontological security and human identity:

> We must cultivate institutions and practices that reflect the irreducible plurality of human sense-making ... where meaning, value and agency are not byproducts but foundational.

Meaning is perhaps the most consistently affirmed non-material dimension. As George Ellis and Markus Gabriel argue, meaning is not reducible to neurological patterns or utilitarian outcomes but is ontologically real and existentially central. It orients our lives and gives coherence to our actions. It is central not only to metaphysical and philosophical reorientation of the sciences but also to the individual and its place in the world from a psychological perspective as Bouman argues (Chapter 4). Richard Hecht (Chapter 7) and the Epilogue by Mary McAleese underscore how meaning, often tied to sacred narratives and religious traditions, provides moral depth and resilience even in the face of suffering.

Closely linked to meaning is the presence of **hope**. It is a forward-looking sense of possibility that sustains agency amid uncertainty. Ian Hughes (Chapter 14) describes hope as integral to any viable project of institutional or societal transformation. Hecht (Chapter 7), drawing on Viktor Frankl and Ernst Bloch, treats hope as an ontological stance that animates the human spirit. Far from passive optimism, hope in these contexts is a form of active, imaginative engagement with the future. It is, as Zwitter (Chapter 5) argues, a spiritual virtue with teleological power that aligns our vision of a better future with a transcendental goal and our immanent life as a process of flourishing.

Intuition also emerges as a vital mode of knowing. Harald Atmanspacher (Chapter 6) reclaims intuitive cognition not as irrational, but as a legitimate, even essential, aspect of insight. In many traditions represented in this volume – from Confucian ethics (Chapter 9) to Indigenous knowledge systems (Chapter 8) – intuition is a means of grasping interconnectedness and moral truth that cannot be accessed through analytical reasoning alone.

Spirituality and **transcendence** form another axis of non-material flourishing. Karma Ura's account of Bhutanese Gross National Happiness (Chapter 11) is grounded in Buddhist values of compassion and non-attachment. Wakanyi Hoffman (Chapter 8) and Victoria Sukhomlinova (Chapter 9) describe flourishing as embedded in sacred cosmologies and rituals that bind communities to each other and to the Earth. Mary McAleese's Epilogue calls for reclaiming the sacred, not necessarily as dogma, but as a lived category of reverence and ethical attunement.

Finally, the **ethic of care**, as articulated by Ariel Hernandez (Chapter 13), Maggie O'Neill (Chapter 10), and others, is a non-material stance towards life and responsibility. Care cannot be measured in GDP or performance metrics, but it is foundational to human interdependence. It also repositions the moral subject not as a self-interested actor, but as a relational being responsive to vulnerability and obligation.

Together, these non-material foundations reveal that flourishing is not merely a state of well-being or access to resources, but a cultivated orientation towards life that integrates meaning, hope, intuition, care and the sacred. If global governance frameworks and scientific paradigms are to take flourishing seriously, they must embrace these inner dimensions with the same urgency as material needs.

5 Perennial insights from diverse traditions

Despite their differences in cultural background, disciplinary orientation and ontological commitments, many of the contributions in this volume echo one another in affirming several shared, perhaps perennial, dimensions of flourishing. These recurring themes of relationality, meaning, care and transformation suggest a converging grammar of well-being that transcends context without erasing it.

First, **relationality** is a foundational condition across the board. Whether in Indigenous cosmologies (Chapter 8), Confucian ethics (Chapter 9), or feminist epistemologies (Chapter 10), flourishing emerges not in isolation but through embeddedness in relationships – with others, with nature, with ancestors and with the transcendent. Human beings are never autonomous atoms but nodes in living networks.

Second, the **pursuit of meaning**, often expressed through narrative, ritual or philosophical reflection, appears as a universal human endeavour. As Ellis (Chapter 1), Gabriel (Chapter 2), and Hecht (Chapter 7) argue in different ways, meaning is not a byproduct of cognitive illusion but a constitutive feature of human life. Bouman (Chapter 4) furthermore emphasizes that meaning and purpose play a central role in psychological well-being. Meaning guides action, sustains resilience and orients hope. Without it, material sufficiency is empty.

Third, **care**, understood both as a practice and an ethic, is a recurring motif. Ariel Hernandez's *Homo curans* (Chapter 13), Maggie O'Neill's feminist praxis (Chapter 10) and Karma Ura's Bhutan's Gross National Happiness framework (Chapter 11) all elevate care from the private or emotional realm to the centre of public life and governance. Care here is not merely compassion; it is a normative stance towards interdependence and responsibility.

Finally, **transformation** on personal, collective and planetary levels is posited not as a goal but as a constitutive process of flourishing, whether through the power of virtues as discussed in Bouman's Positive Psychology, (Chapter 4), Zwitter's spiritual virtues (Chapter 5), Hecht's Arc of Hope (Chapter 7), the planetary thinking of Hanusch and Leggewie (Chapter 12), or the systemic reforms proposed in Chapter 16. Flourishing is not an end in itself but a means to a greater harmonic symbiosis between humans and their material and non-material environment with which we are inextricably linked. Flourishing is, thus, seen as dynamic, always in the making, and often emerging from struggle, contradiction or crisis.

These perennial dimensions – relationality, meaning, care and transformation – do not prescribe a single model of flourishing, but they offer a shared horizon. They suggest that amidst the pluralism of perspectives, there are resonant themes that can inform global dialogue and mutual understanding without flattening difference. In recognizing these patterns, we do not claim universality in a prescriptive sense but rather invite a deeper inquiry into the connective tissue of human aspiration across space, time and worldview.

6 Closing reflections

The chapters of this book converge towards a compelling conclusion: that human flourishing must be understood as a deeply integrative concept, grounded in epistemic humility, cultural diversity and spiritual insight. No single tradition, method or worldview can claim ownership of what it means to flourish. Rather, flourishing emerges from dialogue: between disciplines, between cultures, and between the material and the transcendent.

This volume is therefore both an intervention and an invitation. It intervenes against the narrowing forces of reductionism, technocracy, and dogmatic universalism and scientism. At the same time, it invites readers to imagine new ways of being, knowing and governing that are more attuned to the richness of human experience. In doing so, it offers not only a critique of the present but a vision for the future, one in which science and storytelling, care and justice, individuality and community, reason, intuition and reverence all play their part.

If we take seriously the insights gathered here, the task ahead is not merely to implement new policies or metrics, but to reweave the fabric of meaning that sustains our shared humanity as a foundation of flourishing. In this spirit, we hope that the reflections in this book serve as a foundation for further inquiry, collaboration and transformation towards a world in which all beings may truly flourish.

Epilogue

Religions in Transformation to Sustainability: The Role of the Sacred in Human Flourishing

Mary McAleese

It was Plato who said 'The City is what it is because our citizens are what they are' (Pradeau 2002). He could have been talking about today's planet Earth and us the citizens of the Earth. There is growing debate about the Earth's existential crises and how to overcome them. The evidence is in that if we continue to be as we are and do as we do then this City which is our planet Earth will continue to experience the increasingly catastrophic consequences of human-made failures. That statement is utterly ironic in a world where those human-made failures are avoidable and where the rapid technological, scientific and medical advances, mainly of the past century, have brought great benefits to humanity. Those advances fuelled by the massification of education have released the talents and voices of once disempowered masses. Empires have fallen, democracies have grown, globally agreed treaties and institutions have created a web of principled global connectivity and a degree of collegiality unheard of before the mid-twentieth century. None of it is yet enough to stem and turn the tide of human stupidity and iniquity.

There is no doubt that our burgeoning technical, scientific and medical power have the capacity to solve or at least mitigate the problems that beset large swathes of the Earth's citizens, whether those problems are human-made or arise from the many cruel ways nature has of reminding us of our individual and communal vulnerability. But despite all the amazing gamechangers that are reshaping our everyday lives and despite that fact that this generation of

---

Mary McAleese was President of Ireland 1997-2011 and is a Professor of Children, Law & Religion at the University of Glasgow, and Chancellor of Trinity College Dublin.

Earth's inhabitants have the resources and opportunity to collaboratively lift all of humanity to new heights of widespread well-being our shoulders are not yet fully to the wheels that need to grind to give every child the education, healthcare, nutrition, food security, housing, welfare, empowerment, protection and peace he or she needs to flourish.

Instead, the old gravitational pull of superficiality, elitism, war-mongering, rivalry, exploitation, greed, negligence, short-termism and downright selfishness conspire to ensure that not only are opportunities wasted on a global scale but the very technologies and sciences that could be life enhancing are instead menacingly capable of extinguishing life on a global scale, extinguishing even the Earth itself.

Why are we in this mess? Could it be as some like Harari (2015; 2017; 2019) and Legrand (2021) suggest that the world of science and technology has far outrun the development of robust moral, spiritual and philosophical engines needed to provide us with the depth of human wisdom and the sound moral compass demanded by these times of churn. If they are right then looking for answers in better and bigger technologies and in the political, economic and social policies that focus on them may simply be kicking a lot of cans down a shortening, land-mined road. Sustainable answers that embrace the Earth and all its inhabitants lie in profoundly changed hearts and minds, in remaking the City to a new vision for humanity so that it works for all its citizens, a new vision that leads to a new praxis and not just a welter of worthy words. But while we know we are capable of change, even rapid change as Covid showed, are we capable of the scale of change these times and that vision demand?

Despite years of high-level international diplomacy discussions, agreements and treaties how trustworthy are the fingers that can press the buttons on devastating nuclear weapons and how effective have been the actions taken to turn back the tide of menacing climate change? Not very! The Earth's existential crises exist, are real, are immediate and they are our problem, but in the eyes of Earth's citizenry, are they big enough, credible enough, near enough, compelling enough to galvanize a thoroughgoing common human effort focused on promoting the best interests of all humanity? Can we see any realistic way in which a sense of shared responsibility and of active good neighbourliness can take hold globally in place of the aggressive, hostile, mutually suspicious, mutually ignorant, competitive finger-pointing and othering, the obsessions with things that ignore our mortality and keep us distracted from the urgent problems mounting around us? Are we naive to imagine humanity is capable of hitting a reset button of such an order of magnitude that we can collaborate to finally end deadly conflict and climate damage so that the future will be made in the likeness of hope from lessons learnt and not from mistakes repeated over and over again?

Today's reality is that the miserable past keeps repeating itself over and over again. Many who inhabit this Earth, despite huge advances in communication and even because of those huge advances, live in embedded and persistent ignorance of one another, fear of one another, contempt for one another, at war with one another, at odds with one another or just in simple but treacherous indifference towards one another. Some have too much, most have too little. A few care a lot, most are not bothered.

The Earth's resources as in the past are still exploited to the detriment of the many, the aggrandizement of the few, the appeasement of some with crumbs from the rich man's table, the distracting entertainment of the masses with the trivial and frivolous. Political power structures even in the most stable liberal democracies are susceptible to economic and political aneurysms that can weaken their moral core. The rise of the right is capable of compromising the resolve of the centre, shifting its language, hardening its heart.

Political and civic leadership at national and international levels can and do often work within unified structures but it often seems that in reality their mutual engagement results in maximum publicity but minimal progress. They have littered our threatened world with photo ops, worthy words, broken promises, unmet targets, careful compromises and dates for the next meeting where hands will be wrung and ominous warnings issued but in all likelihood nothing much will have improved enough to reassure us that we are making the future work well for a blossoming humanity in a way that the past has not.

Should we just watch hard-won liberal democracy disintegrate, cower before yet another generation of self-righteous dysfunctional leaders and their bullying followers puffed up with notions that money makes them masters of the universe? Is it acceptable to watch such civilization as we have managed to muster get washed away in the tsunamis of war, poverty, racism, fire, water and fake narratives that conduce to hatred and othering or do we do as we are doing here – insist on fighting for a culture of human decency and mutual care strong enough to resist the worst in humanity and convincing enough to promote the best that humanity is capable of? Can we change the trajectory of long and deeply embedded thinking and actions that have made bitter enemies and strangers even and especially of near neighbours?

Can the hugely diverse peoples of the Earth be rallied purposefully in a collaborative endeavour that could transform life on Earth for the benefit of all, becoming single-minded about what really matters, humanity and the Earth, elevating them to our highest priorities, as sacred core values hallowed by all.

I use the word *sacred* here not in a particularly or exclusively religious sense though I accept it has been used and manipulated by religions and may invoke understandable scepticism. But it has not been consumed to the point of exhaustion by religion and is a concept capable of being understood in the secular as well as the religious spheres (Grzymała-Busse 2023) as embracing the notions of reverence and respect, of inviolability and precedence demanded by the inalienable dignity of every human being and by the Earth itself. It is a word that helps us prioritize putting the Earth and its resources at the service of humanity. It is not the opposite of the secular. It is not even at odds with the secular for like the strands in a skein of wool the secular and religious have been woven together for centuries. Rather, a sense of the sacred is potentially a powerful and accessible, knowable resource already present within our secular as well as our religious spheres.

The sacred simply points up the things that demand to be revered, made sacrosanct, because they secure the best interests of humanity and the planet. Placed at the centre of an advanced human wisdom a developed politics of the sacred would insist on the protection of the Earth from violation, desecration and damage by us and the protection of the human person from tyranny, cruelty, subjugation imposed by us, so that each can flourish, have a roof, water, food, education, sufficient unto their day and with a careful stewardship of the legacy handed to the next generation.

The sacred understands that nature gives and takes away life; that nature can be cruel, that damage and disease lie unannounced, uninvited around every corner. But a new culture and politics infused with a deep sense of what it is that is or should be sacred to us would insist that these very realities not of our human making provide us with enough challenges to overcome without the additional overwhelming challenge of preventable human-made messes. What is sacred to us should be capable of healing the world, not hurting it. What is sacred to us should not be vanities that desecrate but ideals that consecrate, though we have to acknowledge history, past and present, is littered with sacred vanities that claim divine authorship while wreaking havoc. Slavery, imperialism, manipulation, misogyny, homophobia, anti-semitism, religious wars, to name just a few.

When Thomas LeGrand (2021) calls for a new politics of being, I recognize what we here might more colourfully call for a not-unrelated politics of the sacred. Both are about the business of redefining what is in the best interests of all humanity and the best way to serve those interests in the here and now.

We may be a little nervous around the word sacred for we have to acknowledge that people of faith, who are the vast majority of inhabitants of the world, are already familiar with this concept. The word sacred already has a place within their world view. It may place limits on that view; it may be more sword than ploughshare. It may even be an obstacle in its current understanding to the development of a new politics of the sacred but also it may not. It could be just the word we need. A word and idea that can be retrieved and retooled to mend a broken world.

Nervousness around discussion of the sacred is a natural consequence of the dismissal of religion because of its at times baleful contribution to human development. Who can doubt its formidable braking effect on the advancement of the human rights of all, especially women and LGBTIQ+ and its toxic dalliance with politics throughout global history including today.

But if we dismiss religion, are we missing out on a transcendent dimension embedded widely in every continent and culture and particularly already present in hearts and minds where there is a familiarity with the notion that some things are of their nature sacred, bigger than anything else, more important even than the biggest of our tribal preoccupations. Could that transcendent dimension if properly harnessed, fed with updated heroic ideals, build dialogical bridges across the many cavernous differences to a miraculous realization of our common vested interest in elevating to radical new levels of consciousness and action, the sacredness of the Earth, of the individual and of creation.

The religions of the world have until recently been largely missing in action on this intriguing front except to the extent that they flatter themselves by preaching to the secular world from the lofty heights of their self-reverential pulpits. They are great at telling governments or indeed each other what to do but not so good, in fact resistant to internal self-critique, to critically pondering their own great givens. But what if their adherents and leaders were invited to consider the possibilities of a new politics of the sacred capable of saving our world and indeed its religions from perdition, for of course if the Earth disappears its faith systems disappear too. Sad to say that many religions seem more concerned with their own dwindling numbers than the bigger existential picture.

Could the world's religions and their adherents have a significant role to play in building 'with other religious and secular actors on the basis of a common ethical foundation of care, love and respect'[1] a wholly new, wholesome, culture of embrace not disgrace.

Let us look at what potential exists, what possible roadmaps can be devised for making the world's religions and religious actors both active and lapsed, generators of and collaborators with each other and with non-religious actors, in the seminal watershed that can move us out of bunkers and trenches that ironically reinforce instability and global precariousness and help make us instead wherever we are in the world, open shelters of care for all the Earth, shepherds of all its peoples.

The story is mixed and complex. It is to a large extent also overlooked, even dismissed but we dismiss the pervasive, ingrained, but also shifting role of religion, at our peril. It is a reality worth understanding. It is a resource worth reevaluating for the evidence is in, particularly here in the Western world that religion itself is in a process of transformation, some might even say of disintegrating as numbers across all the major Christian denominations in the West plummet, finances become parlous, schism, dissent, scepticism and disinterest conduce to internal existential crises.

That situation creates a real opportunity for radical rethinking. By far the most important characteristic of religion is the elementary fact that according to the PEW/Templeton report, 84 per cent of the world's inhabitants are aligned with religion in one form or another and a majority are aligned (Nadeem 2022) with the three largest faith systems, Christianity, Islam and Hinduism. If the entire world is heading towards falling off a cliff, then religions and religious adherents will be the biggest losers by far.

The process evident currently in 'Christian' Europe, once so strong and hierarchically dominated, so conformist and controlling of members, is of seismic change and a struggle for both credibility and relevance in the face of growing challenge, distrust and cynicism, especially among the educated young. It is best summarized by a stanza from Seamus Heaney's poem 'From the Canon of Expectation' (Heaney 1987): 'What looks the strongest has outlived its term/ the future lies with what's affirmed from under'.

Voices and views once silenced within hierarchical systems are now being heard. Many people today walk away from formal religions because they can, because they are human rights literate and know that they have the inalienable right to freedom of conscience, opinion, belief, religion including the right to change religion (cf. Universal Declaration of Human Rights). They refuse to be life-long religious conscripts by birth or baptism or cultural initiation as demanded by significant numbers of influential world religions. The prevailing normative notion of faith membership by conscription rather than personal conviction is the source of much of today's crisis within faith systems.

Take the one I know best, the Latin Catholic Church, with a membership of one in six of the world's population, the biggest NGO in the world, a key influencer on five continents, an unequalled global provider of education and healthcare, especially to the poor, the only faith system to have permanent representative status at the United Nations. It has its own extensive legal system, called canon law, which insists on the right of its governing hierarchy (Magisterium) to limit the inalienable intellectual freedoms of Church members[2] set out in the Universal Declaration of Human Rights. It also insists that membership begins at baptism with promises made at baptism, that the obligations of membership are life-long and cannot be resiled from. Yet 84 per cent of all Catholics are baptized as babies as canon law demands, in other words when non-sentient and incapable of evaluating the nature of the obligations imposed by baptism and manifestly incapable of making any kind of promises. The myths upon which Church membership and magisterial control are founded are just that, fictions and incapable of sustaining serious scrutiny. And that scrutiny is growing exponentially.

Until recently the 1.3 billion members of the Catholic Church had no internal forum in which to state their views. But the murmurings of an educated laity, a collapse in vocations to the priesthood, a collapse in income, the sound of feet retreating from the Church, created a pressure-cooker context in which Pope Francis resorted to opening a global synodal process. Hard as he tried to keep controversy off the agenda, the genie was already out of the bottle and the debate which opened up in almost every diocese in the world showed clear support for gender equality, for an end to demonization of LGBTIQ+ people, for accountable governance, protection of children and penalization of bishops who betray the trust of abuse victims. The final report of the Synod (2024), a drab, timid document that caused widespread disappointment, made no mention of LGBTIQ+ and took the issue of female ordination off the agenda (Pope Francis 2024).

While progress seems grindingly slow and papal as well as magisterial pushback is strong, the reality is that it is no longer possible to ignore the huge subterranean undertow driven by people of faith in the direction of serious reform. '*The future lies with what is affirmed from under*' (Heaney 1987: n. 8).

Recent referenda in Ireland on same sex marriage (2015) and abortion (2018) showed a voting population nominally at least majority Christian and Catholic, at odds with and overwhelmingly defying Church teaching on both subjects. These and other contested issues repeated particularly in other Christian countries are indicative of the human capacity to change, to be persuaded by compelling arguments, to think again, to act differently, to be transformed by information from within and without. Without faith in that possibility I would not be here, I could easily be persuaded to let this dismal Earth and all who share it disappear in self-inflicted ignominy, to let nature take its course, as it did with the dinosaurs, to concede that after all we humans may not be worth the candle, this Earth and all in it may not be worth preserving.

I have reason to know the possibility and the price of transformation. I was born in Belfast shortly after the Second World War had made a bloodbath of Europe. I lived as a Catholic in a sectarian quagmire in a corner of the United Kingdom, Northern Ireland, which called itself a Christian democracy, but which was neither. It was created as a Protestant government for a Protestant people capable of manipulation of laws, policing, jobs, housing and voting rights in order to exclude the Catholic minority from civic and political equality – with baleful consequences.

Ironically I grew up just as an embryonic idealistic brave new post-war world was emerging with the Universal Declaration of Human Rights, the Council of Europe, the United Nations, the European Union, a multiplicity of global rights-based treaties with global accountability mechanisms, International Courts, a burgeoning industry of human rights NGOs, free media, social media, massified education and the onward march it seemed of liberal democracy. None of it unfortunately was enough to stop the streets I grew up in from becoming a sectarian war zone. In Northern Ireland, where you can encounter a Christian Church on almost every street corner, the Christmas image of peace on Earth and good will to all was daily dishonoured in the breach with an intensity that caused thousands of deaths and injuries. The Churches in Northern Ireland were singularly ineffective for decades in countering the intercommunal tensions. A tepid minimalism was the order of the day. But the patient endeavours of a small cohort of courageous peacemakers drawn from all sides, over decades, eventually led to the Good Friday Agreement of 1998[3] which provided for peace through shared governance and parity of esteem.

What that Peace Process taught me was the remarkable transformative power of courageous people, working individually and collectively who believed change was possible, who analysed all sides of the problem with a generosity of spirit, stood in the shoes of the other, refused to be blinded by othering the other and who constructed a vision for a shared future and developed it insistently against a wall of hostility and scepticism from their friends and enemies alike. Their work revealed the existence of a sacred space, a space for the sacred within all perspectives where contemplation of a different kind of future could with persuasion, take root. The old paradigm of winners and losers gave way to the politics of compromise and win-win.

However, the political and legal and cultural institutions designed to underpin peaceful coexistence are in and of themselves not enough to secure a peaceful, just and equitable future without the heft of mobilizing the occupants of that sacred space and sense of the sacred. They had to become and remain the persuaders to the future captured in the Good Friday Agreement. Luckily it transpired they are legion.

Today we can say that whatever their differences they are all part of the twenty-first century savvy generations that are living with common existential threats right on their doorsteps no matter what continent or country. They have access to vastly effective global communications systems, burgeoning international and intranational formal and informal groupings dedicated to the advancement of human rights and care for the Earth. There is huge current and future potential though there is still only meagre evidence of the purposive use of their collective power to be the agents of change we need if this Earth is to be nurturing mother and father and family to all its inhabitants.

Once the possibility of change arises and once the imperative of change becomes desperately obvious as it now is then the extent of change, the content of change, the purpose, the how and why become our call to dialogue, discernment, creative thinking and action. Engaging with people and institutions we have little time for is precisely how problems get resolved. Bunkers, no matter how comfortable, need to fling open their doors to let the future in.

The sheer extent of the undertaking is as exciting as it is ambitious, as undermined by pessimism as it is by the realization that the power of persuasive ideas is a double-edged sword. History, even recent and indeed contemporary discourse, shows how toxic ideas, myths, perceptions and ideologies can run amok, can overwhelm and wreak bullying chaos. The danger is that the vast and rapid change of culture and consciousness needed to hold back the surging tide of human stupidity that is continuously fuelling the existential crises will just seem too unlikely, too difficult. The rise of the political right makes it essential that the tolerant, egalitarian centre must hold. How?

Once we focus on the mobilization of the sense of the sacred as an ally in the task, we can see how it meets our requirements, whether we are people of faith or not. The process of radical transformation has to be global in its reach. It has to appeal across physical and cultural borders. It has to reach into hearts and minds already formed along particular and highly differentiated even oppositional pathways and perspectives. It has to reach across the current generations and time is of the essence for no matter how long our lives, in the great scheme that is humanity, life is always very short and the time available to make an impact is short in the short term and shorter still in the medium- to long-term.

Evil has swamped our world, century after century. Bloodbaths of epic and industrial proportions have written the history of the world, right up to and including now. We are a sobered European and Western citizenry, educated and confident in ways our forebearers were not, seeking to develop the radical answers, ideas and structures that take us far beyond those developed in the post-war era and which despite their integrity, with the best will in the world, have not yet cooled the magma that courses beneath the surface of humanity and which is always in danger of erupting in the zero sum games of tyranny or worse still apathy. And all the while the existential crisis grows exponentially, and we are forced to be spectators as we waste time we do not have. So, we are forced to turn our minds to fresh possibilities we have overlooked or dismissed, including the mobilization of a sense of the sacredness of the Earth and human life which is a feature of all major religions. Could its synthesis become a new problem-solving, epoch changing power, a global politics of the sacred? The obstacles are formidable but far from insurmountable despite the reality that the world's great faith systems have historically worked independently with their backs to one another rather than collaboratively side by side. Not only that but human dominion over the Earth including the asserted right to plunder its resources, to conquer a neighbour's territory or even his or her mind, has sometimes relied on divine authority gleaned from the great sacred texts and from religious backing. New theologies of the sacred, new scholarly interpretations and challenges are needed to push the idea of the sacred to new and incorruptible limits.

The historic relationship between the world's religions is largely one of historic enmity, mutual suspicion, fear of mutual conversion, hubristic claims of exclusive divinely ordained superiority, religious wars, petty vanities. Collaborations and partnerships have been relatively rare. Ecumenical relationships between the major Christian denominations even after decades of late twentieth and early twenty-first century endeavour have barely managed to progress beyond photo opportunities and aspirational language. Somehow the focus on doctrinal and dogmatic differences, the unhealed wounds inflicted by history that allows each to see themselves as victims and martyrs at the hands of the others, have conspired to keep the individual religious bunkers hermetically sealed. But even as the world's children by the tens of million are being educated in their respective faith systems and left largely ignorant of if not indifferent or hostile to the faith systems of others, even as internal disputes tear at the cohesion within denominations and faith institutions and entities, there is a sacred space where all can comfortably stand, where voices argue for mutual respect, mutual understanding, dialogue, tolerance of diversity and co-responsibility for each other's welfare and that of our planet.

Green shoots of hope from advanced thinkers within and without the major faith systems and political systems are bubbling although relatively long-standing organizations like the World Parliament of Religions and Religions for Peace in reality have made little impact to date apart from keeping the space for tepid dialogue warm.

But green shoots began to break through the surface most visibly in 2015. It was in the lead up to the Paris Climate Accords negotiated by 196 member states of the United Nations and today ratified by almost all. It was the year of significant interventions by leaders of faith systems on the subject of climate change.

There was the Rabbinic Letter on Climate Change (Earthday.org 2015) signed by hundreds of rabbis insisting that all share responsibility for the Earth and the well-being of humanity; the Buddhist Declaration on Climate Change (Various Signatories 2015); the second Hindu Declaration on Climate Change (Hindu American Foundation 2012), which noted that despite great diversity within Hinduism there was a unity of faith through which Hindus 'will be able to make the sort of inner and outer transitions that addressing climate change requires'. There was the Islamic Declaration on Global Climate Change issued by Islamic scholars (IFEES 2015), reconciling climate science with the Quran and calling on Muslims 'wherever they may be [...] to tackle habits, mindsets and the root causes of climate change, environmental degradation and the loss of biodiversity in their particular spheres of influence, following the example of the Prophet Muhammad (peace and blessings be upon him) and bring about a resolution to the challenges that now face us'.

The Church of England Synod (Church of England 2015) in that year urged world leaders to work for the global consensus needed to provoke action on carbon reduction and it pledged to look internally at what it could do in practical ways to promote actions that were consonant with sustainability.

Of all the faith interventions in 2015 probably the most comprehensive and impactful was that of Pope Francis' encyclical Laudato Si' (Pope Francis 2015) which may yet prove to be his most enduring legacy. Repeating and building on concerns which had been expressed by his predecessors John XXIII, Paul VI, John Paul II, Benedict XVI and Ecumenical Patriarch Bartholomew and acknowledging the concerns expressed by other religions he addressed not just Catholics but 'every person living on this planet' saying:

> The urgent challenge to protect our common home includes a concern to bring the whole human family together to seek a sustainable and integral development, for we know that things can change. I urgently appeal, then, for a new dialogue about how we are shaping the future of our planet. We need a conversation which includes everyone, since the environmental challenge we are undergoing, and its human roots, concern and affect us all. The worldwide ecological movement has already made considerable progress and led to the establishment of numerous organizations committed to raising awareness of these challenges. Regrettably, many efforts to seek concrete solutions to the environmental crisis have proved ineffective, not only because of powerful opposition but also because of a more general lack of interest. Obstructionist attitudes, even on the part of believers, can range from denial of the problem to indifference, nonchalant resignation or blind confidence in technical solutions. We require a new and universal solidarity (...) A healthy politics is sorely needed, capable of reforming and coordinating institutions, promoting best practices and overcoming undue pressure and bureaucratic inertia. It should be added though, that even the best mechanisms can break down when there are no worthy goals and values, or a genuine and profound humanism to serve as the basis of a noble and generous society.
>
> (Pope Francis 2015, paras. 13; 14; 181)

Within the text of Laudato Si' was a strong theological advocacy for individual and collective responsibility. It was followed in 2019 by the Abu Dhabi Declaration (Fraternal Declaration) between Pope Francis and Sheikh Ahmed el-Tayeb, Grand Imam of Al-Azhar (Vatican 2019) which invites 'all persons who have faith in God and faith in human fraternity to unite and work together (...) to advance a culture of mutual respect in the awareness of the great divine grace that makes all human beings brothers and sisters'. This is by far the most important document in terms of inter-faith dialogue for it embraces the world's two biggest religions accounting for virtually half the world's population and exhorts a future of ongoing dialogue designed to break down mutual fear and ignorance and to 'declare the adoption of a culture of dialogue as the path; mutual cooperation as the code of conduct; reciprocal understanding as the method and standard'. They called on world leaders 'to work strenuously to spread the culture of tolerance and of living together in peace; to intervene at the earliest opportunity to stop the shedding of innocent blood and bring an end to wars, conflicts, environmental decay and the moral and cultural decline that the world is presently experiencing'. They asked leaders and would-be influencers 'to rediscover the values of peace, justice, goodness, beauty, human fraternity and coexistence in order to confirm the importance of these values as anchors of salvation for all, and to promote them everywhere'.

In recent years more conscious debate on the issues covered in the Abu Dhabi Declaration including dialogue concerning peacebuilding and climate sustainability is evident within faith systems and inter-faith groups that have built networks of common endeavour particularly in the lead up to COP intergovernmental meetings since Paris. So, an emerging nucleus of effort and focus is evident. An emerging though still inchoate theology of the role of the sacred in care for the Earth is also evident. But given that 84 per cent of the world's citizens have a religious affiliation somehow there is a real deficit in terms of political impact and identifiable progress. One wonders why.

The answer lies in how major faith systems self-identify as the teachers and not as those who have something to learn. That is their modus operandi, to preach, to teach, to convert, to rely on old worn texts, to resist updating, to assume they know best and have everything to offer others and little to learn from them. The Latin Catholic Church to which I belong is good at telling the rest of the world how to behave, what to believe. It has very poor self-critiquing skills, is quick to reprimand and penalize internal criticism no matter how deserved or well-intentioned. It is still operating a governance and decision-making structure that is unelected, unaccountable, hierarchical, exclusively male and self-perpetuating. It is increasingly irrelevant to its own constituency. The gender decision-making gap is of gargantuan proportions. When it preaches human rights and dialogue it can easily be accused of hypocrisy and populist minimalism. Yet still it has created important conduits for discussion, mutual discernment, challenge and change, the impetus for which more often comes from below and is not driven top-down but puts pressure for top-down responses. A global web of politics of the sacred capable of embracing the world's religious leaders and faithful could be their redemption and their challenge. In alliance with the secular, political world it could be something almost miraculous. A new politics of the sacred could be the Hail Mary pass the world has been waiting for. What benign power that alignment of secular, spiritual, religious and non-religious interests and actions could unleash into our world.

Now we just need the coherent vision and wisdom to inspire the people of the City with new ideas, respectful but relentless dialogue, persuasive arguments, a message of hope, a call to action and faith in the capacity of the heart and mind to be changed and courage to face down the nay-sayers, the bigots, the climate deniers. The people of the City are reachable as never before. Many are erudite, cultivated and equipped as never before to reappraise the old systems, the givens and structures that have dragged them down cul de sacs of sheer waste and worst practice. The politics of the sacred for those of real faith and those of none could prove to be nothing less than the fulness of Nature/God's roadmap to peace on a sustainable Earth and goodwill to all, at last.

Notes

1. Extract from the introduction to the Programme for The New Institute Hamburg, Workshop, Religions in transformation to Sustainability-the role of the sacred in human flourishing, 12-13 June 2024, The New Institute, Hamburg.
2. Cf. John Paul II, *Codex Iuris Canonici*, 25 January 1983, in *AAS* 75/2 (1983) 1–324. The Code was promulgated by John Paul II, ap. const. *Sacrae disciplinae leges*, 25 January 1983, in *AAS* 75/2 (1983) vii-xiv. It came into effect on 27 November 1983. It abrogated the 1917 Pio-Benedictine *Codex Iuris Canonici* (CIC/1917). The official language of the Code is Latin. Cf. canons 209, 212 §1 and 3, 223 §1 and 227. Cf. Congregation for the Doctrine of the Faith Instruction on infant baptism, 20 October 1980, *Pastoralis actio*, 22, which describes such rights as 'an illusion'.
3. The Good Friday Agreement is an International Peace Agreement signed on 10 April 1998, also known as the Belfast Agreement.

Bibliography

Abdallah, S., A. Hoffman and L. Akenji. (2024). *The 2024 Happy Planet Index*. Hot or Cool Institute.

Aberle, D. F., A. K. Cohen, A. K. Davis, M. J. Levy and F. X. Sutton. (1950). 'The Functional Prerequisites of a Society', *Ethics* 60 (2): 100–11. https://doi.org/10.1086/290705.

Abram, David. (1996). *The Spell of the Sensuous: Perception and Language in a More-Than-Human World*. Pantheon Books.

Achebe, Chinua. (1958). *Things Fall Apart*. William Heinemann.

Acker, Emma, Sue Canterbury, Adrian Daub and Lauren Palmor. (2018). *Cult of the Machine: Precisionism and American Art*. 1st edition. Yale University Press.

Acquaviva, Graziella. (2019). 'Cultural Values of Trees in the East African Context', *Kervan. International Journal of African and Asian Studies* 23 (1). https://doi.org/10.13135/1825-263X/3290.

Adloff, Frank and Margaret Clarke. (2014). 'Convivialist Manifesto: A Declaration of Interdependence'. *Global Dialogues*, no. 3. https://doi.org/10.14282/2198-0403-GD-3.

Adonis, Digna Lipa-od. (2011). 'The Community Development Concepts of the Igorot Indigenous Peoples in Benguet, Philippines', Thesis, Australian Catholic University. https://doi.org/10.4226/66/5a961d63c6866.

Adorno, Theodor W. (1984). *Aesthetic Theory*. Routledge & Kegan Paul.

Agamben, Giorgio. (1998). *Homo Sacer: Sovereign Power and Bare Life*. Translated by Daniel Heller-Roazen. 1st edition. Stanford University Press.

Akiwowo, Akinsola A. (1983). *Ajobi and Ajogbe: Variations of the Theme of Sociation*. Inaugural Lecture Series. Ife-Ife: University of Ife Press. https://123pdf.org/document/yevr89xr-ajobi-ajogbe-variations-theme-sociation.html (accessed 11 October 2025).

Alaimo, Stacy. (2016). *Exposed: Environmental Politics and Pleasures in Posthuman Times*. University of Minnesota Press.

Alcoff, Linda Martín. (2022). 'Extractivist Epistemologies', *Tapuya: Latin American Science, Technology and Society* 5 (1): 2127231. https://doi.org/10.1080/25729861.(2022).2127231.

Alkire, Sabina. (2002). 'Conceptual Framework for Human Security', *Centre for Research on Inequality, Human Security and Ethnicity, CRISE,* Queen Elizabeth House, University of Oxford, CRISE Working Paper, Working Paper 2. https://assets.publishing.service.gov.uk/media/57a08cf740f0b652dd001694/wp2.pdf (accessed 11 October 2025).

Allen, Catherine J. (2002). *The Hold Life Has: Coca and Cultural Identity in an Andean Community*. 2nd edition. Smithsonian Books.

Altizer, Thomas J. J. and William Hamilton. (1966). *Radical Theology and the Death of God*. First Edition. Bobbs-Merrill Co.

Anālayo, Bhikkhu. (2018). *Satipaṭṭhāna Meditation: A Practice Guide*. Windhorse Publications.

Andreotti, V. (2011). Engaging with Other Knowledge Systems: The Through Other Eyes Initiative, in *Actionable Postcolonial Theory in Education*. Postcolonial Studies in Education. Palgrave Macmillan. https://doi.org/10.1057/9780230337794_14.

Angle, Stephen C. (2006). 'A Fresh Look at Knowledge and Action: Wang Yangming in Comparative Perspective', *Journal of Chinese Philosophy* 33 (2): 287–98. https://doi.org/10.1111/j.1540-6253.2006.00354.x.

Anglés-Alcázar, Daniel, Claude-André Faucher-Giguère, Dušan Kereš, Philip F. Hopkins, Eliot Quataert and Norman Murray. (2017). 'The Cosmic Baryon Cycle and Galaxy Mass Assembly in the FIRE Simulations', *Monthly Notices of the Royal Astronomical Society* 470 (4): 4698–719. https://doi.org/10.1093/mnras/stx1517.

Antonovsky A. (1987). *Unraveling the Mystery of Health*. Jossey-Bass.

Aquinas, Saint Thomas. (2005). *Disputed Questions on the Virtues*. Edited by E. Margaret Atkins and Thomas Williams. Cambridge University Press.

Archer, Margaret S. (2015). *The Relational Subject*. Cambridge University Press.

Archibald, Jo-Ann. (2008). *Indigenous Storywork: Educating the Heart, Mind, Body, and Spirit*. University of British Columbia Press.

Arendt, Hannah. 2002. 'Reflections of Violence', in Catherine Besteman (ed.), *Violence: A Reader*, 19–34. New York University Press.

Aristotle. (2000). *Nicomachean Ethics*. Translated by Roger Crisp. Cambridge University Press.

Aristotle. (2009). *The Nicomachean Ethics*. Edited by Lesley Brown. Translated by W. D. Ross. Oxford University Press.

Armstrong, Paul B. (2013). *How Literature Plays with the Brain: The Neuroscience of Reading and Art*. Johns Hopkins University Press.

Armstrong, Paul B. (2019). 'Neuroscience, Narrative, and Narratology', *Poetics Today* 40 (3): 395–428. https://doi.org/10.1215/03335372-7558052.

Arthur, W. Brian. (2009). *The Nature of Technology: What It Is and How It Evolves*. Simon and Schuster.

Atkinson, Anthony B. (2015). *Inequality: What Can Be Done?* 1st edition. Harvard University Press.

Atmanspacher, Harald. (2024). 'Psychophysical Neutrality and Its Descendants: A Brief Primer for Dual-Aspect Monism', *Synthese* 203 (1): 25. https://doi.org/10.1007/s11229-023-04449-z.

Atmanspacher, Harald and Dean Rickles. (2022). *Dual-Aspect Monism and the Deep Structure of Meaning*. Routledge.

'Atmospheric Physics.' (2022). ECMWF. 11 May 2022. https://www.ecmwf.int/en/research/modelling-and-prediction/atmospheric-physics (accessed 11 October 2025).

Aydin, Ciano. (2017). 'The Posthuman as Hollow Idol: A Nietzschean Critique of Human Enhancement', *The Journal of Medicine and Philosophy: A Forum for Bioethics and Philosophy of Medicine* 42 (3): 304–27. https://doi.org/10.1093/jmp/jhx002.

Backhaus, Julia, Audley Genus, Sylvia Lorek, Edina Vadovics and Julia M. Wittmayer, (eds). (2017). *Social Innovation and Sustainable Consumption: Research and Action for Societal Transformation*. Routledge-SCORAI Studies in Sustainable Consumption. Routledge.

Badiou, Alain. (2022). *A New Dawn for Politics*. Translated by Robin Mackay. English edition. Polity Press.

Bajohr, Hannes. (2019). 'Anthropocene and Negative Anthropology', Public Seminar, 29 July 2019. https://publicseminar.org/essays/anthropocene-and-negative-anthropology (accessed 11 October 2025).

Baker, Maria, Kataraina Pipi and Terri Cassidy. (2015). 'Kaupapa Māori Action Research in a Whānau Ora Collective: An Exemplar of Māori Evaluative Practice and the Findings', *Evaluation Matters—He Take Tō Te Aromatawai*, July, 113–36. https://doi.org/10.18296/em.0006.

Ball, Philip. (2022). *The Book of Minds: How to Understand Ourselves and Other Beings, from Animals to AI to Aliens*. University of Chicago Press. https://doi.org/10.7208/chicago/9780226822044.

Ball, Philip. (2023). *How Life Works: A User's Guide to the New Biology*. Picador.

Bannink, Fredrike and Nicole Geschwind. (2021). *Positive CBT: Individual and Group Treatment Protocols for Positive Cognitive Behavioral Therapy*. 1st edition. Hogrefe Publishing. https://doi.org/10.1027/00578-000.

Barad, Karen. (2007). *Meeting the Universe Halfway: Quantum Physics and the Entanglement of Matter and Meaning*. Duke University Press.

Bartlett, Albert A. (1994). 'Reflections on Sustainability, Population Growth, and the Environment', *Population and Environment* 16 (1): 5–35. https://doi.org/10.1007/BF02208001.

Bauer, Keith A. 2010. 'Transhumanism and Its Critics: Five Arguments against a Posthuman Future', *International Journal of Technoethics* 1 (3): 1–10. https://doi.org/10.4018/jte.2010070101.

Beck, Ulrich. (2016). *The Metamorphosis of the World*. Polity Press.

Bell, Brian. (2020). 'Microorganisms in Parched Regions Extract Needed Water from Colonized Rocks', *UC Irvine News*, 4 May 2020. https://news.uci.edu/2020/05/04/microorganisms-in-parched-regions-extract-needed-water-from-colonized-rocks (accessed 11 October 2025).

Bellacasa, Maria Puig de la. (2017). *Matters of Care: Speculative Ethics in More than Human Worlds*. University of Minnesota Press.

Bello-Bravo, Julia. (2019). 'When Is Indigeneity: Closing a Legal and Sociocultural Gap in a Contested Domestic/International Term', *AlterNative* 15 (2): 111–20. https://doi.org/10.1177/1177180119828380.

Benatar, David. (2008). *Better Never to Have Been: The Harm of Coming into Existence*. Oxford University Press.

Bennett, Jane. (1987). *Unthinking Faith and Enlightenment: Nature and Politics in a Post-Hegelian Era*. 1st edition. New York University Press.

Bennett, Jane. (2001). *The Enchantment of Modern Life: Attachments, Crossings, and Ethics*. Princeton University Press.

Bennett, Jane. (2002). *Thoreau's Nature: Ethics, Politics, and the Wild*. Rowman & Littlefield.

Bennett, Jane. (2010). *Vibrant Matter: A Political Ecology of Things*. Duke University Press.

Benton-Banai, Edward. (1988). *The Mishomis Book: The Voice of the Ojibway*. Red School House.

Beraldo, Sergio and Luigino Bruni. (2020). 'Special Issue: Happiness, Capabilities, and Opportunities', *International Review of Economics* 67 (1): 1–3. https://doi.org/10.1007/s12232-020-00346-w.

Berkes, Fikret. (2012). *Sacred Ecology*. 3rd edition. Routledge.

Beukelaer, Christiaan De, M. Pyykkönen and J. Singh. (2014). *Globalization, Culture, and Development: The UNESCO Convention on Cultural Diversity*. 2015 edition. Palgrave Macmillan.

Bieri, Peter. (2017). *Wie wäre es, gebildet zu sein?: Extra: Die Vielfalt des Verstehens*. Komplett Media GmbH.

Biermann, Frank, Thomas Hickmann, Carole-Anne Sénit, Marianne Beisheim, Steven Bernstein, Pamela Chasek, Leonie Grob, et al. (2022). 'Scientific Evidence on the Political Impact of the Sustainable Development Goals', *Nature Sustainability* 5 (9): 795–800. https://doi.org/10.1038/s41893-022-00909-5.

Billi, Daniela, Clelia Staibano, Cyprien Verseux, Claudia Fagliarone, Claudia Mosca, Mickael Baqué, Elke Rabbow and Petra Rettberg. (2019). 'Dried Biofilms of Desert Strains of *Chroococcidiopsis* Survived Prolonged Exposure to Space and Mars-Like Conditions in Low Earth Orbit', *Astrobiology* 19 (8): 1008–17. https://doi.org/10.1089/ast.(2018).1900.

Blitstein, Pablo. (2021). 'Confucianism in Late Nineteenth-Early Twentieth Century China', in David Elstein (ed.), *Dao Companion to Contemporary Confucian Philosophy*, 15: 27–46. Dao Companions to Chinese Philosophy. Springer International Publishing. https://doi.org/10.1007/978-3-030-56475-9_2.

Bloch, Ernst. (1959). *Das Prinzip Hoffnung*. Vol. 1. Suhrkamp. http://archive.org/details/dasprinziphoffnu0000bloc (accessed 12 October 2025).

Bloch, Ernst. (1995). *The Principle of Hope*. Translated by Neville Plaice, Stephen Plaice and Paul Knight. Reprint edition. The MIT Press.

Bloch, Ernst. (2006). *Traces*. Translated by Anthony A. Nassar. Stanford University Press.

Blumenberg, Hans. (1987). *The Genesis of the Copernican World*. Translated by Robert M. Wallace. MIT Press. http://archive.org/details/hans-blumenberg-robert-m.-wallace-the-genesis-of-the-copernican-world-studies-in (accessed 12 October 2025).

Blumenberg, Hans. (1988). *Work on Myth*. MIT Press.

Bohlmeijer, Ernst and Gerben Westerhof. (2021). 'The Model for Sustainable Mental Health: Future Directions for Integrating Positive Psychology Into Mental Health Care', *Frontiers in Psychology* 12 (October): 747999. https://doi.org/10.3389/fpsyg.2021.747999.

Bolier, Linda, Merel Haverman, Gerben J. Westerhof, Heleen Riper, Filip Smit and Ernst Bohlmeijer. (2013). 'Positive Psychology Interventions: A Meta-Analysis of Randomized Controlled Studies', *BMC Public Health* 13 (1): 119. https://doi.org/10.1186/1471-2458-13-119.

Bothe, Michael and Jennifer Ernst. (2023). 'Humanitäre Hilfe – ein Beitrag zum Recht der Solidarität,' in Philipp B. Donath, Alexander Heger, Moritz Malkmus and Orhan Bayrak (eds), *Der Schutz des Individuums durch das Recht: Festschrift für Rainer Hofmann zum 70. Geburtstag*, 381–98. Springer. https://doi.org/10.1007/978-3-662-66978-5_26.

Brand, Fridolin Simon and Kurt Jax. (2007). 'Focusing the Meaning(s) of Resilience: Resilience as a Descriptive Concept and a Boundary Object', *Ecology and Society* 12 (1). https://hdl.handle.net/10535/3371 (accessed 12 October 2025).

Breithaupt, Fritz Alwin. (2025). *The Narrative Brain: The Stories Our Neurons Tell*. Yale University Press.

Brewer, Marilynn B. (1999). 'The Psychology of Prejudice: Ingroup Love and Outgroup Hate?', *Journal of Social Issues* 55 (3): 429–44. https://doi.org/10.1111/0022-4537.00126.

Bronowski, Jacob. (2011). *The Ascent of Man*. Random House.

Brooks, Harvey. (1980). 'Technology, Evolution, and Purpose', *Daedalus* 109 (1): 65–81.

Brooks, Samantha, R. Amlôt, G. J. Rubin and N. Greenberg. (2020). 'Psychological Resilience and Post-Traumatic Growth in Disaster-Exposed Organisations: Overview of the Literature', *BMJ Military Health* 166 (1): 52–6. https://doi.org/10.1136/jramc-2017-000876.

Bruton, A. and J. Sorin. (2022). 'Reimagining Public Value: Our Learning Journey in King County, Washington', *Centre for Public Impact* (blog), 2022. https://centreforpublicimpact.org/resource-hub/2076 (accessed 12 October 2025).

Brutschin, Elina, Silvia Pianta, Massimo Tavoni, Keywan Riahi, Valentina Bosetti, Giacomo Marangoni and Bas J. van Ruijven. (2021). 'A Multidimensional Feasibility Evaluation of Low-Carbon Scenarios', *Environmental Research Letters* 16 (6): 064069. https://doi.org/10.1088/1748-9326/abf0ce.

Burtynsky, Edward, Jennifer Baichwal and Nick De Pencier. (2018). *Anthropocene*. Steidl.

Cabinet Affairs and Strategic Coordination. (2024). 'Thirteenth Five Year Plan of Bhutan, 2024–2029', Cabinet Secretariat, Royal Government of Bhutan.

Cabinet Secretariat. (2015). 'Protocol for Policy Formulation', Cabinet Secretariat, Royal Government of Bhutan.

Cabrera, Victoria and Stewart I. Donaldson. (2024). 'PERMA to PERMA+4 Building Blocks of Well-Being: A Systematic Review of the Empirical Literature', *The Journal of Positive Psychology* 19 (3): 510–29. https://doi.org/10.1080/17439760.2023.2208099.

Cadena, Marisol de la. (2015). *Earth Beings: Ecologies of Practice across Andean Worlds.* Duke University Press.

Cajete, Gregory and Leroy Little Bear. (2000). *Native Science: Natural Laws of Interdependence.* 1st edition. Clear Light Publishers.

Canonici, Noverino N. (1996). 'The Hare and the Lion', in Noverino N. Canonici (ed.), *Zulu Oral Traditions*, 123–6. University of Natal Press.

Capps, Walter H. (1970). *The Future of Hope; [Essays by Bloch, Fackenheim, Moltmann, Metz, Capps].* Fortress Press. http://archive.org/details/futureofhopeessa0000walt (accessed 12 October 2025).

Capps, Walter H. (1972). *Time Invades the Cathedral: Tension in the School of Hope.* Fortress Press.

Capps, Walter H. (1976). *Hope against Hope: Molton [i.e. Moltmann] to Merton in One Decade.* Fortress Press. http://archive.org/details/hopeagainsthopem0000capp (accessed 12 October 2025).

Capra, Fritjof and Pier Luigi Luisi. (2014). *The Systems View of Life: A Unifying Vision.* Cambridge University Press.

Carneiro, Robert L. (1970). 'A Theory of the Origin of the State', *Science* 169 (3947): 733–8.

Carpenter, Amber D. (2021). 'Why Do Bad Things Happen to Good People? "And None of Us Deserving the Cruelty or the Grace": Buddhism and the Problem of Evil', in Steven M. Emmanuel (ed.), *Philosophy's Big Questions: Comparing Buddhist and Western Approaches*, 164–204. Columbia University Press.

Carr, Alan, Laura Finneran, Christine Boyd, Claire Shirey, Ciaran Canning, Owen Stafford, James Lyons, et al. (2024). 'The Evidence-Base for Positive Psychology Interventions: A Mega-Analysis of Meta-Analyses', *The Journal of Positive Psychology* 19 (2): 191–205. https://doi.org/10.1080/17439760.2023.2168564.

Carroll, Sean. (2017). *The Big Picture: On the Origins of Life, Meaning, and the Universe Itself.* Penguin.

Cashdan, Elizabeth and Stephen M. Downes. (2012). 'Evolutionary Perspectives on Human Aggression: Introduction to the Special Issue', *Human Nature (Hawthorne, N.Y.)* 23 (1): 1–4. https://doi.org/10.1007/s12110-012-9133-0.

Castillo, Fides A. del. (2024). 'Valuing Informal Care and Recognizing Care as Foundational', *Journal of Public Health* 46 (2): e344–5. https://doi.org/10.1093/pubmed/fdad264.

Castro, Maria. (2022). 'Diego Rivera's New American Art: San Francisco, 1930-31', in James Oles (ed.), *Diego Rivera's America*, 1st edition. University of California Press.

Cavallar, Georg. (2002). *The Rights of Strangers: Theories of International Hospitality, the Global Community, and Political Justice Since Vitoria.* Ashgate.

Cavell, Stanley. (1999). *The Claim of Reason: Wittgenstein, Skepticism, Morality, and Tragedy.* New edition. Oxford University Press.

Cebral-Loureda, Manuel, Enrique Tamés-Muñoz and Alberto Hernández-Baqueiro. (2022). 'The Fertility of a Concept: A Bibliometric Review of Human Flourishing', *International Journal of Environmental Research and Public Health* 19 (5): 2586. https://doi.org/10.3390/ijerph19052586.

Cedervall Lauta, Kristian. (2016). 'Human Rights and Natural Disasters', in Susan C. Breau and Katja L. H. Samuel (eds), *Research Handbook on Disasters and International Law*, 91–110. Edward Elgar Publishing.

Chakrabarty, Dipesh. (2019). 'The Planet: An Emergent Humanist Category', *Critical Inquiry* 46 (1): 1–31. https://doi.org/10.1086/705298.

Chan, Wing-Tsit. (1963). *A Source Book in Chinese Philosophy*. Princeton University Press.

Charry, Ellen T. and Russell D. Kosits. (2017). 'Christian Theology and Positive Psychology: An Exchange of Gifts', *The Journal of Positive Psychology* 12 (5): 468–79. https://doi.org/10.1080/17439760.2016.1228010.

Chazan, G. (2020). 'Tesla's German Gigafactory Held up by Sand Lizards', *Financial Times*, 22 December 2020. https://www.ft.com/content/d6d51776-76af-45cf-9cfa-d557412f037b (accessed 12 October 2025).

Chigangaidze, Robert. (2021). 'Utilising Ubuntu in Social Work Practice: Ubuntu in the Eyes of the Multimodal Approach', *Journal of Social Work Practice* 36 (3): 291–301. https://doi.org/10.1080/02650533.2021.1981276.

Chik, HinMingFrankie. (2021). 'Liji 禮記', *UBC Community and Partner Publications, Database of Religious History*. https://doi.org/10.14288/1.0404466.

Church of England. (2015). 'Urgent Action Needed on Climate Change Urges Synod', The Church of England, 13 July 2015. https://www.churchofengland.org/media/press-releases/urgent-action-needed-climate-change-urges-synod (accessed 12 October 2025).

Cilliers, Paul. (2001). 'Boundaries, Hierarchies and Networks in Complex Systems', *International Journal of Innovation Management* 5 (2): 135–47. https://doi.org/10.1142/S1363919601000312.

Clark, Andy. (2013). 'Whatever Next? Predictive Brains, Situated Agents, and the Future of Cognitive Science', *Behavioral and Brain Sciences* 36 (3): 181–204. https://doi.org/10.1017/S0140525X12000477.

Coccia, Emanuele. (2019). *The Life of Plants: A Metaphysics of Mixture*. Polity Press.

Coccia, Emanuele. (2021). *Metamorphoses*. Translated by Robin Mackay. Polity Press.

Cockell, C. S., T. Bush, C. Bryce, S. Direito, M. Fox-Powell, J. P. Harrison, H. Lammer, et al. (2016). 'Habitability: A Review', *Astrobiology* 16 (1): 89–117. https://doi.org/10.1089/ast.2015.1295.

Cohen, Dara Kay and Sabrina M. Karim. (2022). 'Does More Equality for Women Mean Less War? Rethinking Sex and Gender Inequality and Political Violence', *International Organization* 76 (2): 414–44. https://doi.org/10.1017/S0020818321000333.

Collins, Patricia Hill. (2000). *Black Feminist Thought: Knowledge, Consciousness, and the Politics of Empowerment*. Routledge. https://doi.org/10.4324/9780203900055.

Collins, Patricia Hill, and Sirma Bilge. (2020). *Intersectionality*. Polity Press.

'Conceptions of Human Flourishing - Programme.' (2024). THE NEW INSTITUTE, Hamburg, 12 June 2024. https://thenew.institute/en/programs/conceptions-of-human-flourishing (accessed 12 October 2025).

Confucius. (1979). *The Analects*. Translated by D. C. Lau. 1st edition. Penguin Classics.

Conklin, Carli N. (2014). 'The Origins of the Pursuit of Happiness', *Washington University Jurisprudence Review* 7: 195.

Connes, Alain. (2001). *Triangle of Thoughts*. American Mathematical Society. http://archive.org/details/triangleofthough0000conn (accessed 12 October 2025).

Conze, Edward. (ed.). (1975). *The Large Sutra on Perfect Wisdom: With the Divisions of the Abhisamayālaṅkāra*. University of California Press.

Coole, Diana. (2018). *Should We Control World Population?* Polity Press.

Cooper, Gregory S., Simon Willcock and John A. Dearing. (2020). 'Regime Shifts Occur Disproportionately Faster in Larger Ecosystems', *Nature Communications* 11 (1): 1175. https://doi.org/10.1038/s41467-020-15029-x.

Cornell, Drucilla. (2016). *The Imaginary Domain: Abortion, Pornography and Sexual Harassment*. Routledge. https://doi.org/10.4324/9780203760383.

Corntassel, Jeff. (2008). 'Toward Sustainable Self-Determination: Rethinking the Contemporary Indigenous-Rights Discourse', *Alternatives: Global, Local, Political* 33 (1): 105–32.

Cote, Muriel and Andrea J. Nightingale. (2012). 'Resilience Thinking Meets Social Theory: Situating Social Change in Socio-Ecological Systems (SES) Research', *Progress in Human Geography* 36 (4): 475–89. https://doi.org/10.1177/0309132511425708.

Courlander, Harold. (1971). *The Fourth World of the Hopis*. University of New Mexico Press http://archive.org/details/fourthworldofhop0000cour (accessed 12 October 2025).

Crenshaw, Kimberlé. (1989). 'Demarginalizing the Intersection of Race and Sex: A Black Feminist Critique of Antidiscrimination Doctrine, Feminist Theory and Antiracist Politics', *University of Chiago Legal Forum* 1989 (11): 139.

Crick, Francis. (1994). *Astonishing Hypothesis: The Scientific Search for the Soul*. Charles Scribner's Sons.

Crocetti, Elisabetta, Flavia Albarello, Wim Meeus and Monica Rubini. (2023). 'Identities: A Developmental Social-Psychological Perspective.' *European Review of Social Psychology* 34 (1): 161–201. https://doi.org/10.1080/10463283.2022.2104987.

Csikszentmihalyi, Mihaly, (ed.). (2014). *Flow and the Foundations of Positive Psychology: The Collected Works of Mihaly Csikszentmihalyi*. Springer Netherlands. https://doi.org/10.1007/978-94-017-9088-8.

Cuijpers, Pim, Clara Miguel, Mathias Harrer, Constantin Yves Plessen, Marketa Ciharova, David Ebert and Eirini Karyotaki. (2023). 'Cognitive Behavior Therapy vs. Control Conditions, Other Psychotherapies, Pharmacotherapies and Combined Treatment for Depression: A Comprehensive Meta-Analysis Including 409 Trials with 52,702 Patients', *World Psychiatry* 22 (1): 105–15. https://doi.org/10.1002/wps.21069.

Curry, Oliver Scott, Lee A. Rowland, Caspar J. Van Lissa, Sally Zlotowitz, John McAlaney and Harvey Whitehouse. (2018). 'Happy to Help? A Systematic Review and Meta-Analysis of the Effects of Performing Acts of Kindness on the Well-Being of the Actor', *Journal of Experimental Social Psychology* 76 (May): 320–9. https://doi.org/10.1016/j.jesp.2018.02.014.

Cusicanqui, Silvia Rivera. (2012). 'Ch'ixinakax Utxiwa: A Reflection on the Practices and Discourses of Decolonization', *South Atlantic Quarterly* 111 (1): 95–109. https://doi.org/10.1215/00382876-1472612.

Czekierda, Katarzyna, Anna Banik, Crystal L. Park and Aleksandra Luszczynska. (2017). 'Meaning in Life and Physical Health: Systematic Review and Meta-Analysis', *Health Psychology Review* 11 (4): 387–418. https://doi.org/10.1080/17437199.2017.1327325.

Dahl, Cortland J., Antoine Lutz and Richard J. Davidson. (2015). 'Reconstructing and Deconstructing the Self: Cognitive Mechanisms in Meditation Practice', *Trends in Cognitive Sciences* 19 (9): 515–23. https://doi.org/10.1016/j.tics.2015.07.001.

Dahlbeck, Johan. (2016). *Spinoza and Education: Freedom, Understanding and Empowerment*. Routledge. https://doi.org/10.4324/9781315679495.

Dakwar, Elias. (2024). *The Captive Imagination: Addiction, Reality, and Our Search for Meaning*. HarperCollins.

Dalai Lama, and Howard Cuttler. (1999). *The Art of Happiness: A Handbook for Living*. New edition. MOBIUS.

Daly, Mary. (2021). 'The Concept of Care: Insights, Challenges and Research Avenues in Covid-19 Times', *Journal of European Social Policy* 31 (1): 108–18. https://doi.org/10.1177/0958928720973923.

Dauenhauer, Nora Marks and Richard Dauenhauer. (1987). *Haa Shuká, Our Ancestors: Tlingit Oral Narratives*. University of Washington Press.

Davidson, Richard J. and Cortland J. Dahl. (2017). 'Varieties of Contemplative Practice', *JAMA Psychiatry* 74 (2): 121–3. https://doi.org/10.1001/jamapsychiatry.2016.3469.

Davis, Angela. (2014). 'Angela Davis, Live at Southern Illinois University Carbondale.' Q&A post lecture as noted by Dr Jonathan Flowers in a post he contemporaneously shared on Tumblr, February 13.

Dawkins, Richard. (2016). *The Selfish Gene: 40th Anniversary Edition*. 44th edition. Oxford Landmark Science. Oxford University Press.

De Kirchner, Beatriz Bossi. (1986). 'Aquinas as an Interpreter of Aristotle on the End of Human Life', *The Review of Metaphysics* 40 (1): 41–54.

Deacon, Terrence W. (1997). *The Symbolic Species: The Co-Evolution of Language and the Brain*. W. W. Norton.

Deci, Edward L. and Richard M. Ryan. (1985). 'Conceptualizations of Intrinsic Motivation and Self-Determination', in Edward L. Deci and Richard M. Ryan (eds), *Intrinsic Motivation and Self-Determination in Human Behavior*, 11–40. Springer. https://doi.org/10.1007/978-1-4899-2271-7_2.

Denoual, Fabienne. (2020). 'Le Designer de l'Anthropocène: Vers une Éthique de l'Habitabilité Élargie', *Sciences du Design* 11 (1): 42–50. https://doi.org/10.3917/sdd.011.0042.

Derrida, Jacques and Anne Dufourmantelle. (2000). *Of Hospitality: Anne Dufourmantelle Invites Jacques Derrida to Respond*. Translated by Rachel Bowlby. Stanford University Press.

Deutsch, David. (2012). *The Beginning of Infinity: Explanations That Transform the World*. Penguin Press.

Dickerman, Leah and Anna Indyck-López. (eds). (2011). *Diego Rivera: Murals for the Museum of Modern Art*. Museum of Modern Art.

Diener, Ed and Martin E. P. Seligman. (2002). 'Very Happy People', *Psychological Science* 13 (1): 81–4. https://doi.org/10.1111/1467-9280.00415.

Dierendonck, Dirk van. (2012). 'Spirituality as an Essential Determinant for the Good Life, Its Importance Relative to Self-Determinant Psychological Needs', *Journal of Happiness Studies* 13 (4): 685–700. https://doi.org/10.1007/s10902-011-9286-2.

Dieye, Adama. (2020). *An Islamic Model for Stabilization and Growth*. Springer International Publishing. https://doi.org/10.1007/978-3-030-48763-8.

Dikeç, Mustafa, Nigel Clark and Clive Barnett. (2009). 'Extending Hospitality: Giving Space, Taking Time', *Paragraph* 32 (1): 1–14. https://doi.org/10.3366/E0264833409000376.

Dirac, P. A. M. (1977). 'Recollections of an Exciting Era', in C. Weiner (ed.), *History of Twentieth Century Physics*, 109–46. Academic Press.

Dobrokhotov, Alexander. (2008). *Selected Works [Избранное]*. Territoriya Budushchego.

Dodaro, Robert. (2004). 'Political and Theological Virtues in Augustine, Letter 155 to Macedonius', *Augustiniana* 54 (1/4): 431–74.

Doerfler, Jill. (2015). *Those Who Belong: Identity, Family, Blood, and Citizenship among the White Earth Anishinaabeg*. 1st edition. Michigan State University Press.

Dolphijn, Rick and Iris van der Tuin. (2012). *New Materialism: Interviews & Cartographies*. Open Humanities Press. https://www.openhumanitiespress.org/books/titles/new-materialism (accessed 12 October 2025).

Domenico, Manlio De and Hiroki Sayama. (2019). 'Complexity Explained' Booklet of the Complexity Explained project (June). https://doi.org/10.17605/OSF.IO/TQGNW.

Donaldson, Stewart I., Llewellyn Ellardus Van Zyl and Scott I. Donaldson. (2022). 'PERMA+4: A Framework for Work-Related Wellbeing, Performance and Positive Organizational Psychology 2.0', *Frontiers in Psychology* 12 (January): 817244. https://doi.org/10.3389/fpsyg.2021.817244.

Dong, Minglai. (2019). 'Correcting Things as Correcting Feelings: A Phenomenological Study of Wang Yang-Ming's Doctrine of Ge-Wu', *Comparative Philosophy: An International Journal of Constructive Engagement of Distinct Approaches toward World Philosophy* 10 (1). https://doi.org/10.31979/2151-6014(2019).100106.

Downs, Linda Bank. (1999). *Diego Rivera: The Detroit Industry Murals*. Illustrated edition. W. W. Norton.

Dowson, Martin, Stuart Devenish and Maureen Miner. (2012). *Beyond Well-Being: Spirituality and Human Flourishing*. Information Age Publishing.

Dryzek, John S. and Jonathan Pickering. (2019). *The Politics of the Anthropocene*. Oxford University Press.

Dunbar, Robin. (2010). *How Many Friends Does One Person Need?: Dunbar's Number and Other Evolutionary Quirks*. Harvard University Press.

Dunbar, Robin. (2012). *The Science of Love*. John Wiley & Sons.

Dunlap, Richard A. (1997). *The Golden Ratio and Fibonacci Numbers*. WORLD SCIENTIFIC. https://doi.org/10.1142/3595.

Dupuy, Pierre-Marie. (2011). 'International Law and Domestic (Municipal) Law', in *Max Planck Encyclopedia of International Law, Online Version*. https://opil.ouplaw.com/display/10.1093/law:epil/9780199231690/law-9780199231690-e1056?prd=MPIL (accessed 12 October 2025).

Duru, Michel, Olivier Therond and M'hand Fares. (2015). 'Designing Agroecological Transitions; A Review.' *Agronomy for Sustainable Development* 35 (4): 1237–57. https://doi.org/10.1007/s13593-015-0318-x.

Earthday.org. (2015). 'Rabbinic Letter on the Climate Crisis Calls for Action', Earth Day, 12 May 2015. https://www.earthday.org/rabbinic-letter-climate-crisis-calls-action (accessed 12 October 2025).

Ehrenfeld, John. (2019). *The Right Way to Flourish: Reconnecting to the Real World*. Routledge.

Einstein, Albert. (1954). 'Appendix II', in Jacques Hadamard (ed.), *The Psychology of Invention in the Mathematical Field*. Dover Publications Inc.

Eisler, Riane and Douglas P. Fry. (2019). *Nurturing Our Humanity: How Domination and Partnership Shape Our Brains, Lives, and Future*. Oxford University Press.

Eldred, Michael. (2024). *On Human Temporality: Recasting Whoness Da Capo*. De Gruyter.

Elhacham, Emily, Liad Ben-Uri, Jonathan Grozovski, Yinon M. Bar-On and Ron Milo. (2020). 'Global Human-Made Mass Exceeds All Living Biomass', *Nature* 588 (7838): 442–4. https://doi.org/10.1038/s41586-020-3010-5.

Ellis, George. (2017). *How Can Physics Underlie the Mind?: Top-Down Causation in the Human Context*. Springer.

Ellis, George F. R. (2005). 'Physics, Complexity and Causality', *Nature* 435 743 (2005). https://doi.org/10.1038/435743a.

Ellis, George F. R. (2023). 'Efficient, Formal, Material, and Final Causes in Biology and Technology', *Entropy* 25 (9): 1301. https://doi.org/10.3390/e25091301.

Ellis, George F. R. and Markus Gabriel. (2021). 'Physical, Logical, and Mental Top-Down Effects', in Jan Voosholz and Markus Gabriel (eds), *Top-Down Causation and Emergence* 439: 3–37. Synthese Library. Springer International Publishing. https://doi.org/10.1007/978-3-030-71899-2_1.

Elman, Benjamin A. (1984). *From Philosophy to Philology: Intellectual and Social Aspects of Change in Late Imperial China*. Harvard East Asian Monographs 110. Council on East Asian Studies, Harvard University.

Elman, Benjamin A. (2013). *Civil Examinations and Meritocracy in Late Imperial China*. Harvard University Press.

Emejulu, Akwugo. (2022). *Fugitive Feminism*. Silver Press.

Emmons, Robert A. (2020). 'Joy: An Introduction to This Special Issue', *The Journal of Positive Psychology* 15 (1): 1–4. https://doi.org/10.1080/17439760.(2019).1685580.

Emmons, Robert A. and Michael E. McCullough. (2003). 'Counting Blessings versus Burdens: An Experimental Investigation of Gratitude and Subjective Well-Being in Daily Life', *Journal of Personality and Social Psychology* 84 (2): 377–89. https://doi.org/10.1037/0022-3514.84.2.377.

Erdoes, Richard and Alfonso Ortiz. (1999). *American Indian Trickster Tales*. Penguin Books.

Erel, Umut, Jin Haritaworn, Encarnación Gutiérrez Rodríguez and Christian Klesse. (2010). 'On the Depoliticisation of Intersectionality Talk: Conceptualising Multiple Oppressions in Critical Sexuality Studies', in Yvette Taylor, Sally Hines and Mark E. Casey (eds), *Theorizing Intersectionality and Sexuality*, 56–77. Palgrave Macmillan UK. https://doi.org/10.1057/9780230304093_4.

Erel, Umut, Erene Kaptani, Maggie O'Neill and Tracey Reynolds. (2022). 'PAR: Resistance to Racist Migration Policies in the UK', in Azril Bacal Roij (ed.), Transformative Research and Higher Education, 93–106. Emerald Publishing.

Escobar, Arturo. (2009). 'Beyond the Third World: Imperial Globality, Global Coloniality and Anti-Globalisation Social Movements', in Mark Berger (ed.), *After the Third World?*, 194–217. Routledge.

Escobar, Arturo. (2014). 'Other Worlds Are (Already) Possible: Self-Organization, Complexity and Post-Capitalist Cultures', in Ravi Savyasaachi Kumar (ed.), *Social Movements*, 289–303. Routledge India.

Escobar, Arturo. (2018a). '3. Transition Discourses and the Politics of Relationality: Toward Designs for the Pluriverse', in Bernd Reiter (ed.), *Constructing the Pluriverse*, 63–89. Duke University Press.

Escobar, Arturo. (2018b). *Designs for the Pluriverse: Radical Interdependence, Autonomy, and the Making of Worlds*. New Ecologies for the Twenty-First Century. Duke University Press.

Esposito, John L. (2011). *What Everyone Needs to Know about Islam*. 2nd edition. Oxford University Press.

European Court of Justice. (2018). 'C-266/16.' https://curia.europa.eu/juris/liste.jsf?language=en&num=C-266/16 (accessed 12 October 2025).

Falkner, Robert and Barry Buzan. (2019). 'The Emergence of Environmental Stewardship as a Primary Institution of Global International Society', *European Journal of International Relations* 25 (1): 131–55. https://doi.org/10.1177/1354066117741948.

Faust, Drew Gilpin. (2008). *This Republic of Suffering: Death and the American Civil War*. Knopf.

Fava, G. A. (1999). 'Well-Being Therapy: Conceptual and Technical Issues', *Psychotherapy and Psychosomatics* 68 (4): 171–79. https://doi.org/10.1159/000012329.

Fei, Xiaotong. (2017). *From the Soil [Xiangtu Zhongguo]*. Beijing Daxue Chubanshe.

Fernández-Llamazares, Álvaro, Julien Terraube, Michael C. Gavin, Aili Pyhälä, Sacha M. O. Siani, Mar Cabeza and Eduardo S. Brondizio. (2020). 'Reframing the Wilderness Concept Can Bolster Collaborative Conservation', *Trends in Ecology & Evolution* 35 (9): 750–3. https://doi.org/10.1016/j.tree.(2020).06.005.

Fiksel, Joseph. (2006). 'Sustainability and Resilience: Toward a Systems Approach', *Sustainability: Science, Practice and Policy* 2 (2): 14–21. https://doi.org/10.1080/15487733.2006.11907980.

Fine, Michael. (2024). 'Care Consequences of the COVID-19 Pandemic', *International Journal of Care and Caring* 8 (2): 1–7. https://doi.org/10.1332/23978821Y2024D000000031.

Fisch, Jörg. (2009). 'Die Geschichte Des Selbstbestimmungsrechts Der Völker, Oder Der Versuch, Einem Menschenrecht Die Zähne Zu Ziehen', in Peter Hilpold (ed.), *Das Selbstbestimmungsrecht Der Völker*. Peter Lang D. https://doi.org/10.3726/978-3-653-00219-5.

Fischer, John Martin. (2023). Review of Review of 'Determined: A Science of Life Without Free Will,' by Robert M. Sapolsky. *Notre Dame Philosophical Reviews*, November. https://ndpr.nd.edu/reviews/determined-a-science-of-life-without-free-will (accessed 12 October 2025).

Florea, Alin I. and Steven B. Caudill. (2014). 'Happiness, Religion and Economic Transition', *The Economics of Transition* 22 (1): 1–12.

Fox, Claire F. (2022). 'Diego Rivera's Pan-America', in James Oles (ed.), *Diego Rivera's America*, 1st edition. University of California Press.

Fox, Nick J. and Pam Alldred. (2015). 'New Materialist Social Inquiry: Designs, Methods and the Research-Assemblage', *International Journal of Social Research Methodology* 18 (4): 399–414. https://doi.org/10.1080/13645579.(2014).921458.

Frankl, Viktor E. (2004). *Man's Search for Meaning: The Classic Tribute to Hope from the Holocaust*. Simon & Schuster.

Frankl, Viktor E. (2006). *Man's Search for Meaning*. Beacon Press.

Fredrickson, Barbara L. (2001). 'The Role of Positive Emotions in Positive Psychology: The Broaden-and-Build Theory of Positive Emotions', *American Psychologist* 56 (3): 218–26. https://doi.org/10.1037/0003-066X.56.3.218.

Fredrickson, Barbara L. (2013). 'Positive Emotions Broaden and Build', in *Advances in Experimental Social Psychology* 47: 1–53. Elsevier. https://doi.org/10.1016/B978-0-12-407236-7.00001-2.

Frege, Gottlob. (1892). 'Über Sinn und Bedeutung', *Zeitschrift für Philosophie und philosophische Kritik* NF 100/1: 25–50.

Frege, Gottlob. (1918). 'Der Gedanke. Eine Logische Untersuchung'. Edited by A. Roser and F. Börnke. *Beiträge Zur Philosophie Des Deutschen Idealismus* 1 (2): 58–77.

Freistein, Katja, (ed.) (2022). *Imagining Pathways for Global Cooperation*. Edward Elgar Publishing.

Frith, Chris and Uta Frith. (2005). 'Theory of Mind', *Current Biology* 15 (17): R644–6. https://doi.org/10.1016/j.cub.2005.08.041.

Fuchs, Thomas. (2020). 'Our Future on Earth: Science Insights into Our Planet and Society', Google Docs, 2020. https://drive.google.com/file/d/1chEx2Aewehp1_0nXYnERwUViJI6qR2hi/view?usp=embed_facebook (accessed 14 October 2025).

Fuchs, Thomas. (2021). *In Defence of the Human Being: Foundational Questions of an Embodied Anthropology*. Oxford University Press.

Funnell, Antony. (2024). 'Fear and Anger – the Complicated Emotions That Govern Our World.' Future Tense. ABC.Net.Au. https://www.abc.net.au/listen/programs/futuretense/fear-and-anger-the-complicated-emotions-that-govern-our-world/103544982 (accessed 10 April 2025).

Fuso Nerini, Francesco, Mariana Mazzucato, Johan Rockström, Harro van Asselt, Jim W. Hall, Stelvia Matos, Åsa Persson, Benjamin Sovacool, Ricardo Vinuesa and Jeffrey Sachs. (2024). 'Extending the Sustainable Development Goals to 2050 — a Road Map', *Nature* 630 (8017): 555–8. https://doi.org/10.1038/d41586-024-01754-6.

Gable, Shelly L. and Jonathan Haidt. (2005). 'What (and Why) Is Positive Psychology?', *Review of General Psychology* 9 (2): 103–10. https://doi.org/10.1037/1089-2680.9.2.103.

Gabriel, Markus. (2011). *Transcendental Ontology: Essays in German Idealism*. Continuum.

Gabriel, Markus. (2015). *Fields of Sense*. Edinburgh University Press.

Gabriel, Markus. (2017). *I Am Not a Brain: Philosophy of Mind for the 21st Century*. Polity Press.

Gabriel, Markus. (2018). *Neo-Existentialism: How to Conceive of the Human Mind after Naturalism's Failure*. Edited by Jocelyn Maclure. Polity Press.

Gabriel, Markus. (2020a). 'Intelligence and Understanding – Limits of Artificial Intelligence', *Mind and Matter* 18 (1): 39–60.

Gabriel, Markus. (2020b). *The Meaning of Thought*. Polity Press.

Gabriel, Markus. (2022). *Moral Progress in Dark Times: Universal Values for the Twenty-First Century*. Translated by Wieland Hoban. English edition. Polity Press.

Gabriel, Markus. (2024a). *Fictions*. Translated by Wieland Hoban. 1st edition. Polity.

Gabriel, Markus. (2024b). 'Responses', in Jan Voosholz (ed.), *Markus Gabriel's New Realism*, 323–73. Springer Nature.

Gabriel, Markus. (2024c). *Sense, Nonsense, and Subjectivity*. Harvard University Press.

Gabriel, Markus. (2024d). *The Human Animal: Why We Still Don't Fit into Nature*. Translated by Karl von der Luft. Polity Press.

Gabriel, Markus. (2025). *Moralische Tatsachen: Warum Sie Existieren Und Wie Wir Sie Erkennen Können*. München: C.H.Beck.

Gabriel, Markus, Christoph Horn, Anna Katsman, Wilhelm Krull, Anna Luisa Lippold, Corine Pelluchon and Ingo Venzke. (2022). *Towards a New Enlightenment - The Case for Future-Oriented Humanities*. transcript Verlag. https://doi.org/10.1515/9783839465707.

Gabriel, Markus and Graham Priest. (2022). *Everything and Nothing*. Polity Press.

Galesic, Mirta, Daniel Barkoczi, Andrew M. Berdahl, Dora Biro, Giuseppe Carbone, Ilaria Giannoccaro, Robert L. Goldstone, et al. (2023). 'Beyond Collective Intelligence: Collective Adaptation', *Journal of the Royal Society Interface* 20 (200): 20220736. https://doi.org/10.1098/rsif.2022.0736.

Gane, Nicholas. (2006). 'When We Have Never Been Human, What Is to Be Done?: Interview with Donna Haraway', *Theory, Culture & Society* 23 (7–8): 135–58. https://doi.org/10.1177/0263276406069228.

Garfield, Jay L. (2022). *Buddhist Ethics: A Philosophical Exploration*. Buddhist Philosophy for Philosophers. Oxford University Press.

Garmestani, Ahjond S. and Melinda Harm Benson. (2013). 'A Framework for Resilience-Based Governance of Social-Ecological Systems', *Ecology and Society* 18 (1). https://www.jstor.org/stable/26269259 (accessed 14 October 2025).

Garnett, Stephen T., Neil D. Burgess, Julia E. Fa, Álvaro Fernández-Llamazares, Zsolt Molnár, Cathy J. Robinson, James E. M. Watson, et al. (2018). 'A Spatial Overview of the Global Importance of Indigenous Lands for Conservation', *Nature Sustainability* 1 (7): 369–74. https://doi.org/10.1038/s41893-018-0100-6.

Garske, Pia. 2014. 'What's the "Matter?" Der Materialitätsbegriff des "New Materialism" und dessen Konsequenzen für feministisch-politische Handlungsfähigkeit', *PROKLA. Zeitschrift für kritische Sozialwissenschaft* 44 (174): 111–29. https://doi.org/10.32387/prokla.v44i174.194.

Geertz, Clifford and Robert Darnton. (2017). *The Interpretation of Cultures: Selected Essays*. 3rd edition. Basic Books.

Geinitz, Christian. (2024). 'Bericht von Expertenrat: Nur ein Scheinerfolg für den Klimaschutz', *Frankfurter Allgemeine Zeitung - FAZ.NET*, 15 April 2024. https://www.faz.net/aktuell/wirtschaft/bericht-von-expertenrat-nur-ein-scheinerfolg-fuer-den-klimaschutz-19655034.html (accessed 14 October 2025).

Gentile, Emilio. (2006). *Politics as Religion*. Translated by George Staunton. Princeton University Press.

Gesang, Bernward. (2013). 'What Climate Policy Can a Utilitarian Justify?', *Journal of Agricultural and Environmental Ethics* 26 (2): 377–92. https://doi.org/10.1007/s10806-012-9380-4.

Gilligan, Carol. (2008). 'Moral Orientation and Moral Development', in Alison Bailey and Chris J. Cuomo (eds), *The Feminist Philosophy Reader*. McGraw-Hill.

Gilmore, Jane. (2019). *Fixed It: Violence and the Representation of Women in the Media*. Viking.

Giraud, Eva. (2019). 'The Planetary Is Political', *BioSocieties* 14 (3): 472–81. https://doi.org/10.1057/s41292-019-00169-1.

Glass, Lisa-Maria, Jens Newig and Simon Ruf. (2023). 'MSPs for the SDGs – Assessing the Collaborative Governance Architecture of Multi-Stakeholder Partnerships for Implementing the Sustainable Development Goals', *Earth System Governance* 17 (August): 100182. https://doi.org/10.1016/j.esg.2023.100182.

Glenn, E. N. (2021). 'Reimagining Care and Care Work', in Nadya Araujo Guimarães and Helena Hirata (eds), *Care and Care Workers: A Latin American Perspective*, 1st edition Springer.

Gloster, Andrew T., Noemi Walder, Michael E. Levin, Michael P. Twohig and Maria Karekla. (2020). 'The Empirical Status of Acceptance and Commitment Therapy: A Review of Meta-Analyses', *Journal of Contextual Behavioral Science* 18 (October): 181–92. https://doi.org/10.1016/j.jcbs.2020.09.009.

Godden, Lee and Ray Ison. (2019). 'Community Participation: Exploring Legitimacy in Socio-Ecological Systems for Environmental Water Governance', *Australasian Journal of Water Resources* 23 (1): 45–57. https://doi.org/10.1080/13241583.2019.1608688.

Godoy, Emilio. (2021). 'Indigenous Communities in Mexico Fight Energy Projects', *Global Issues*, 19 March 2021. https://www.globalissues.org/news/2021/03/19/27403 (accessed 14 October 2025).

Godrej, Farah. (2016). 'Culture and Difference', in Teena Gabrielson, Cheryl Hall, John M. Meyer and David Schlosberg (eds), *The Oxford Handbook of Environmental Political Theory*. Oxford University Press.

Goethe, Johann Wolfgang von. (2009). *The Metamorphosis of Plants*. Translated by Gordon L. Miller. The MIT Press.

Gough, Ian. (2017). 'Recomposing Consumption: Defining Necessities for Sustainable and Equitable Well-Being', *Philosophical Transactions of the Royal Society A: Mathematical, Physical and Engineering Sciences* 375 (2095): 20160379. https://doi.org/10.1098/rsta.2016.0379.

Gowan, Richard. (2024). 'The Double Standards Debate at the UN'. *International Crisis Group*, 7 March 2024. https://www.crisisgroup.org/middle-east-north-africa/east-mediterranean-mena/israelpalestine/double-standards-debate-un (accessed 14 October 2025).

Grace, Sharon, Maggie O'Neill, Tammi Walker, Hannah King, Lucy Baldwin, Alison Jobe, Orla Lynch, Fiona Measham, Kate O'Brien and Vicky Seaman. (2022). *Criminal Women: Gender Matters*. Bristol Policy Press.

Graczyk, Alicja Małgorzata. (2021). 'Households Behaviour towards Sustainable Energy Management in Poland—The Homo Energeticus Concept as a New Behaviour Pattern in Sustainable Economics', *Energies* 14 (11): 3142. https://doi.org/10.3390/en14113142.

Graham, Liam. (2023). *Molecular Storms: The Physics of Stars, Cells and the Origin of Life*. Springer.

Granderson, Ainka A. (2017). 'The Role of Traditional Knowledge in Building Adaptive Capacity for Climate Change: Perspectives from Vanuatu', *Weather, Climate, and Society* 9 (3): 545–61. https://doi.org/10.1175/WCAS-D-16-0094.1.

Gray, Robert H. (2020). 'The Extended Kardashev Scale', *The Astronomical Journal* 159 (5): 228. https://doi.org/10.3847/1538-3881/ab792b.

Green, Rhys E., Stephen J. Cornell, Jörn P. W. Scharlemann and Andrew Balmford. (2005). 'Farming and the Fate of Wild Nature', *Science* 307 (5709): 550–5. https://doi.org/10.1126/science.1106049.

Greene, Brian. (2021). *Until the End of Time: Mind, Matter, and Our Search for Meaning in an Evolving Universe*. Vintage.

Grosfoguel, Ramón. (2011). 'Decolonizing Post-Colonial Studies and Paradigms of Political-Economy: Transmodernity, Decolonial Thinking, and Global Coloniality', *TRANSMODERNITY: Journal of Peripheral Cultural Production of the Luso-Hispanic World* 1 (1). https://doi.org/10.5070/T411000004.

Grzymała-Busse, Anna Maria. (2023). *Sacred Foundations: The Religious and Medieval Roots of the European State*. Princeton University Press.

Gudynas, Eduardo. (2011). 'Buen Vivir: Today's Tomorrow', *Development* 54 (4): 441–7. https://doi.org/10.1057/dev.2011.86.

Guidi, Jenny and Giovanni A. Fava. (2021). 'Conceptual and Clinical Innovations of Well-Being Therapy', *International Journal of Cognitive Therapy* 14 (1): 196–208. https://doi.org/10.1007/s41811-021-00101-1.

Gyekye, Kwame. (1996). *African Cultural Values: An Introduction*. Sankofa Publishing Company.

Gyekye, Kwame. (1997). *Tradition and Modernity: Philosophical Reflections on the African Experience*. Oxford University Press.

Haaken, Janice K. and Maggie O'Neill. (2014). 'Moving Images: Psychoanalytically Informed Visual Methods in Documenting the Lives of Women Migrants and Asylum Seekers', *Journal of Health Psychology* 19 (1): 79–89. https://doi.org/10.1177/1359105313500248.

Habermas, Jurgen. (1969). 'Ernst Bloch—A Marxist Romantic', *Salmagundi*, no. 10/11, 311–25.

Hadamard, Jacques. (1954). *Essay on the Psychology of Invention in the Mathematical Field*. Princeton University Press.

Haidt, Jonathan. (2006). *The Happiness Hypothesis: Finding Modern Truth in Ancient Wisdom*. Basic Books.

Hall, David L. and Roger T. Ames. (1987). *Thinking Through Confucius*. SUNY Press.

Hall, Ralph P., Barbara Van Koppen and Emily Van Houweling. (2014). 'The Human Right to Water: The Importance of Domestic and Productive Water Rights', *Science and Engineering Ethics* 20 (4): 849–68. https://doi.org/10.1007/s11948-013-9499-3.

Harari, Yuval Noah. (2015). *Sapiens: A Brief History of Humankind*. Translated by John Purcell and Haim Watzman. Popular Science. Vintage Books.

Harari, Yuval Noah. (2017). *Homo Deus: A Brief History of Tomorrow*. Revised edition. Vintage.

Harari, Yuval Noah. (2019). *21 Lessons for the 21st Century*. Vintage.

Haraway, Donna J. (2003). *The Companion Species Manifesto: Dogs, People, and Significant Otherness*. University of Chicago Press.

Haraway, Donna J. (2016). *Staying with the Trouble: Making Kin in the Chthulucene*. Duke University Press. https://doi.org/10.1215/9780822373780.

Harman, Graham. (2018). *Object-Oriented Ontology: A New Theory of Everything*. Pelican.

Harris, David and Sarah Harris. (2013). *Digital Design and Computer Architecture*. Elsevier.

Harrison, Peter. (2017). *The Territories of Science and Religion*. University of Chicago Press.

Harrow, K. (2015). 'Ngoma ya Kimya', in A. Drury (ed.), *Stray Truths: Selected Poems of Euphrase Kezilahabi,* pp. 42–3). Michigan State University Press.

Haugen, Gary A. and Victor Boutros. (2015). *The Locust Effect: Why the End of Poverty Requires the End of Violence*. Oxford University Press.

Haybron, Daniel M. (2008). 'Happiness, the Self and Human Flourishing', *Utilitas* 20 (1): 21–49. https://doi.org/10.1017/S0953820807002889.

Hayward, Matt W., Robert J. Scanlon, Alexandra Callen, Lachlan G. Howell, Kaya L. Klop-Toker, Yamil Di Blanco, Niko Balkenhol, et al. (2019). 'Reintroducing Rewilding to Restoration – Rejecting the Search for Novelty', *Biological Conservation* 233 (May): 255–9. https://doi.org/10.1016/j.biocon.2019.03.011.

Heaney, Seamus. (1961). *The Cure at Troy: A Version of Sophocles' Philoctetes*. Farrar, Straus and Giroux.

Heaney, Seamus. (1987). *The Haw Lantern*. Faber & Faber.

Heitler, W. (1998). 'Goethian Science', in David Seamon and Arthur Zajonc (eds), *Goethe's Way of Science: A Phenomenology of Nature*. State University of New York Press.

Held, Barbara S. (2004). 'The Negative Side of Positive Psychology', *Journal of Humanistic Psychology* 44 (1): 9–46. https://doi.org/10.1177/0022167803259645.

Held, Virginia. (2006). *The Ethics of Care: Personal, Political, and Global*. Oxford University Press.

Helliwell, John F., Richard Layard, Jeffrey D. Sachs, Jan-Emmanuel De Neve, Lara B. Aknin and Shun Wang. 2024. 'World Happiness Report 2024'. Oxford University: Wellbeing Research Centre. https://worldhappiness.report/ed/2024 (accessed 14 October 2025).

Herget, Lauren. (2007). 'Country of Contradictions: Mexico's Transition to Modernity', *Cornell International Affairs Review* 1 (1). http://www.inquiriesjournal.com/articles/1296/country-of-contradictions-mexicos-transition-to-modernity (accessed 14 October 2025).

Hernández, Ariel. (2014a). *Nation-Building and Identity Conflicts: Facilitating the Mediation Process in Southern Philippines*. Springer Fachmedien. https://doi.org/10.1007/978-3-658-05215-7.

Hernández, Ariel. (2014b). *Strategic Facilitation of Complex Decision-Making: How Process and Context Matter in Global Climate Change Negotiations*. Springer.

Hernández, Ariel. (2017). 'Facilitar La Toma De Decisiones - Escenarios De Transformación De Las Economías Bajas En Carbon', in Helena Cabal, Efraín Gómez-Arias and Antonio Rodríguez Martínez (eds), *Perspectivas De Sustentabilidad En México*, 93–107. Universidad Autónoma del Estado de Morelos.

Hernández, Ariel. (2021a). 'SDG-Aligned Futures and the Governance of Transformation to Sustainability: Reconsidering Governance Perspectives on the Futures We Aspire To', *Discussion Paper*, 30/2021. https://doi.org/10.23661/DP30.2021.

Hernández, Ariel. (2021b). *Taming the Big Green Elephant. Setting in Motion the Transformation Towards Sustainability*. Springer. https://doi.org/10.1007/978-3-658-31821-5.

Hernández, Ariel. (2022a). 'The Phases Model of the Transformation to Sustainability (T2S)—Structuring through the Negotiation Perspective', *Sustainability* 14 (9): 5024. https://doi.org/10.3390/su14095024.

Hernández, Ariel. (2022b). 'When Cooperation Meets Negotiations - An Approach to Address the Complexity of Transformation to Sustainability', in Imme Scholz, Lilian Busse, and Thomas Fues (eds), *Transboundary Cooperation and Global Governance for Inclusive Sustainable Development: Contributions in Honour of Dirk Messner's 60th Birthday*, 199–205. Nomos Verlagsgesellschaft mbH & Co. KG. https://doi.org/10.5771/9783748930099.

Hernández, Ariel. (2024). 'Reclaiming the "Resilient Filipino": From Romanticised Climate Narratives Towards Resilience-Focused Climate Policy in the Philippines'. *FULCRUM*. 10 December 2024. https://fulcrum.sg/reclaiming-the-resilient-filipino-from-romanticised-climate-narratives-towards-resilience-focused-climate-policy-in-the-philippines (accessed 14 October 2025).

Hernández, Ariel and Yudhi Timor Bimo Prakoso. (2021). 'The Learning Activation Approach—Understanding Indonesia's Energy Transition by Teaching It', *Energies* 14 (17): 5224. https://doi.org/10.3390/en14175224.

Heyes, Cecilia M. and Chris D. Frith. (2014). 'The Cultural Evolution of Mind Reading', *Science* 344 (6190): 1243091. https://doi.org/10.1126/science.1243091.

Hickel, Jason. (2021). *Less Is More: How Degrowth Will Save the World*. 1st edition. Random House.

Hindu American Foundation. (2012). 'Hindu Declaration on Climate Change 2015', *Hindu American Foundation* (blog). 1 December 2012. https://www.hinduamerican.org/hindu-declaration-climate-change-2015 (accessed 14 October 2025).

Hodgkin, Luke. (2005). *A History of Mathematics: From Mesopotamia to Modernity*. Oxford University Press.

Höffe, Otfried. (2020). 'Von der humanitären Forschung kann es nie genug geben', *Frankfurter Rundschau*, 13 May 2020. https://www.fr.de/kultur/gesellschaft/humanitaerenforschung-kann-genug-geben-13759160.html (accessed 14 October 2025).

Hoffman, Donald. (2019). *The Case Against Reality: Why Evolution Hid the Truth from Our Eyes*. W. W. Norton.

Hoffmann, Peter M. (2012). *Life's Ratchet: How Molecular Machines Extract Order from Chaos*. Basic Books.

Høivik, Tord. (1977). 'The Demography of Structural Violence', *Journal of Peace Research* 14 (1): 59–73.

Honneth, Axel. (2009). *Pathologies of Reason: On the Legacy of Critical Theory*. Translated by James Ingram. Columbia University Press.

Horton, Richard, Robert Beaglehole, Ruth Bonita, John Raeburn, Martin McKee and Stig Wall. (2014). 'From Public to Planetary Health: A Manifesto', *Lancet* 383 (9920): 847. https://doi.org/10.1016/S0140-6736(14)60409-8.

Hossenfelder, Sabine. (2022). *Existential Physics: A Scientist's Guide to Life's Biggest Questions*. Penguin Publishing Group.

Hountondji, Paulin J. (1996). *African Philosophy: Myth and Reality*. 2nd edition. Indiana University Press.

Howarth, Richard B. (2015). 'Sustainability, Well-Being, and Economic Growth', *Center for Humans & Nature* (blog). 30 December 2015. https://humansandnature.org/sustainability-well-being-and-economic-growth (accessed 14 October 2025).

Huang, Wei, Emine Ertekin, Taifeng Wang, Luz Cruz, Micah Dailey, Jocelyne DiRuggiero and David Kisailus. (2020). 'Mechanism of Water Extraction from Gypsum Rock by Desert Colonizing Microorganisms', *Proceedings of the National Academy of Sciences* 117 (20): 10681–7. https://doi.org/10.1073/pnas.2001613117.

Hudson, Barbara. (2006). 'Beyond White Man's Justice: Race, Gender and Justice in Late Modernity', *Theoretical Criminology* 10 (1): 29–47. https://doi.org/10.1177/1362480606059981.

Hughes, Ian. (2018). *Disordered Minds: How Dangerous Personalities Are Destroying Democracy*. John Hunt Publishing.

Hughes, Ian, Edmond Byrne, Markus Glatz-Schmallegger, Clodagh Harris, William Hynes, Kieran Keohane and Brian Ó. Gallachóir. (2021). 'Deep Institutional Innovation for Sustainability and Human Development', *World Futures* 77 (5): 371–94. https://doi.org/10.1080/02604027.2021.1929013.

Hughes, Ian, Ariel Hernandez, James Glynn, William Hynes and Brian Ó Gallachóir. (2024). 'Conceptualising Global Cultural Transformation—Developing Deep Institutional Scenarios for Whole of Society Change', *Environmental Research Letters* 19 (9): 094050. https://doi.org/10.1088/1748-9326/ad6d7f.

'Human Sacrifice.' (2025). In *Wikipedia*. https://en.wikipedia.org/w/index.php?title=Human_sacrifice&oldid=1282230502 (accessed 14 October 2025).

Hursthouse, Rosalind. (2002). *On Virtue Ethics*. Oxford University Press.

Husserl, Edmund. (1954). *Die Krisis Der Europäischen Wissenschaften Und Die Transzendentale Phänomenologie*. Husserliana - Ergänzungsband Texte Aus Dem Nachlass 1934—1937, VI. Den Haag: Martinus Nijhoff.

Hyde, Lewis. (1998). *Trickster Makes This World: Mischief, Myth, and Art*. Farrar Straus & Giroux.

ICCA Consortium. (2021). 'Territories of Life 2021 Report: A Global Overview of Indigenous Peoples' and Local Communities' Lands, Territories, and Areas Conserved by Them', *Gland: ICCA Consortium*. https://www.iccaconsortium.org/tag/territories-of-life-2021-report (accessed 14 October 2025).

IDG. (2023). 'Inner Development Goals: Background, Method and the IDG Framework'. https://innerdevelopmentgoals.org/framework (accessed 14 October 2025).

IFEES. (2015). 'Islamic Declaration on Global Climate Change', ifees.ecoislam(blog), 18 August 2015. https://www.ifees.org.uk/about/islamicdeclaration (accessed 14 October 2025).

IISD - SDG Knowledge Hub. (2021). 'Statistical Commission's 52nd Session Advances Beyond GDP Approach', March 10, 2021. http://sdg.iisd.org/news/statistical-commissions-52nd-session-advances-beyond-gdp-approach (accessed 14 October 2025).

'Inner Development Goals.' (2024). https://innerdevelopmentgoals.org (accessed 14 October 2025).

International Seabed Authority. (2022a). 'The Mining Code', 17 March 2022. https://www.isa.org.jm/the-mining-code/official-documents (accessed 14 October 2025).

International Seabed Authority. (2022b). 'The Mining Code: Draft Exploitation Regulations', 17 March 2022. https://www.isa.org.jm/the-mining-code/draft-exploitation-regulations-2 (accessed 14 October 2025).

Internet Encyclopedia of Philosophy.' (n.d.). 'Plotinus: Virtue Ethics' Accessed February 8, 2024. https://iep.utm.edu/plot-v-e (accessed 14 October 2025).

Inwood, Brad and Raphael Woolf, (eds). (2012). *Aristotle: Eudemian Ethics*. Translated by Brad Inwood and Raphael Woolf. Cambridge Texts in the History of Philosophy. Cambridge University Press. https://doi.org/10.1017/CBO9781139043281.

IPCC. (2022). 'Summary for Policymakers', in Hans-Otto Pörtner, Debra Cynthia Roberts, Melinda M. B. Tignor, Elvira S. Poloczanska, Katja Mintenbeck, Andrès Alegría, Marlies Craig, et al. (eds), *Climate Change 2022: Impacts, Adaptation and Vulnerability. Contribution of Working Group II to the Sixth Assessment Report of the Intergovernmental Panel on Climate Change.* Cambridge University Press.

Ipsen, Knut. (2024). *Völkerrecht: ein Studienbuch*. Edited by Volker Epping and Wolff Heintschel von Heinegg. 8. Völlig neu bearbeitete Auflage. Kurzlehrbücher für das Juristische Studium. C.H. Beck. https://doi.org/10.17104/9783406799273.

Irni, Sari. (2013). 'The Politics of Materiality: Affective Encounters in a Transdisciplinary Debate', *European Journal of Women's Studies* 20 (4): 347–60. https://doi.org/10.1177/1350506812472669.

Jacques, Paul. (2024). 'Call to Action to Tackle "Epidemic Scale" of VAWG', *Police Professional* (blog), 23 July 2024. https://policeprofessional.com/news/call-to-action-to-tackle-epidemic-scale-of-vawg (accessed 14 October 2025).

James, P. D. (2000). *Time to Be in Earnest: A Fragment of Autobiography*. Faber and Faber.

James, William. (1895). 'Is Life Worth Living?', *International Journal of Ethics* 6 (1): 1–24.

James, William. (1912). *Is Life Worth Living?* S. Burns Weston. http://archive.org/details/islifeworthlivin00jameuoft (accessed 14 October 2025).

Jamieson, Lynn and Roona Simpson. (2013). *Living Alone: Globalization, Identity and Belonging*. Palgrave Macmillan UK. https://doi.org/10.1057/9781137318527.

Jampalyang, Chim. (2018). *Ornament of Abhidharma: A Commentary on Vasubandhu's Abhidharmakoa*. Translated by Ian James Coghlan. Vol. 23. The Library of Tibetan Classics. Wisdom.

Jaspers, Karl. (1966). *The Great Philosophers, Vol. 2: The Original Thinkers- Anaximander, Heraclitus, Parmenides, Plotinus, Anselm, Nicholas of Cusa, Spinoza, Lao-Tzu, Nagarjuna*. Harcourt, Brace & World, Inc.

Jaspers, Karl. (1967). *Philosophical Faith and Revelation*. 1st edition. Harper & Row.

Jewell, Jessica and Aleh Cherp. (2020). 'On the Political Feasibility of Climate Change Mitigation Pathways: Is It Too Late to Keep Warming below 1.5°C?', *WIREs Climate Change* 11 (1): e621. https://doi.org/10.1002/wcc.621.

Jimenez, Alberto Corsin. (2008). *Culture and Well-Being: Anthropological Approaches to Freedom and Political Ethics*. Pluto Press.

Johnson, Samuel G. B., Avri Bilovich and David Tuckett. (2023). 'Conviction Narrative Theory: A Theory of Choice Under Radical Uncertainty', *Behavioral and Brain Sciences* 46 (January): e82. https://doi.org/10.1017/S0140525X22001157.

Johnston, Basil. (1976). *Ojibway Heritage*. 1st edition. McClelland and Stewart.

Johnston, Ian and Ping Wang, (eds). (2012). *Daxue and Zhongyong*. Bilingual edition. Chinese University of Hong Kong.

Jonaitis, Aldona, (ed.). (1991). *Chiefly Feasts: The Enduring Kwakiutl Potlatch*. American 1st edition. University of Washington Press.

Jonas, Hans. (1985). *The Imperative of Responsibility: In Search of an Ethics for the Technological Age*. University of Chicago Press.

Joseph, Stephen, (ed.). (2015). *Positive Psychology in Practice: Promoting Human Flourishing in Work, Health, Education, and Everyday Life*. Wiley. https://doi.org/10.1002/9781118996874.

Joyce, Sindy, Olive O'Reilly, Margaret O'Brien, David Joyce, Jennifer Schweppe and Amanda Haynes. (2022). 'Irish Travellers' Access to Justice.' Report. University of Limerick. https://researchrepository.ul.ie/server/api/core/bitstreams/e4237fbb-3384-44aa-b17a-0a2f93dc5198/content (accessed 14 October 2025).

Jung, Carl Gustav. (1933). *Modern Man in Search of a Soul*. Harvest.

Jung, Carl Gustav. (1989). *Memories, Dreams, Reflections*. Edited by Aniela Jaffe, Clara Winston and Richard Winston. Vintage.

Kallis, Giorgos. (2018). *Degrowth*. The Economy Key Ideas. Agenda Publishing.

Kallis, Giorgos, Susan Paulson, Giacomo D'Alisa and Federico Demaria. (2020). *The Case for Degrowth*. The Case for Series. Polity.

Kanger, Laur and Johan Schot. (2019). 'Deep Transitions: Theorizing the Long-Term Patterns of Socio-Technical Change', *Environmental Innovation and Societal Transitions* 32 (September): 7–21. https://doi.org/10.1016/j.eist.2018.07.006.

Kant, Immanuel. (2002). *Groundwork of the Metaphysics of Morals, in Focus*. Edited by Lawrence Pasternack. Routledge Philosophers in Focus Series. Routledge.

Kant, Immanuel. (2003). *To Perpetual Peace: A Philosophical Sketch*. Translated by Ted Humphrey. Hackett Publishing Co, Inc.

Kant, Immanuel. (2007). *Critique of Pure Reason*. Translated by Marcus Weigelt and Max Muller. Penguin Classics.

Kaptani, Erene, Erel, Umut, O'Neill, Maggie and Tracey and Reynolds. (2021). 'Methodological Innovation in Research: Participatory Theater with Migrant Families on Conflicts and Transformations over the Politics of Belonging', *Journal of Immigrant & Refugee Studies* 19 (1): 68–81. https://doi.org/10.1080/15562948.2020.1843748.

Kardashev, Nikolai S. (1964). 'Transmission of Information by Extraterrestrial Civilizations', *Soviet Astronomy* 8 (2): 217–21.

Karthikeya, R., E. Antonacopoulou, B. Keating, E. Mountbatten-O'Malley, A. Nevin, E. Neill, J. Ritchie-Dunham, and M. T. Lee. (2024). 'The Global Flourishing Goals: An Invitation', *Flourishing Network Resources*. https://projects.iq.harvard.edu/fnr/home (accessed 14 October 2025).

Keating, AnaLouise. (2012). *Transformation Now!: Toward a Post-Oppositional Politics of Change*. University of Illinois Press.

Kenyatta, Jomo. (1938). *Facing Mount Kenya: The Tribal Life of the Gikuyu*. Secker and Warburg. https://ehrafworldcultures.yale.edu/cultures/fl10/documents/004 (accessed 14 October 2025).

Kernberg, O. F. (2003). 'Sanctioned Social Violence: A Psychoanalytic View-Part I', *The International Journal of Psychoanalysis* 84 (3): 683–98.

Kiknadze, Nona C. and Blaine J. Fowers. (2023). 'Cultural Variation in Flourishing', *Journal of Happiness Studies* 24 (7): 2223–44. https://doi.org/10.1007/s10902-023-00677-9.

Kim, Julia C., Julie A. Richardson and Tsoki Tenzin. (2023). 'Weaving Wellbeing into the Fabric of the Economy: Lessons from Bhutan's Journey Toward Gross National Happiness' in Elizabeth Rieger, Robert Costanza, Ida Kubiszewski and Paul Dugdale (eds), *Toward an Integrated Science of Wellbeing*. Oxford University Press. https://doi.org/10.1093/oso/9780197567579.003.0015.

Kimmerer, Robin Wall. (2013). *Braiding Sweetgrass: Indigenous Wisdom, Scientific Knowledge and the Teachings of Plants*. 111st edition. Milkweed Editions.

King, Martin Luther Jr. (1991). 'Remaining Awake Through a Great Revolution', in James Melvin Washington (ed.) *A Testament of Hope: The Essential Writings and Speeches of Martin Luther King Jr.*, 268–78. HarperCollins Publishers.

Kingdon, John W. (1984). *Agendas, Alternatives, and Public Policies*. Little, Brown. http://archive.org/details/agendasalternati0000king_o1l9 (accessed 14 October 2025).

Klein, Melanie. (1996). 'Notes on Some Schizoid Mechanisms', *The Journal of Psychotherapy Practice and Research* 5 (2): 160–79.

Kleinig, John and Nicholas G. Evans. (2013). 'Human Flourishing, Human Dignity, and Human Rights', *Law and Philosophy* 32 (5): 539–64. https://doi.org/10.1007/s10982-012-9153-2.

Kobsev, Artem. (2002). *Philosophy of Chinese Neo-Confucianism [Философия Китайского Неоконфуцианства]*. Vostochnaya literatura RAN.

Koehrsen, Jens and Fabian Huber. (2021). 'A Field Perspective on Sustainability Transitions: The Case of Religious Organizations', *Environmental Innovation and Societal Transitions* 40 (September): 408–20. https://doi.org/10.1016/j.eist.2021.09.005.

Koenigsberger, Leo. (1906). *Hermann von Helmholtz*. Clarendon Press.

Kovach, Margaret. (2010). *Indigenous Methodologies: Characteristics, Conversations, and Contexts*. Reprint edition. University of Toronto Press.

Kyabgon, Traleg. (2010). *Influence of Yogacara on Mahamudra*. KTD Publications.

Ladner, Gerhart B. (1967). 'Homo Viator: Mediaeval Ideas on Alienation and Order', *Speculum* 42 (2): 233–59. https://doi.org/10.2307/2854675.

LaDuke, Winona. (2005). *Recovering the Sacred: The Power of Naming and Claiming*. South End Press.

Latour, Bruno. (1993). *We Have Never Been Modern*. Translated by Catherine Porter. Harvard University Press.

Latour, Bruno. (2009). *Das Parlament der Dinge: Für eine politische Ökologie*. Translated by Gustav Roßler. 6th edition. Suhrkamp Verlag.

Lau, D. C. (1967). 'A Note on Ke Wu', *Bulletin of the School of Oriental and African Studies* 30: 353–7. https://doi.org/10.1017/s0041977x0006225x.

Lawrence, Michael, Thomas Homer-Dixon, Scott Janzwood, Johan Rockstöm, Ortwin Renn and Jonathan F. Donges. (2024). 'Global Polycrisis: The Causal Mechanisms of Crisis Entanglement', *Global Sustainability* 7 (January): 6. https://doi.org/10.1017/sus.2024.1.

Lawson, Andrew. (2019). 'A Conceptual Framework for Exploring Voluntary Stewardship Programs for Land Managers as a Tool of New Environmental Governance', *Australasian Journal of Environmental Management* 26 (2): 137–55. https://doi.org/10.1080/14486563.2019.1599741.

Layard, Richard. (2006). *Happiness: Lessons from a New Science*. Reprint edition. Penguin Books.

Layton, Lynne. (2020). *Toward a Social Psychoanalysis: Culture, Character, and Normative Unconscious Processes*. 1st edition. Routledge.

Leaney, Gregory John. (2012). 'Flourishing in the Flesh of the Interworld: Ecophenomenological Intertwining and Environmental Virtue Ethics', Thesis, UNSW Sydney. https://doi.org/10.26190/unsworks/16113.

Lederman, Harvey. (2022). 'What Is the "Unity" in the "Unity of Knowledge and Action"?', *Dao* 21 (4): 569–603. https://doi.org/10.1007/s11712-022-09853-9.

Lee, Anthony W. (1999). *Painting on the Left: Diego Rivera, Radical Politics, and San Francisco's Public Murals*. 1st edition. University of California Press.

Lee, Bandy. (2019a). *The Dangerous Case of Donald Trump: 37 Psychiatrists and Mental Health Experts Assess a President - Updated and Expanded with New Essays*. Thomas Dunne Books.

Lee, Bandy. (2019b). *Violence: An Interdisciplinary Approach to Causes, Consequences, and Cures*. Wiley-Blackwell.

Legge, James. (1893). *Confucian Analects: The Great Learning, and the Doctrine of the Mean*. Clarendon Press. http://archive.org/details/confuciananalect00conf (accessed 14 October 2025).

Legrand, Thomas. (2021). *Politics of Being: Wisdom and Science for a New Development Paradigm*. Ocean of Wisdom Press.

Leinen, Jo and Andreas Bummel. (2018). *A World Parliament: Governance and Democracy in the 21st Century*. Democracy Without Borders.

Leinfelder, Reinhold. (2018). 'The Future of Reefs in the Anthropocene. Integrated High-Resolution Stratigraphy as a Monitoring, Assisting and Predictive Tool', *Geophysical Research Abstracts* 20: 7256.

Lenton, Timothy M., Hermann Held, Elmar Kriegler, Jim W. Hall, Wolfgang Lucht, Stefan Rahmstorf and Hans Joachim Schellnhuber. (2008). 'Tipping Elements in the Earth's Climate System', *Proceedings of the National Academy of Sciences* 105 (6): 1786–93. https://doi.org/10.1073/pnas.0705414105.

Levine, Robert V. (2006). *A Geography of Time: Temporal Misadventures of a Social Psychologist, or How Every Culture Keeps Time Just a Little Bit Differently*. 1st edition. One World Publications.

Lewis, David K. (1986). *On the Plurality of Worlds*. Blackwell.

Ley, Madelaine. (2023). 'Care Ethics and the Future of Work: A Different Voice', *Philosophy & Technology* 36 (1): 7. https://doi.org/10.1007/s13347-022-00604-5.

Li, Jian-Bin, Kai Dou and Yue Liang. (2021). 'The Relationship Between Presence of Meaning, Search for Meaning, and Subjective Well-Being: A Three-Level Meta-Analysis Based on the Meaning in Life Questionnaire', *Journal of Happiness Studies* 22 (1): 467–89. https://doi.org/10.1007/s10902-020-00230-y.

Lifton, Robert Jay. (2017). 'Foreword', in Bandy X. Lee (ed.), *The Dangerous Case of Donald Trump: 27 Psychiatrists and Mental Health Experts Assess a President*, 1st edition. Thomas Dunne Books.

Lintsen, H. W., F. C. A. Veraart, J. P. H. Smits and John Grin. (2018). *Well-Being, Sustainability and Social Development: The Netherlands 1850–2050*. Philosophy of Nature. Springer. https://doi.org/10.1007/978-3-319-76696-6.

Little Bear, Leroy. (2000). 'Jagged Worldviews Colliding', in Marie Ann Battiste (ed.), *Reclaiming Indigenous Voice and Vision*, 77–85. UBC Press.

Liu, Jenny Jing Wen, Natalie Ein, Julia Gervasio, Mira Battaion and Kenneth Fung. (2022). 'The Pursuit of Resilience: A Meta-Analysis and Systematic Review of Resilience-Promoting Interventions', *Journal of Happiness Studies* 23 (4): 1771–91. https://doi.org/10.1007/s10902-021-00452-8.

Lockwood, Michael, Julie Davidson, Allan Curtis, Elaine Stratford and Rod Griffith. (2010). 'Governance Principles for Natural Resource Management', *Society and Natural Resources* 23 (10): 1–16. https://doi.org/10.1080/08941920802178214.

Lomas, Tim, R. Noah Padgett, Alden Yuanhong Lai, James O. Pawelski and Tyler J. VanderWeele. (2025). 'A Multidimensional Assessment of Global Flourishing: Differential Rankings of 145 Countries on 38 Wellbeing Indicators in the Gallup World Poll, with an Accompanying Principal Components Analyses of the Structure of Flourishing', *The Journal of Positive Psychology* 20 (3): 397–421. https://doi.org/10.1080/17439760.2024.2370538.

Lomas, Tim, Lea Waters, Paige Williams, Lindsay G. Oades and Margaret L. Kern. (2021). 'Third Wave Positive Psychology: Broadening Towards Complexity', *The*

Journal of Positive Psychology 16 (5): 660–74. https://doi.org/10.1080/17439760.2020.1805501.

Lorde, Audre. (2018). *The Master's Tools Will Never Dismantle the Master's House*. Penguin Classics.

Losch, Andreas. (2019). 'The Need of an Ethics of Planetary Sustainability', *International Journal of Astrobiology* 18 (3): 259–66. https://doi.org/10.1017/S1473550417000490.

Louv, Richard. (2019). *Our Wild Calling: How Connecting with Animals Can Transform Our Lives—and Save Theirs*. Algonquin Books.

Lovelock, James. (2020). *Novacene: The Coming Age of Hyperintelligence*. MIT Press.

Ludwig, Paul W. (2002). *Eros and Polis: Desire and Community in Greek Political Theory*. Cambridge University Press. https://doi.org/10.1017/CBO9780511497773.

Lynch, Kathleen. (2022). *Care and Capitalism: Why Affective Equality Matters for Social Justice*. Polity Press.

Lynch, Tony and Stephen Norris. (2016). 'On the Enduring Importance of Deep Ecology', *Environmental Ethics* 38 (1): 63–75. https://doi.org/10.5840/enviroethics20163815.

Lyubomirsky, Sonja. (2008). *The How of Happiness: A New Approach to Getting the Life You Want*. Reprint edition. Penguin Books.

Machan, Tibor R. (2004). *Putting Humans First: Why We Are Nature's Favorite*. Rowman & Littlefield Publishers.

Machiavelli, Niccolo. (2003). *The Prince*. Edited by Anthony Grafton. Translated by George Bull. Reissue edition. Penguin Classics.

Mahoney, Kathleen E. and Paul Mahoney, (eds). (1993). *Human Rights in the Twenty-First Century: A Global Challenge*. M. Nijhoff.

Makeham, John. (2008). *Lost Soul: 'Confucianism' in Contemporary Chinese Academic Discourse*. Harvard-Yenching Institute Monograph Series 64. Harvard University Press.

Makeham, John., (ed.). (2012). *Learning to Emulate the Wise: The Genesis of Chinese Philosophy as an Academic Discipline in Twentieth-Century China*. Formation of Disciplines Series. The Chinese University Press.

Manca, Anna Rita. (2014). 'Social Cohesion', in Alex C. Michalos (ed.), *Encyclopedia of Quality of Life and Well-Being Research*, 6026–8. Springer. https://doi.org/10.1007/978-94-007-0753-5_2739.

Mann, Barbara Alice. (2000). *Iroquoian Women: The Gantowisas*. Peter Lang.

Mansuri, Ghazala and Vijayendra Rao. (2013). 'Localizing Development : Does Participation Work?' Policy Research Report. Washington DC: World Bank. https://openknowledge.worldbank.org/entities/publication/7ec97d6d-602c-5525-858a-0bfc6cafff7f (accessed 14 October 2025).

Manwell, Laurie A., Merelle Tadros, Tiana M. Ciccarelli and Roelof Eikelboom. (2022). 'Digital Dementia in the Internet Generation: Excessive Screen Time During Brain Development Will Increase the Risk of Alzheimer's Disease and Related Dementias in Adulthood', *Journal of Integrative Neuroscience* 21 (1): 28. https://doi.org/10.31083/j.jin2101028.

Marcel, Gabriel. (2010). *Homo Viator: Introduction to the Metaphysic of Hope.* Translated by Emma Craufurd and Paul Seaton. St. Augustine's Press.

Margulis, L. and D. Sagan. (1999). *What is Life?* Simon & Schuster.

Marx, Axel, Charline Depoorter, Santiago Fernandez de Cordoba, Rupal Verma, Mercedes Araoz, Graeme Auld, Janne Bemelmans, et al. (2024). 'Global Governance through Voluntary Sustainability Standards: Developments, Trends and Challenges', *Global Policy* 15 (4): 708–28. https://doi.org/10.1111/1758-5899.13401.

Maslow, Abraham H. (1943). 'A Theory of Human Motivation', *Psychological Review* 50 (4): 370–96. https://doi.org/10.1037/h0054346.

Massingham, Peter. (2019). 'An Aristotelian Interpretation of Practical Wisdom: The Case of Retirees', *Palgrave Communications* 5 (1): 1–13. https://doi.org/10.1057/s41599-019-0331-9.

Mau, Steffen. (2019). *The Metric Society: On the Quantification of the Social.* Translated by Sharon Howe. Polity Press.

Mayer, Colin. (2024). *Capitalism and Crises: How to Fix Them.* Oxford University Press.

Mbiti, John S. (2010). *African Religions & Philosophy.* Second revised and enlarged edition. Heinemann.

Mbow, Cheikh and Cynthia Rosenzweig. (2019). 'Chapter 5 : Food Security.' Special Report on Climate Change and Land. IPCC. https://www.ipcc.ch/srccl/chapter/chapter-5 (accessed 14 October 2025).

Mc Groarty, Kieran. (1994). 'Plotinus on Eudaimonia', *Hermathena* 157, 103–15.

McCloskey, Deirdre. (2008). 'Adam Smith, the Last of the Former Virtue Ethicists', *History of Political Economy* 40 (1): 43–71. https://doi.org/10.1215/00182702-2007-046.

McCloskey, Deirdre Nansen. (2010). *The Bourgeois Virtues: Ethics for an Age of Commerce.* University of Chicago Press.

McDaniel, Justin Thomas and Hector Kilgoe, (eds). (2024). *Religious Studies, Theology, and Human Flourishing.* Oxford University Press.

McDermott, Ronan and Tommaso Natoli. (2018). 'Disaster Law', in Hans-Joachim Heintze and Pierre Thielbörger (eds), *International Humanitarian Action.* Springer.

McGrath, Alister E. (2001). *Christian Theology: An Introduction.* Wiley-Blackwell.

McGregor, Deborah. (2018). 'Reconciliation and Environmental Justice', *Journal of Global Ethics* 14 (2): 222–31. https://doi.org/10.1080/17449626.2018.1507005.

McMahon, Darrin M., (ed.). (2023). *History and Human Flourishing.* Oxford University Press.

Mead, Margaret. (2015). 'Warfare Is Only an Invention—Not a Biological Necessity', in Richard K. Betts (ed.), *Conflict After the Cold War: Arguments on Causes of War and Peace,* 6th edition, 254–8. Routledge.

Means, Alexander J. (2014). 'Educational Commons and the New Radical Democratic Imaginary', *Critical Studies in Education* 55 (2): 122–37. https://doi.org/10.1080/17508487.2014.903502.

Menkiti, Ifeanyi. (1984). 'Person and Community in African Traditional Thought', in Richard A. Wright (ed.), *African Philosophy: An Introduction,* 171–81. University Press of America.

Metz, Thaddeus. (2011). 'Ubuntu as a Moral Theory and Human Rights in South Africa', *African Human Rights Law Journal* 11 (2): 532–59.

Meunier, Corinne. (2020). *Verkehrswende für ALLE*. Umwelt Bundesamt. https://www.umweltbundesamt.de/publikationen/verkehrswende-fuer-alle (accessed 14 October 2025).

Milanovic, Marko. (2024). 'ICJ Delivers Advisory Opinion on the Legality of Israel's Occupation of Palestinian Territories', *EJIL: Talk!* (blog), 20 July 2024. https://www.ejiltalk.org/icj-delivers-advisory-opinion-on-the-legality-of-israels-occupation-of-palestinian-territories (accessed 14 October 2025).

Mill, John Stuart. (1863). *Utilitarianism*. Parker, Son, and Bourn.

Miller, Michelle Ann, Middleton, Carl, Rigg, Jonathan and Taylor, David. (2020). 'Hybrid Governance of Transboundary Commons: Insights from Southeast Asia', *Annals of the American Association of Geographers* 110 (1): 297–313. https://doi.org/10.1080/24694452.2019.1624148.

Miller, Robert. (2000). *Researching Life Stories and Family Histories*. SAGE Publications Ltd. https://doi.org/10.4135/9781849209830.

Miner, Maureen, Martin Dowson and Stuart Devenish. (2012). *Beyond Well-Being: Spirituality and Human Flourishing*. Information Age Publishing.

Mistry, Jayalaxshmi and Andrea Berardi. (2016). 'Bridging Indigenous and Scientific Knowledge', *Science* 352 (6291): 1274–75. https://doi.org/10.1126/science.aaf1160.

Mitchell, Kevin J. (2023). *Free Agents: How Evolution Gave Us Free Will*. 1st edition. Princeton University Press.

Mittelmark, Maurice B., Shifra Sagy, Monica Eriksson, Georg F. Bauer, Jürgen M. Pelikan, Bengt Lindström and Geir Arild Espnes, (eds). (2017). *The Handbook of Salutogenesis*. Springer. http://www.ncbi.nlm.nih.gov/books/NBK435831 (accessed 14 October 2025).

Mogi, Ken. (2022). *Ikigai: Die japanische Lebenskunst*. Translated by Sofia Blind. DuMont Buchverlag GmbH & Co. KG.

Momaday, N. Scott. (1976). 'Native American Attitudes to the Environment', in Åke Hultkrantz and Walter H. Capps (eds), *Seeing with a Native Eye: Essays on Native American Religion*, 79–85. Harper & Row.

Moody, Robert V. and Ming-Dao Deng. (2024). *The Pattern of Change*. Cambridge University Press.

Moore, Jason W. (2015). *Capitalism in the Web of Life: Ecology and the Accumulation of Capital*. Verso.

Moraga, Cherríe and Gloria Anzaldúa, (eds). (1981). *This Bridge Called My Back, Fourth Edition: Writings by Radical Women of Color*. Persephone Press.

Morgan, Kara, Zachary A. Collier, Elisabeth Gilmore and Ketra Schmitt. (2022). 'Decision-First Modeling Should Guide Decision Making for Emerging Risks', *Risk Analysis: An Official Publication of the Society for Risk Analysis* 42 (12): 2613–19. https://doi.org/10.1111/risa.13888.

Moriggi, Angela, Katriina Soini, Bettina B. Bock and Dirk Roep. (2020). 'Caring in, for, and with Nature: An Integrative Framework to Understand Green Care Practices', *Sustainability* 12 (8): 3361. https://doi.org/10.3390/su12083361.

Morrison, Toni. (1987). *Beloved*. Alfred A. Knopf.

Morton, Timothy. (2013). *Hyperobjects: Philosophy and Ecology after the End of the World*. 1st edition. University of Minnesota Press.

Mouffe, Chantal. (2019). 'Critical Artistic Practices: An Agonistic Approach', *The Large Glass Journal of Contemporary Art, Culture and Theory* 27 (28): 117–20.

Mountbatten-O'Malley, Eri. (2024). *Human Flourishing*. Bloomsbury Publishing.

Mumford, Lewis. (1922). *The Story of Utopias*. Boni and Liveright.

Murphy, Nancey and Warren S. Brown. (2009). *Did My Neurons Make Me Do It?: Philosophical and Neurobiological Perspectives on Moral Responsibility and Free Will*. 1st edition. Oxford University Press.

Museka, Godfrey and Manasa Munashe Madondo. (2012). 'The Quest for a Relevant Environmental Pedagogy in the African Context: Insights from Unhu/Ubuntu Philosophy', *Journal of Ecology and the Natural Environment* 4 (10): 258–65. https://doi.org/10.5897/JENE12.052.

Musikanski, Laura, Rhonda Phillips, James Bradbury, John De Graaf and Clinton L. Bliss. (2021). *Happiness, Well-Being and Sustainability: A Course in Systems Change*. Taylor & Francis.

Nadeem, Reem. (2022). 'Key Findings From the Global Religious Futures Project', *Pew Research Center* (blog), 21 December 2022. https://www.pewresearch.org/religion/2022/12/21/key-findings-from-the-global-religious-futures-project (accessed 14 October 2025).

Naess, Arne. (1977). 'Spinoza and Ecology', *Philosophia* 7 (1): 45–54. https://doi.org/10.1007/BF02379991.

Nagel, Thomas. (1972). 'Aristotle on Eudaimonia', *Phronesis* 17 (3): 252–9.

Nagel, Thomas. (1979). *Mortal Questions*. Cambridge University Press.

Nagel, Thomas. (2012). *Mind and Cosmos: Why the Materialist Neo-Darwinian Conception of Nature Is Almost Certainly False*. Oxford University Press. https://doi.org/10.1093/acprof:oso/9780199919758.001.0001.

Nakashima, Douglas, Kirsty Galloway McLean, Hans Thulstrup, Ameyali Ramos Castillo and Jennifer Rubis. (2012). *Weathering Uncertainty: Traditional Knowledge for Climate Change Assessment and Adaptation*. UNESCO Digital Library. UNU. https://unesdoc.unesco.org/ark:/48223/pf0000216613 (accessed 14 October 2025).

Nasr, Vali. (2007). *The Shia Revival: How Conflicts Within Islam Will Shape the Future*. W. W. Norton & Co Inc.

Nassar, Dalia. (2022). *Romantic Empiricism: Nature, Art, and Ecology from Herder to Humboldt*. Oxford University Press.

National Police Chiefs' Council. (2024). 'VAWG Is a National Emergency, Say Police Chiefs', End Violence Against Women, 23 July 2024. https://www.

endviolenceagainstwomen.org.uk/vawg-is-a-national-emergency-say-police-chiefs (accessed 14 October 2025).

Neiman, Susan. (2015). *Why Grow Up?: Subversive Thoughts for an Infantile Age*. Macmillan.

Ng, Weiting and Kuei Rong Ong. (2022). 'Using Positive Psychological Interventions to Improve Well-Being: Are They Effective Across Cultures, for Clinical and Non-Clinical Samples?', *Journal of Contemporary Psychotherapy* 52 (1): 45–53. https://doi.org/10.1007/s10879-021-09513-8.

Nicholsen, Shierry Weber. (2016). *Dance to the Tune of Life: Biological Relativity*. Cambridge University Press. https://doi.org/10.1017/9781316771488.

Nicholsen, Shierry Weber. (2018). 'The Impact of Extreme Historical Situations on the Psyche: On the Work of Robert Jay Lifton'. Presentation at SPSI session / personal communication.

Nicholson, Shierry Weber. (2018a). Malignant Normality and the Dilemmas of Resistance (personal communication from the author).

Nicholson, Shierry Weber. (2018b). The Impact of Extreme Historical Situations on the Psyche: On the Work of Robert Jay Lifton. Presentation at Seattle Psychoanalytic Society and Institute (SPSI) session (personal communication from the author).

Nicholsen, Shierry Weber. (2019). 'After Wrong Doing: On Moral Injury, Apology and the Case for Reparation'. Personal communication.

Nicholsen, Shierry Weber. (2021). 'Malignant Normality and the Dilemma of Resistance: Honoring Minima Moralia', *Krisis | Journal for Contemporary Philosophy* 41 (2): 93–4. https://doi.org/10.21827/krisis.41.2.38250.

Nicholsen, Shierry Weber. (2021). 'The Role of Stochasticity in Biological Communication Processes', *Progress in Biophysics and Molecular Biology* 162 (July): 122–8. https://doi.org/10.1016/j.pbiomolbio.2020.09.008.

Noble, Denis. (2012). 'A Theory of Biological Relativity: No Privileged Level of Causation', *Interface Focus* 2 (1): 55–64. https://doi.org/10.1098/rsfs.2011.0067.

Noble, Denis and George Ellis. (2021). 'Biological Relativity Revisited: The Pre-Eminent Role of Values', *Theoretical Biology Forum* 115 (1/2): 45–69. https://doi.org/10.19272/202211402004.

Noble, Raymond and Denis Noble. (2021). 'Can Reasons and Values Influence Action: How Might Intentional Agency Work Physiologically?', *Journal for General Philosophy of Science* 52 (2): 277–95. https://doi.org/10.1007/s10838-020-09525-3.

Noddings, Nel. (2013). *Caring: A Relational Approach to Ethics and Moral Education*. University of California Press.

Nolan, Richard L. and David C. Croson. (1995). *Creative Destruction: A Six-Stage Process for Transforming the Organization*. Harvard Business Review Press.

Norren, Dorine E. van. (2022). 'African Ubuntu and Sustainable Development Goals: Seeking Human Mutual Relations and Service in Development', *Third World Quarterly* 43 (12): 2791–810. https://doi.org/10.1080/01436597.2022.2109458.

Novak, Gregor, and August Reinisch. (2012). 'Article 49', in Nikolai Wessendorf, Bruno Simma, Daniel-Erasmus Khan, Georg Nolte and Andreas Paulus (eds), *The Charter of the United Nations: A Commentary, Volume II*. Oxford University Press. https://doi.org/10.1093/law/9780199639779.003.0018.

Novak, Gregor and August Reinisch. (2024). 'Article 50', in Bruno Simma, Daniel-Erasmus Khan, Georg Nolte and Andreas Paulus (eds), *The Charter of the United Nations: A Commentary*. Oxford University Press. https://doi.org/10.1093/law/9780192864536.003.0066.

Nussbaum, Martha C. (2016). *Anger and Forgiveness: Resentment, Generosity, and Justice*. Oxford University Press.

Nussbaum, Martha. (2023). *Justice for Animals*. Simon & Schuster.

'ODA Levels in 2022 – Preliminary Data Detailed Summary Note.' (2023). OECD, April 2023. https://one.oecd.org/document/DCD(2024)32/en/pdf#:~:text=Preliminary%20ODA%20levels%20in%202022&text=Total%20ODA%20in%202022%20rose,compared%20to%200.33%25%20in%202021 (accessed 14 October 2025).

O'Donnell, Erin L. and Julia Talbot-Jones. (2018). 'Creating Legal Rights for Rivers: Lessons from Australia, New Zealand, and India', *Ecology and Society* 23 (1). https://doi.org/10.5751/ES-09854-230107.

OECD. (2001). *The Well-Being of Nations: The Role of Human and Social Capital*. Organisation for Economic Co-operation and Development. https://www.oecd-ilibrary.org/education/the-well-being-of-nations_9789264189515-en (accessed 14 October 2025).

OECD. (2020). 'Beyond Growth: Towards a New Economic Approach', Organisation for Economic Co-operation and Development. https://www.oecd.org/en/publications/beyond-growth_33a25ba3-en.html (accessed 14 October 2025).

Olendzki, Andrew. (2003). 'Buddhist Psychology' in Seth Zuihō Segall (ed.), *Encountering Buddhism: Western Psychology and Buddhist Teachings*, 9–14. SUNY Series in Transpersonal and Humanistic Psychology. State University of New York Press.

Oliver, Mary. (1986). *Dream Work*. Atlantic Monthly Press.

Öllinger, Michael and Günther Knoblich. (2009). 'Psychological Research on Insight Problem Solving', in Harald Atmanspacher and Hans Primas (eds), *Recasting Reality: Wolfgang Pauli's Philosophical Ideas and Contemporary Science*, 275–300. Springer. https://doi.org/10.1007/978-3-540-85198-1_14.

O'Mahony, Tadhg. (2022). 'Toward Sustainable Wellbeing: Advances in Contemporary Concepts', *Frontiers in Sustainability* 3 (March). https://doi.org/10.3389/frsus.2022.807984.

O'Neill, Maggie. (2018). 'Walking, Well-Being and Community: Racialized Mothers Building Cultural Citizenship Using Participatory Arts and Participatory Action Research', *Ethnic and Racial Studies* 41 (1): 73–97. https://doi.org/10.1080/01419870.2017.1313439.

O'Neill, Maggie. (2001). *Prostitution and Feminism: Towards a Politics of Feeling*. Polity Press.

O'Neill, Maggie. (2024). 'Journeying with Adorno: Passion, Pathways and Ethno-Mimesis', *Irish Journal of Sociology* 32 (3): 252–71. https://doi.org/10.1177/07916035241295891.

O'Neill, Maggie, Bea Giaquinto and Fahira Hasedžic. (2019). 'Migration, Memory and Place: Arts and Walking as Convivial Methodologies in Participatory Research – A Visual Essay' in Mette Louise Berg and Magdalena Nowicka (eds), *Studying Diversity, Migration and Urban Multiculture*, 96–120. Convivial Tools for Research and Practice. UCL Press. https://doi.org/10.2307/j.ctvfrxs30.11.

O'Neill, Maggie, Sara Giddens, Patricia Breatnach, Carl Bagley, Darren Bourne and Tony Judge. (2002). 'Renewed Methodologies for Social Research: Ethno-Mimesis as Performative Praxis', *The Sociological Review* 50 (1): 69–88. https://doi.org/10.1111/1467-954X.00355.

O'Neill, Maggie and Mary Laing. (2018). 'Rights, Recognition and Resistance: Analysing Legal Challenges, Sex Workers Rights and Citizenship', in Sharron A. Fitzgerald and Kathryn McGarry (eds), *Realising Justice for Sex Workers: An Agenda for Change,*, 161–82. Global Political Economies of Gender and Sexuality. Rowman & Littlefield Publishers.

O'Neill, Maggie and Brian Roberts. (2019). *Walking Methods: Research on the Move*. Routledge. https://doi.org/10.4324/9781315646442.

O'Neill, Maggie and John Perivolaris. (2020). 'Critical Theory in Practice: Walking, Art and Narrative as Conjunctural Analysis in The Large Glass', *Museum of Contemporary Art Skopje, Macedonia. No 29/30* (blog), 2020. https://blogs.kent.ac.uk/edgesmargins/files/2021/01/The-Large-Glass-web-Maggie-ONeill-and-John-Perivolaris-Critical-Theory-in.pdf (accessed 14 October 2025).

Oral, Nilüfer. (2023). 'The Common Heritage of Mankind Under International Law: An Overview', in Virginie Tassin (ed.), *Routledge Handbook of Seabed Mining and the Law of the Sea*. Routledge.

Overy, Richard. (2024). *Why War?* Penguin Books.

Padmanabhan, Martina. (2018). *Transdisciplinary Research and Sustainability: Collaboration, Innovation and Transformation*. Routledge.

Park, Stephen M. (2014). *The Pan American Imagination: Contested Visions of the Hemisphere in Twentieth-Century Literature*. Kindle edition. University of Virginia Press.

Pauli, Wolfgang. (1952). 'Der Einfluss Archetypischer Vorstellungen Auf Die Bildung Naturwissenschaftlicher Theorien Bei Kepler', in Carl Gustav Jung and Wolfgang Pauli (eds), *Naturerklärung Und Psyche*. Studien Aus Dem C. G. Jung-Institut Zürich. Rascher.

Peckham, Robert. (2023). *Fear: An Alternative History of the World*. Profile Books.

Penetrante, Ariel. (2012). 'Negotiating Memories and Justice in the Philippines', in I. William Zartman, Mark Anstey, and Paul Meerts (eds), *The Slippery Slope to*

Genocide: Reducing Identity Conflicts and Preventing Mass Murder, 85–109. Oxford University Press. https://doi.org/10.1093/acprof:oso/9780199791743.001.0001.

Peng, Wei, Gokul Iyer, Valentina Bosetti, Vaibhav Chaturvedi, James Edmonds, Allen Fawcett, Stephane Hallegatte, David G. Victor, Detlef van Vuuren and John Weyant. (2021). 'Climate Policy Models Need to Get Real about People — Here's How', *Nature* 594 (7862): 174–6. https://doi.org/10.1038/d41586-021-01500-2.

Perelomov, L. S. (1998). *Confucius. Lunyu: Research, Translation from Chinese, Commentary [Конфуций. Лунь Юй: Исследование, Перевод с Китайского, Комментарии]*. Moscow: Vostochnaya Literatura.

Phalan, Benjamin T. (2018). 'What Have We Learned from the Land Sparing-Sharing Model?' *Sustainability* 10 (6): 1760. https://doi.org/10.3390/su10061760.

Plaks, Andrew H. (2014). 'The Daxue (Great Learning) and the Zhongyong (Doctrine of the Mean)', in Vincent Shen (ed.), *Dao Companion to Classical Confucian Philosophy*,, 3: 139–52. Dao Companions to Chinese Philosophy. Springer. https://doi.org/10.1007/978-90-481-2936-2_6.

Plotinus. (n.d.). *The Six Enneads*. Translated by Stephen Mackenna and B. S. Page. Internet Classics Archive. MIT Internet Classics Archive. Accessed February 8, 2024. https://classics.mit.edu/Plotinus/enneads.mb.txt (accessed 16 October 2025).

Plumwood, Val. (2002). *Feminism and the Mastery of Nature*. Routledge. https://doi.org/10.4324/9780203006757.

Poincaré, Henri. (1910). 'Mathematical Creation', *The Monist* 20 (3): 321–35.

Poincaré, Henri. (2003). *Science and Method*. Translated by Francis Maitland. Courier Corporation.

Pope Francis. (2015). 'Laudato Si', Encyclical letter. The Holy See. https://www.vatican.va/content/francesco/en/encyclicals/documents/papa-francesco_20150524_enciclica-laudato-si.html (accessed 16 October 2025).

Pope Francis. (2024). 'For a Synodal Church: Communion, Participation, Mission, Final Document', XVI Ordinary General Assembly of the Synod of Bishops. Vatican. https://www.synod.va/content/dam/synod/news/2024-10-26_final-document/ENG---Documento-finale.pdf (accessed 16 October 2025).

Porter, Jean (1955- [1990]). *The Recovery of Virtue: The Relevance of Aquinas for Christian Ethics*. 1st edition. J. Knox Press.

Pradeau, Jean-François. (2002). *Plato and the City: A New Introduction to Plato's Political Thought*. Translated by Janet Lloyd. Liverpool University Press.

Pratt, Mary Louise. (2002). 'Modernity and Periphery', in Elisabeth Mudimbe-Boyi (ed.), *Beyond Dichotomies: Histories, Identities, Cultures, and the Challenge of Globalization*, 21–47. State University of New York Press.

Price, Huw. (2011). *Naturalism Without Mirrors*. Oxford University Press.

Prigogine, Ilya. (2004). *Is Future Given?* World Scientific Publishing.

Protevi, John. (2006). 'Katrina', *Symposium* 10 (1): 363–81. https://doi.org/10.5840/symposium200610123.

'Protocol Additional to the Geneva Conventions of 12 August 1949, and Relating to the Protection of Victims of International Armed Conflicts (Protocol I).' (1977). 8 June 1977. https://ihl-databases.icrc.org/en/ihl-treaties/api-1977 (accessed 16 October 2025).

Puar, J. (2012) 'I would rather be a cyborg than a goddess'. Becoming-Intersectional in Assemblage Theory. Transversal Texts available at: https://transversal.at/transversal/0811/puar/en.

Putnam, Hilary. (1981). *Reason, Truth and History*. Reprinted edition 1998. Cambridge University Press.

Quammen, David. (2013). *Spillover: Animal Infections and the Next Human Pandemic*. 1st edition. W. W. Norton & Co Inc.

Quammen, David. (2020). 'Opinion | We Made the Coronavirus Epidemic', *The New York Times*, 28 January 2020, sec. Opinion. https://www.nytimes.com/2020/01/28/opinion/coronavirus-china.html (accessed 16 October 2025).

Quinlan, Denise, Nicola Swain and Dianne A. Vella-Brodrick. (2012). 'Character Strengths Interventions: Building on What We Know for Improved Outcomes', *Journal of Happiness Studies* 13 (6): 1145–63. https://doi.org/10.1007/s10902-011-9311-5.

Radhakrishna, Sindhu and Asmita Sengupta. (2020). 'What Does Human-Animal Studies Have to Offer Ethology?', *Acta Ethologica* 23 (3): 193–9. https://doi.org/10.1007/s10211-020-00349-4.

Ramose, Mogobe B. (1999). *African Philosophy Through Ubuntu*. Mond Books.

Ramvi, Ellen, Ingvil Hellstrand, Ida Bruheim Jensen, Birgitta Haga Gripsrud and Brita Gjerstad. (2023). 'Ethics of Care in Technology-Mediated Healthcare Practices: A Scoping Review', *Scandinavian Journal of Caring Sciences* 37 (4): 1123–35. https://doi.org/10.1111/scs.13186.

Rao, Narasimha D., Jihoon Min and Alessio Mastrucci. (2019). 'Energy Requirements for Decent Living in India, Brazil and South Africa', *Nature Energy* 4 (12): 1025–32. https://doi.org/10.1038/s41560-019-0497-9.

Rassool, Naz, Viv Edwards and Carole Bloch. (2006). 'Language and Development in Multilingual Settings: A Case Study of Knowledge Exchange and Teacher Education in South Africa', *International Review of Education* 52 (6): 533–52. https://doi.org/10.1007/s11159-006-9008-x.

Raworth, Kate. (2017). *Doughnut Economics: Seven Ways to Think Like a 21st-Century Economist*. Random House Business.

Rees, William E. (1995). 'Achieving Sustainability: Reform or Transformation?', *Journal of Planning Literature* 9 (4): 343–61. https://doi.org/10.1177/088541229500900402.

Reichenbach, H. (1938). *Experience and Prediction*. University of Chicago Press.

Reid, Bill and Robert Bringhurst. (1996). *The Raven Steals the Light*. University of Washington Press.

Riahi, Keywan, Detlef P. van Vuuren, Elmar Kriegler, Jae Edmonds, Brian C. O'Neill, Shinichiro Fujimori, Nico Bauer, et al. (2017). 'The Shared Socioeconomic Pathways and Their Energy, Land Use, and Greenhouse Gas Emissions Implications: An Overview'. *Global Environmental Change* 42 (January): 153–68. https://doi.org/10.1016/j.gloenvcha.2016.05.009.

Ricard, Matthieu. (2016). *Altruism: The Power of Compassion to Change Yourself and the World*. Translation edition. Little, Brown.

Richter, Matthias Ludwig. (2013). *The Embodied Text: Establishing Textual Identity in Early Chinese Manuscripts*. Studies in the History of Chinese Texts, volume 3. Brill.

Rickles, Dean. (2016). 'A Participatory Future of Humanity', in A. Aguirre, B. Foster and Z. Merali (eds), *How Should Humanity Steer the Future?*, 49. Springer.

Rickles, Dean. (2022). *Life Is Short: An Appropriately Brief Guide to Making It More Meaningful*. Princeton University Press.

Rivera, Diego and Gladis Marsh. (1960). *My Art, My Life: An Autobiography*. The Citadel Press.

Rivera, Juan Rafael Coronel and Luis-Martín Lozano. (2008). *Diego Rivera. The Complete Murals*. 1st edition. TASCHEN.

Rockström, Johan, Joyeeta Gupta, Dahe Qin, Steven J. Lade, Jesse F. Abrams, Lauren S. Andersen, David I. Armstrong Mckay, et al. (2023). 'Safe and Just Earth System Boundaries', *Nature* 619 (7968): 102–11. https://doi.org/10.1038/s41586-023-06083-8.

Rockström, Johan, Will Steffen, Kevin Noone, Åsa Persson, F. Stuart Chapin, Eric Lambin, Timothy M. Lenton, et al. (2009). 'Planetary Boundaries: Exploring the Safe Operating Space for Humanity', *Ecology and Society* 14 (2). https://doi.org/10.5751/ES-03180-140232.

Rohrer, Julia M., David Richter, Martin Brümmer, Gert G. Wagner and Stefan C. Schmukle. (2018). 'Successfully Striving for Happiness: Socially Engaged Pursuits Predict Increases in Life Satisfaction', *Psychological Science* 29 (8): 1291–8. https://doi.org/10.1177/0956797618761660.

Rosenberg, Daniel. (2023). 'The Legal Fight Over Deep-Sea Resources Enters a New and Uncertain Phase', *EJIL: Talk!* (blog). 22 August 2023. https://www.ejiltalk.org/the-legal-fight-over-deep-sea-resources-enters-a-new-and-uncertain-phase (accessed 16 October 2025).

Rosenbloom, Daniel. (2017). 'Pathways: An Emerging Concept for the Theory and Governance of Low-Carbon Transitions', *Global Environmental Change* 43 (March): 37–50. https://doi.org/10.1016/j.gloenvcha.2016.12.011.

Rosenthal, Mark. (2015). *Diego Rivera and Frida Kahlo in Detroit*. Illustrated edition. Detroit Institute of Arts.

Rošker, Jana S. (2021). *Interpreting Chinese Philosophy: A New Methodology*. 1st edition. Bloomsbury Academic. https://doi.org/10.5040/9781350199897.

Roskies, Adina L. (2021). 'The Neuroscience of Free Will - University of St. Thomas', *University of St. Thomas Journal of Law & Public Policy* 15 (162). https://researchonline.stthomas.edu/esploro/outputs/journalArticle/The-Neuroscience-of-Free-Will/991015131528403691 (accessed 16 October 2025).

Rubin, Jeffrey B. (2003). 'Close Encounter of a New Kind: Towards an Integration of Psychoanalysis and Buddhism', in Seth Zuihō Segall (ed.), *Encountering Buddhism: Western Psychology and Buddhist Teachings*, 42–71. SUNY Series in Transpersonal and Humanistic Psychology. State University of New York Press.

Ruddick, Will and Aude Peronne. (2024). 'Healing Networks.' Substack newsletter. *Grassroots Economist* (blog), 26 April 2024. https://willruddick.substack.com/p/healing-networks (accessed 16 October 2025).

Ruggeri, Kai, Eduardo Garcia-Garzon, Áine Maguire, Sandra Matz and Felicia A. Huppert. (2020). 'Well-Being Is More Than Happiness and Life Satisfaction: A Multidimensional Analysis of 21 Countries', *Health and Quality of Life Outcomes* 18 (1): 192. https://doi.org/10.1186/s12955-020-01423-y.

Rulli, Maria Cristina, Monia Santini, David T. S. Hayman and Paolo D'Odorico. (2017). 'The Nexus Between Forest Fragmentation in Africa and Ebola Virus Disease Outbreaks', *Scientific Reports* 7 (1): 41613. https://doi.org/10.1038/srep41613.

Russel, Dawn. (2014). 'The Real Impact of High Transportation Costs', *The Supply Chain Xchange*, 11 March 2014. https://www.thescxchange.com/articles/838-the-real-impact-of-high-transportation-costs (accessed 26 October, 2025).

Sachs, Jeffrey D. (2015). *The Age of Sustainable Development*. Columbia University Press.

Sackman, Douglas C. (1996). 'Allegories of Life: Gender, Labor and Bio-Technology in Diego Rivera's Allegory of California', *Human Ecology Review* 3 (1): 115–26.

Saint Augustine. (1968). *The Retractions*. Edited by Sister Mary Inez Bogan. Volume 60. The Fathers of the Church. Catholic University of America Press. https://doi.org/10.2307/j.ctt32b3rt.

Saint Thomas of Aquinas. (1947). *Summa Theologica*. Translated by Fathers of the English Dominican Province. Benziger Brothers edition. Ave Maria Press.

Salgado, Sebastião. (2013). *Genesis*. Taschen.

Sandel, Michael. (2012). *What Money Can't Buy: The Moral Limits of Markets*. Allen Lane.

Sapolsky, Robert M. (2021). 'Neuroscience and the Law', *University of St. Thomas Journal of Law and Public Policy (Minnesota)* 15: 138.

Sapolsky, Robert M. (2023). *Determined: A Science of Life without Free Will*. Penguin.

Saritoprak, Seyma N. and Hisham Abu-Raiya. (2023). 'Living the Good Life: An Islamic Perspective on Positive Psychology', in Edward B. Davis, Everett L. Worthington and Sarah A. Schnitker (eds), *Handbook of Positive Psychology, Religion, and Spirituality*, 179–93. Springer International Publishing. https://doi.org/10.1007/978-3-031-10274-5_12.

Sartre, Jean-Paul. (2018). *Being and Nothingness: An Essay in Phenomenological Ontology*. Routledge. https://doi.org/10.4324/9780429434013.

Sass, Hartmut von. (2019). *Between / Beyond / Hybrid: New Essays on Transdisciplinarity*. Diaphanes.

Saxer, Urs. (2010). *Die internationale Steuerung der Selbstbestimmung und der Staatsentstehung: Selbstbestimmung, Konfliktmanagement, Anerkennung und Staatennachfolge in der neueren Völkerrechtspraxis*. Beiträge zum ausländischen öffentlichen Recht und Völkerrecht 214. Springer. https://doi.org/10.1007/978-3-642-10271-4.

Schlesinger, Arthur M. (1964). 'The Lost Meaning of "The Pursuit of Happiness"', *The William and Mary Quarterly* 21 (3): 326–7. https://doi.org/10.2307/1918449.

Schmithausen, Lambert. (2009). *Plants in Early Buddhism and the Far Eastern Idea of the Buddha-Nature of Grasses and Trees*. Lumbini International Research Institute.

Scholz, Sally J. (2011). 'Solidarity', in Deen K. Chatterjee (ed.), *Encyclopedia of Global Justice*, 1022–5. Springer. https://doi.org/10.1007/978-1-4020-9160-5_85.

Schotanus-Dijkstra, Marijke, M. E. Pieterse, C. H. C. Drossaert, G. J. Westerhof, R. De Graaf, M. Ten Have, J. A. Walburg and E. T. Bohlmeijer. (2016). 'What Factors Are Associated with Flourishing? Results from a Large Representative National Sample', *Journal of Happiness Studies* 17 (4): 1351–70. https://doi.org/10.1007/s10902-015-9647-3.

Schrijver, Nico. (2011). 'Natural Resources, Permanent Sovereignty Over'. *Max Planck Encyclopedia of International Law, Online Version*. https://opil.ouplaw.com/display/10.1093/law:epil/9780199231690/law-9780199231690-e1442?rskey=DYqIAE&result=1&prd=MPIL (accessed 14 October 2025).

Schutte, Nicola S. and John M. Malouff. (2019). 'The Impact of Signature Character Strengths Interventions: A Meta-Analysis', *Journal of Happiness Studies* 20 (4): 1179–96. https://doi.org/10.1007/s10902-018-9990-2.

Schütze, Fritz. (1992). 'Pressure and Guilt: War Experiences of a Young German Soldier and Their Biographical Implications (Part 1 and 2)', *International Sociology* 7 (2/3): 187–208, 347–67. https://doi.org/10.1177/026858092007002005.

Scruton, Roger. (2001). *Kant: A Very Short Introduction*. Oxford University Press.

Seligman, Martin. 2011. *Flourish: A Visionary New Understanding of Happiness and Well-Being*. Free Press.

Seligman, Martin E. P. and Mihaly Csikszentmihalyi. (2000). 'Positive Psychology: An Introduction', *American Psychologist* 55 (1): 5–14. https://doi.org/10.1037/0003-066X.55.1.5.

Seligman, Martin E. P. and Christopher Peterson. (2004). *Character Strengths and Virtues: A Handbook and Classification*. Oxford University Press.

Sen, Amartya. (2001). *Development as Freedom*. Oxford University Press.

Servigne, Pablo and Raphaël Stevens. (2020). *How Everything Can Collapse: A Manual for Our Times*. Translated by Andrew Brown. Polity Press.

Seth, Anil. (2021). *Being You: A New Science of Consciousness*. Faber & Faber.

Shaughnessy, Edward L. (2006). *Rewriting Early Chinese Texts*. SUNY Series in Chinese Philosophy and Culture. State University of New York Press.

Shay, Jonathan. (1994). *Achilles in Vietnam: Combat Trauma and the Undoing of Character*. Maxwell Macmillan. http://archive.org/details/achillesinvietna00shay (accessed 14 October 2025).

Sheldon, Kennon M. and Andrew J. Elliot. (1999). 'Goal Striving, Need Satisfaction, and Longitudinal Well-Being: The Self-Concordance Model', *Journal of Personality and Social Psychology* 76 (3): 482–97. https://doi.org/10.1037/0022-3514.76.3.482.

Shen, Vincent. (2014). 'Introduction: Classical Confucianism in Historical and Comparative Context', in Vincent Shen (ed.), *Dao Companion to Classical Confucian*

Philosophy, 3: 1–19. Dao Companions to Chinese Philosophy. Springer Netherlands. https://doi.org/10.1007/978-90-481-2936-2_1.

Shipp, Abbie J. and Karen J. Jansen. (2021). 'The "Other" Time: A Review of the Subjective Experience of Time in Organizations', *Academy of Management Annals* 15 (1): 299–334. https://doi.org/10.5465/annals.2018.0142.

Simon, Herbert A. (1993). 'The Human Mind: The Symbolic Level', *Proceedings of the American Philosophical Society* 137 (4): 638–47.

Simonton, Dean Keith. (1988). *Scientific Genius: A Psychology of Science*. Scientific Genius: A Psychology of Science. Cambridge University Press.

Singer, Peter. (2011). *The Expanding Circle: Ethics, Evolution and Moral Progress*. 1st Princeton University Press paperback ed. with new Preface and Afterword. Princeton University Press.

Singer, Tania, Matthieu Ricard and Kate Karius, (eds). (2019). *Power and Care: Toward Balance for Our Common Future-Science, Society, and Spirituality*. The MIT Press.

Smith, Anna Marie. (1998). *Laclau and Mouffe: The Radical Democratic Imaginary*. Taylor & Francis Ltd.

Smith, Dorothy E. (1996). 'The Relations of Ruling: A Feminist Inquiry', *Studies in Cultures, Organizations and Societies* 2 (2): 171–90. https://doi.org/10.1080/10245289608523475.

Smith, Linda Tuhiwai. (2012). *Decolonizing Methodologies: Research and Indigenous Peoples*. Zed Books.

Soergel, Bjoern, Sebastian Rauner, Vassilis Daioglou, Isabelle Weindl, Alessio Mastrucci, Fabio Carrer, Jarmo Kikstra, et al. (2024). 'Multiple Pathways Towards Sustainable Development Goals and Climate Targets', *Environmental Research Letters* 19 (12): 124009. https://doi.org/10.1088/1748-9326/ad80af.

Spain, Anna. (2009). 'Who's Going to Copenhagen?: The Rise of Civil Society in International Treaty-Making.' *ASIL Insights* (blog). 11 December 2009. https://www.asil.org/insights/volume/13/issue/25/who%E2%80%99s-going-copenhagen-rise-civil-society-international-treaty-making (accessed 14 October 2025).

Spieker, Heike. (2013). 'Das Mandat der humanitären Hilfe', in Jürgen Lieser and Dennis Dijkzeul (eds), *Handbuch Humanitäre Hilfe*. Springer. https://doi.org/10.1007/978-3-642-32290-7.

Spinoza, Benedictus de. (1994). *A Spinoza Reader: The Ethics and Other Works*. Edited by Edwin Curley. Princeton University Press.

Spitzer, Manfred. (2014). *Digitale Demenz: Wie wir uns und unsere Kinder um den Verstand bringen*. Droemer HC.

Spivak. (1988). 'Can the Subaltern Speak?', in Cary Nelson and Lawrence Grossberg (eds), *Marxism and the Interpretation of Culture*. University of Illinois Press.

Spreckelsen, Tilman. 2021. 'Ein Gespräch mit Cornelia Funke über ihre Romanserie "Reckless"'. *FAZ.NET*, 11 January 2021. https://www.faz.net/aktuell/feuilleton/buecher/themen/ein-gespraech-mit-cornelia-funke-ueber-ihre-romanserie-reckless-17138772.html (accessed 14 October 2025).

Stace, Walter Terence. (1961). *Mysticism and Philosophy*. Macmillan.

Star, Susan Leigh and James R. Griesemer. (1989). 'Institutional Ecology, 'Translations' and Boundary Objects: Amateurs and Professionals in Berkeley's Museum of Vertebrate Zoology, 1907-39', *Social Studies of Science* 19 (3): 387–420. https://doi.org/10.1177/030631289019003001.

Steane, Andrew. (2018). *Science and Humanity: A Humane Philosophy of Science and Religion*. Oxford University Press.

Steane, Andrew. (2023). *Liberating Science: The Early Universe, Evolution and the Public Voice of Science*. Oxford University Press.

Steffen, Will, Wendy Broadgate, Lisa Deutsch, Owen Gaffney and Cornelia Ludwig. (2015). 'The Trajectory of the Anthropocene: The Great Acceleration', *The Anthropocene Review* 2 (1): 81–98. https://doi.org/10.1177/2053019614564785.

Steffen, Will, Johan Rockström, Katherine Richardson, Timothy M. Lenton, Carl Folke, Diana Liverman, Colin P. Summerhayes, et al. (2018). 'Trajectories of the Earth System in the Anthropocene', *Proceedings of the National Academy of Sciences* 115 (33): 8252–9. https://doi.org/10.1073/pnas.1810141115.

Stengers, Isabelle. (2005). 'The Cosmopolitical Proposal', in Bruno Latour and Peter Weibel (eds), *Making Things Public: Atmospheres of Democracy*, 994–1003. MIT Press.

Stewart, Frances. (2005). 'Horizontal Inequalities: A Neglected Dimension of Development', in Anthony B. Atkinson, Kaushik Basu, Jagdish N. Bhagwati, Douglass C. North, Dani Rodrik, Frances Stewart, Joseph E. Stiglitz, and Jeffrey G. Williamson (eds), *Wider Perspectives on Global Development*, 101–35. Palgrave Macmillan UK. https://doi.org/10.1057/9780230501850_5.

Stewart, Ian. (2013). *Seventeen Equations That Changed the World*. Profile Books.

Stiglitz, Joseph E., Amartya Sen and Jean-Paul Fitoussi. (2010). *Mismeasuring Our Lives: Why GDP Doesn't Add Up*. The New Press.

Stone, Mary. (2024). 'The Andean Mother Pachamama: A Mothering Community Paradigm | Program for the Evolution of Spirituality', in Kanu Ikechukwu (ed.), *Alternative Spiritualities of Celebration, Resistance, and Accountability: Engaging Our Colonial and Decolonial Contexts*. Harvard Divinity School.

Stonebridge, Lyndsey. (2024). *We Are Free to Change the World: Hannah Arendt's Lessons in Love and Disobedience*. Hogarth.

Stork, Juliane and Philipp Öhlmann. (2021). 'Religious Communities as Actors for Ecological Sustainability in Southern Africa and Beyond', *Research Programme on Religious Communities and Sustainable Development*, Humboldt-Universität zu Berlin, November. https://doi.org/10.18452/23587.

Storm, Jason Ananda Josephson. (2021). *Metamodernism: The Future of Theory*. University of Chicago Press.

Sundberg, Juanita. (2014). 'Decolonizing Posthumanist Geographies', *Cultural Geographies* 21 (1): 33–47. https://doi.org/10.1177/1474474013486067.

TallBear, Kim. (2017). 'Beyond the Life/Not-Life Binary: A Feminist-Indigenous Reading of Cryopreservation, Interspecies Thinking, and the New Materialisms', in Joanna Radin and Emma Kowal (eds), *Cryopolitics*, 179–202. The MIT Press. https://doi.org/10.7551/mitpress/10456.003.0015.

Taonui, R. (2006). *Polynesian Oral Traditions. Vaka Moana: Voyages of the Ancestors: The Discovery and Settlement of the Pacific*, 22–53.

Taylor, Charles, (ed.). (1985). 'Self-Interpreting Animals', in *Philosophical Papers: Volume 1: Human Agency and Language*, 1: 45–76. Cambridge University Press. https://doi.org/10.1017/CBO9781139173483.003.

Tedeschi, Richard G., Jane Shakespeare-Finch, Kanako Taku and Lawrence G. Calhoun. (2018). *Posttraumatic Growth: Theory, Research, and Applications*. 1st edition Routledge. https://doi.org/10.4324/9781315527451.

The Long Now Foundation, dir. (2019). *The Role of 80-Million Year-Old Rocks in American Slavery — Lewis Dartnell at the Interval*. https://www.youtube.com/watch?v=xrvEE3QVuXc (accessed 14 October 2025).

Thiele, Torsten, Hans-Peter Damian and Pradeep Singh. (2021). 'A Comprehensive Approach to the Payment Mechanism for Deep Seabed Mining', Institute for Advanced Sustainability Studies Policy Brief, 2001, 1. https://doi.org/10.48440/IASS.2021.004.

Thinley, Jigmi Y. and Janette Hartz-Karp. (2019). 'National Progress, Sustainability and Higher Goals: The Case of Bhutan's Gross National Happiness', *Sustainable Earth* 2 (1): 11. https://doi.org/10.1186/s42055-019-0022-9.

Thorne, Brian. (1992). *Carl Rogers*. Sage Publications.

Thorstensen, Vera, Ariel Macaspac Hernandez, Rogerio De Oliviera Corrêa, Dolores Teixeira De Brito, Mauro Kiithi Arima Junior, Catherine Rebouças Mota, Tiago Matsuoka Megale, Amanda Mitsue Zuchieri and Fabio Jorge Thomazella. (2024). 'Voluntary Sustainability Standards (VSS) and the "Greening" of High-Emitting Industry Sectors in Brazil: Mapping the Sustainability Efforts of the Private Sector', *IDOS Discussion Paper*, 1/2024. https://doi.org/10.23661/IDP1.2024.

Thurner, Stefan, Rudolf A. Hanel and Peter Klimek. (2018). *Introduction to the Theory of Complex Systems*. Oxford University Press.

Tian, Nan, Diego Lopes Da Silva, Xiao Liang and Lorenzo Scarazzato. (2024). 'Sipri Fact Sheet April 2024: Trends in World Military Expenditure, 2023.' *Reliefweb*, 22 April 2024. https://reliefweb.int/report/world/sipri-fact-sheet-april-2024-trends-world-military-expenditure-2023-encasv (accessed 14 October 2025).

Tidball, Keith G. (2014). 'Seeing the Forest for the Trees: Hybridity and Social-Ecological Symbols, Rituals and Resilience in Postdisaster Contexts', *Ecology and Society* 19(4): 25. https://doi.org/10.5751/ES-06903-190425.

Tirosh-Samuelson, Hava. (2020). 'Human Flourishing and History: A Religious Imaginary for the Anthropocene', *Journal of the Philosophy of History* 14 (3): 382–418. https://doi.org/10.1163/18722636-12341449.

Todd, Zoe. (2016). 'An Indigenous Feminist's Take On The Ontological Turn: "Ontology" Is Just Another Word for Colonialism', *Journal of Historical Sociology* 29 (1): 4–22. https://doi.org/10.1111/johs.12124.

Toshio, Kuroda, James C. Dobbins and Suzanne Gay. (1981). 'Shinto in the History of Japanese Religion', *The Journal of Japanese Studies* 7 (1): 1–21. https://doi.org/10.2307/132163.

Trabbic, Joseph G. (2011). 'The Human Body and Human Happiness in Aquinas's "Summa Theologiae"', *New Blackfriars* 92 (1041): 552–64.

'Transforming Our World: The 2030 Agenda for Sustainable Development | Department of Economic and Social Affairs.' (n.d.). https://sdgs.un.org/2030agenda (accessed 14 October 2025).

Tronto, Joan. (1993). *Moral Boundaries: A Political Argument for an Ethic of Care*. Routledge. https://doi.org/10.4324/9781003070672.

Tronto, Joan. (2013). *Caring Democracy: Markets, Equality, and Justice*. New York University Press.

Tronto, Joan. (2017). 'There Is an Alternative: Homines Curans and the Limits of Neoliberalism', *International Journal of Care and Caring* 1 (1): 27–43. https://doi.org/10.1332/239788217X14866281687583.

Trutnevyte, Evelina, Léon F. Hirt, Nico Bauer, Aleh Cherp, Adam Hawkes, Oreane Y. Edelenbosch, Simona Pedde and Detlef P. van Vuuren. (2019). 'Societal Transformations in Models for Energy and Climate Policy: The Ambitious Next Step', *One Earth* 1 (4): 423–33. https://doi.org/10.1016/j.oneear.2019.12.002.

Tsebe, Aubrey T. (2021). 'The Epistemology of (M)Other Tongue(s): What Does This Mean for Language in Education?', *The Journal for Transdisciplinary Research in Southern Africa* 17 (1): 8. https://doi.org/10.4102/td.v17i1.1068.

Tugade, Michele M. and Barbara L. Fredrickson. (2004). 'Resilient Individuals Use Positive Emotions to Bounce Back From Negative Emotional Experiences', *Journal of Personality and Social Psychology* 86 (2): 320–33. https://doi.org/10.1037/0022-3514.86.2.320.

Tutu, Desmond. (1999). *No Future Without Forgiveness*. Image.

Ulanowicz, Robert E. (2009). *A Third Window: Natural Life Beyond Newton and Darwin*. Templeton Foundation Press.

UN Department of Economic and Social Affairs. (n.d.). 'The 17 Goals | Sustainable Development.' https://sdgs.un.org/goals (accessed 14 October 2025).

UN Department of Political and Peacebuilding Affairs. (2022). *Repertoire of the Practice of the Security Council: Supplement 2020*. United Nations. https://digitallibrary.un.org/record/3993455 (accessed 14 October 2025).

UN Permanent Forum on Indigenous Issues. (2018). 'Statement of Ms. Victoria Tauli-Corpuz Special Rapporteur on the Rights of Indigenous Peoples', Agenda Item 10, 17th Session. United Nations. https://social.desa.un.org/sites/default/files/migrated/19/2018/04/Statement-by-SRIP-to-Agenda-Item-10_17th-Session-of-the-United-Nations-Permanent-Forum-on-Indigenous-Issues.pdf (accessed 14 October 2025).

United Nations Department of Political and Peacebuilding Affairs. (n.d.). 'The United Nations and Decolonization'. https://www.un.org/dppa/decolonization/en (accessed 14 October 2025).

'United States Recognizes Morocco's Sovereignty Over Western Sahara.' (2021), *American Journal of International Law* 115 (2): 318–23. https://doi.org/10.1017/ajil.2021.11.

UNStats. (2023). 'The Sustainable Development Goals Report 2023: Special Edition - Sounding the Alarm: SDG Progress at the Midpoint.' https://unstats.un.org/sdgs/report/2023/progress-midpoint (accessed 14 October 2025).

Ura, Dasho Karma. (2017). 'Gross National Happiness, Values Education and Schooling for Sustainability in Bhutan' in Rajeswari Namagiri Gorana and Preeti Rawat Kanaujia (eds), *Reorienting Educational Efforts for Sustainable Development: Experiences from South Asia*, 71–88. Springer. https://doi.org/10.1007/978-94-017-7622-6_5.

Ura, Karma. (2019). 'GNH: Development with Integrity.' Distinguished lecture (unpublished), Sheldonian Theatre, University of Oxford, Centre for Bhutan and GNH Studies.

Ura, Karma, Sabina Alkire, Karma Wangdi and Tshoki Zangmo. (2022). '2022 GNH Report.' Centre for Bhutan & GNH Studies. https://ophi.org.uk/publications/Bhutan-GNH-2022 (accessed 14 October 2025).

Ura, Karma, Sabina Alkire, Tshoki Zangmo and Karma Wangdi. (2012). 'An Extensive Analysis of GNH Index. Centre for Bhutan & GNH Studies.' Bhutan: Centre for Bhutan & GNH Studies. https://ophi.org.uk/publications/Extensive-GNH-2012 (accessed 14 October 2025).

Van Zyl, Llewellyn E., Jaclyn Gaffaney, Leoni Van Der Vaart, Bryan J. Dik and Stewart I. Donaldson. (2024). 'The Critiques and Criticisms of Positive Psychology: A Systematic Review', *The Journal of Positive Psychology* 19 (2): 206–35. https://doi.org/10.1080/17439760.2023.2178956.

VanderWeele, Tyler J. (2017). 'On the Promotion of Human Flourishing', *Proceedings of the National Academy of Sciences* 114 (31): 8148–56. https://doi.org/10.1073/pnas.1702996114.

Various Signatories. (2015). 'The Time to Act Is NOW: Buddhist Climate Change Statement', 12 May 2015. https://fore.yale.edu/files/buddhist_climate_change_statement_5-14-15.pdf (accessed 14 October 2025).

Vatican. (2019). 'Human Fraternity for World Peace and Living Together' Signed by His Holiness Pope Francis and the Grand Imam of Al-Azhar Ahamad al-Tayyib. Abu Dhabi. https://www.vatican.va/content/francesco/en/travels/2019/outside/documents/papa-francesco_20190204_documento-fratellanza-umana.html (accessed 14 October 2025).

Velasquez, German. (2014). 'The Right to Health and Medicines: The Case of Recent Multilateral Negotiations on Public Health, Innovation and Intellectual Property', *Developing World Bioethics* 14 (2): 67–74. https://doi.org/10.1111/dewb.12049.

Vella, Shae-Leigh Cynthia and Nagesh B. Pai. (2019). 'A Theoretical Review of Psychological Resilience: Defining Resilience and Resilience Research Over the Decades', *Archives of Medicine and Health Sciences* 7 (2): 233. https://doi.org/10.4103/amhs.amhs_119_19.

Vermeulen-Oskam, E., C. Franklin, L. P. M. Van'T Hof, G. J. J. M. Stams, E. S. Van Vugt, M. Assink, E. J. Veltman, A. S. Froerer, J. P. C. Staaks and A. Zhang. (2024). 'The Current Evidence of Solution-Focused Brief Therapy: A Meta-Analysis of Psychosocial Outcomes and Moderating Factors', *Clinical Psychology Review* 114 (December): 102512. https://doi.org/10.1016/j.cpr.2024.102512.

Vernadsky, V. I. (1998 [1926]). *The Biosphere*. Copernicus.

Vidal, John. (2020). '"Tip of the Iceberg": Is Our Destruction of Nature Responsible for Covid-19?' *The Guardian*, 18 March 2020. https://www.theguardian.com/environment/2020/mar/18/tip-of-the-iceberg-is-our-destruction-of-nature-responsible-for-covid-19-aoe (accessed 14 October 2025).

Vine Jr., Deloria. (2001). 'American Indian Metaphysics', in Vine Deloria Jr. and Daniel R. Wildcat (eds), *Power and Place: Indian Education in America*. Fulcrum Publishing.

Vogel, Steven. (2016). '"Nature" and the (Built) Environment', in Teena Gabrielson, Cheryl Hall, John M. Meyer and David Schlosberg (eds), *The Oxford Handbook of Environmental Political Theory*. Oxford University Press.

Voosholz, Jan, (ed.) (2024a). *Markus Gabriel's New Realism*. 2024 edition. Springer.

Voosholz, Jan, (ed.). (2024b). 'The Concepts of Nature and the Universe in Markus Gabriel's New Realism', in J. Voosholz (ed.) Markus Gabriel's New Realism, *Synthese Library*, 492: 147–66. Springer Nature.

Vuuren, D. P. van, Marcel Kok, P. L. Lucas, Anne Gerdien Prins, Rob Alkemade, Maurits van den Berg, Lex Bouwman, et al. (2015). 'Pathways to Achieve a Set of Ambitious Global Sustainability Objectives by 2050: Explorations Using the Image Integrated Assessment Model', *Technological Forecasting and Social Change* 98: 303–23. https://doi.org/10.1016/j.techfore.2015.03.005.

Wæver, Ole. (2011). 'Politics, Security, Theory', *Security Dialogue* 42 (4–5): 465–80. https://doi.org/10.1177/0967010611418718.

Walker, Matthew. (2017). *Why We Sleep: Unlocking the Power of Sleep and Dreams*. Scribner.

Walsh, Zack, Jessica Böhme and Christine Wamsler. (2021). 'Towards a Relational Paradigm in Sustainability Research, Practice, and Education', *Ambio* 50 (1): 74–84. https://doi.org/10.1007/s13280-020-01322-y.

Walters, Reece. (2003). *Deviant Knowledge*. Willan. https://doi.org/10.4324/9781843924425.

Wang, Canglong, Shuo Wang and Youjiang Gao. (2023). 'Wenhua Dacai (Great Cultural Talent): Paradoxical Discourses and Practices in the Revival of Confucian Classical Education in Contemporary China', *Asia Pacific Education Review* 24 (4): 695–704. https://doi.org/10.1007/s12564-023-09891-9.

Wang, Feifei, Jia Guo and Guoyu Yang. (2023). 'Study on Positive Psychology from 1999 to 2021: A Bibliometric Analysis', *Frontiers in Psychology* 14 (March): 1101157. https://doi.org/10.3389/fpsyg.2023.1101157.

Wang, Huaiyu. (2007). 'On Ge Wu: Recovering the Way of the Great Learning', *Philosophy East and West* 57 (2): 204–26. https://doi.org/10.1353/pew.2007.0028.

Wang, Qiuping and Hao Sun. (2019). 'Traffic Structure Optimization in Historic Districts Based on Green Transportation and Sustainable Development Concept', *Advances in Civil Engineering* 2019 (1): 9196263. https://doi.org/10.1155/2019/9196263.

Wang, Stephen. (2007). 'Aquinas on Human Happiness and the Natural Desire for God', *New Blackfriars* 88 (1015): 322–34.

Watson, Julia. (2019). *Lo-Tek: Design by Radical Indigenism*. Taschen.

Weber, Max. (1918). 'Science as Vocation', in H. H. Gerth and C. Wright Mills (eds), *From Max Weber*. Free Press, 1946.

Wegner, Daniel M. (2018). *The Illusion of Conscious Will*. MIT Press.

Weinberg, Steven. (1976). 'The Forces of Nature', *Bulletin of the American Academy of Arts and Sciences* 29 (4): 13–29. https://doi.org/10.2307/3823787.

Weinberg, Steven. (1993). *The First Three Minutes: A Modern View of the Origin of the Universe*. Basic Books.

Wellens, Karel. (2010). 'Revisiting Solidarity as a (Re-)Emerging Constitutional Principle: Some Further Reflections', in Rüdiger Wolfrum and Chie Kojima (eds), *Solidarity: A Structural Principle of International Law*, 3–54. Springer. https://doi.org/10.1007/978-3-642-11177-8_2.

Wengraf, Tom. (2001). *Qualitative Research Interviewing: Biographic Narrative and Semi-Structured Methods*. 1st edition. Sage Publications Ltd.

Wengrow, David and David Graeber. (2021). *The Dawn of Everything*. Penguin.

Wertheim, Christine and Margaret Wertheim. (2022). *Christine and Margaret Wertheim: Value and Transformation of Corals: Catalogue for the Exhibition at Museum Frieder Burda 2022*. Edited by Udo Kittelmann. Wienand Verlag.

Wertheim, Margaret. (2009). 'Art - Crochet Coral Reef.' Margaret Wertheim. https://www.margaretwertheim.com/crochet-coral-reef (accessed 14 October 2025).

West, Geoffrey. (2017). *Scale: The Universal Laws of Growth, Innovation, Sustainability, and the Pace of Life in Organisms, Cities, Economies, and Companies*. 111st edition. Penguin Press.

Wet, Erika De. (2006). 'The Emergence of International and Regional Value Systems as a Manifestation of the Emerging International Constitutional Order', *Leiden Journal of International Law* 19 (3): 611–32. https://doi.org/10.1017/S0922156506003499.

Weyl, Hermann. (1989). *The Open World*. Ox Bow Press.

Wheeler, John Archibald. (1973). The Princeton Galaxy Interview by Florence Helitzer. Intellectual Digest 3:10.

Wheeler, John Archibald. (1980). 'Pregeometry: Motivations and Prospects', in A. R. Marlow (ed.), *Quantum Theory and Gravitation*, 1–11. Academic Press.

Wheeler, John Archibald and Wojciech Hubert Zurek. (1983). *Quantum Theory and Measurement*. Princeton University Press.

Whitmee, Sarah, Andy Haines, Chris Beyrer, Frederick Boltz, Anthony G Capon, Braulio Ferreira De Souza Dias, Alex Ezeh, et al. (2015). 'Safeguarding Human Health in the Anthropocene Epoch: Report of the Rockefeller Foundation–Lancet Commission on Planetary Health', *The Lancet* 386 (10007): 1973–2028. https://doi.org/10.1016/S0140-6736(15)60901-1.

Whyte, Kyle Powys. (2013). 'On the Role of Traditional Ecological Knowledge as a Collaborative Concept: A Philosophical Study', *Ecological Processes* 2 (1): 7. https://doi.org/10.1186/2192-1709-2-7.

Wilczek, Frank. (2016). *A Beautiful Question*. Penguin Press.

Wildcat, Daniel R. (2009). *Red Alert! Saving the Planet with Indigenous Knowledge*. Speaker's Corner. Fulcrum.

Willett, Walter, Johan Rockström, Brent Loken, Marco Springmann, Tim Lang, Sonja Vermeulen, Tara Garnett, et al. (2019). 'Food in the Anthropocene: The Eat–Lancet Commission on Healthy Diets from Sustainable Food Systems', *The Lancet* 393 (10170): 447–92. https://doi.org/10.1016/S0140-6736(18)31788-4.

Willey, Angela. (2016). 'A World of Materialisms: Postcolonial Feminist Science Studies and the New Natural', *Science, Technology, & Human Values* 41 (6): 991–1014. https://doi.org/10.1177/0162243916658707.

Willmer, Haddon. (2024). 'Images of the City and the Shaping of Humanity.' *While It Is yet Day* (blog). 23 September 2024. https://haddonwillmer.me.uk/images-of-the-city-and-the-shaping-of-humanity (accessed 14 October 2025).

Wink, Walter. (2010). *The Powers That Be: Theology for a New Millennium*. Harmony.

Winner, Langdon. (2004). 'Resistance Is Futile: The Posthuman Condition and Its Advocates', in Harold W. Baillie and Timothy K. Casey(eds), *Is Human Nature Obsolete?*, 385–412. MIT Press. https://doi.org/10.7551/mitpress/3977.003.0020.

Winnicott, W. D. (2005). *Playing and Reality*. Taylor & Francis.

Wiredu, Kwasi. (1996). *Cultural Universals and Particulars: An African Perspective*. Indiana University Press.

Woermann, Minka and Paul Cilliers. (2012). 'The Ethics of Complexity and the Complexity of Ethics', *South African Journal of Philosophy* 31 (2): 447–63. https://doi.org/10.1080/02580136.2012.10751787.

Wolfrum, Rüdiger. (2006). 'Solidarity amongst States: An Emerging Structural Principle of International Law', in Pierre-Marie Dupuy and Christian Tomuschat (eds), *Völkerrecht als Wertordnung: = common values in international law: Festschrift für Christian Tomuschat*. Engel.

Wolfrum, Rüdiger. (2011). 'Common Heritage of Mankind', in *Max Planck Encyclopedia of International Law, Online Version*. https://opil.ouplaw.com/display/10.1093/law:epil/9780199231690/law-9780199231690-e1149 (accessed 14 October 2025).

Wolfson, Harry Austryn. (1934). *The Philosophy of Spinoza: Unfolding the Latent Processes of His Reasoning*. Harvard University Press.

Wolin, Richard. (1994). *Walter Benjamin: An Aesthetic of Redemption*. University of California Press.

Wood, Richard. (2024). *Psychoanalytic Reflections on Vladimir Putin: The Cost of Malignant Leadership*. Taylor & Francis.

World Health Organization. (2023). 'Universal Health Coverage (UHC)', 2023. https://www.who.int/news-room/fact-sheets/detail/universal-health-coverage-(uhc) (accessed 14 October 2025).

Wright, Dale. (2009). *The Six Perfections: Buddhism and the Cultivation of Character*. Oxford University Press.

Wright, Frank Lloyd. (1994). *Frank Lloyd Wright Collected Writings, Vol. 5: 1931–1939*. Edited by Bruce Brooks Pfeiffer. 1st edition. Rizzoli International Publications.

Wulf, Andrea. (2016). *The Invention of Nature: Alexander Von Humboldt's New World*. First Vintage Books edition. Vintage Books.

Yap, Mandy Li-Ming and Krushil Watene. (2019). 'The Sustainable Development Goals (SDGs) and Indigenous Peoples: Another Missed Opportunity?', *Journal of Human Development and Capabilities* 20 (4): 451–67. https://doi.org/10.1080/19452829.2019.1574725.

Yazzie, Robert. (1994). 'Life Comes from It: Navajo Justice Concepts', *New Mexico Law Review* 24 (2): 175.

Yusoff, Kathryn. (2018). *A Billion Black Anthropocenes or None*. University of Minnesota Press.

Zeitlin, Judith T. (1993). *Historian of the Strange*. Stanford University Press.

Zhang, Dexing, Eric K. P. Lee, Eva C. W. Mak, C. Y. Ho and Samuel Y. S. Wong. (2021). 'Mindfulness-Based Interventions: An Overall Review', *British Medical Bulletin* 138 (1): 41–57. https://doi.org/10.1093/bmb/ldab005.

Zhou, Jiayi, Lisa Maria Dellmuth, Kevin M. Adams, Tina-Simone Neset and Nina von Uexkull. (2020). 'The Geopolitics of Food Security: Barriers to the Sustainable Development Goal of Zero Hunger', SIPRI. https://www.sipri.org/publications/2020/sipri-insights-peace-and-security/geopolitics-food-security-barriers-sustainable-development-goal-zero-hunger (accessed 14 October 2025).

Zipes, Jack. (2019). 'Toward the Realization of Anticipatory Illumination', in Jack Zipes (ed.), *Ernst Bloch: The Pugnacious Philosopher of Hope*, 25–41. Springer. https://doi.org/10.1007/978-3-030-21174-5_2.

Zobel, Katharina. (2006). 'Judge Alejandro Álvarez at the International Court of Justice (1946–1955): His Theory of a "New International Law" and Judicial Lawmaking', *Leiden Journal of International Law* 19 (4): 1017–40. https://doi.org/10.1017/S0922156506003736.

Zolbrod, Paul G. (1984). *Dine Bahane: The Navajo Creation Story*. University of New Mexico Press.

Zuboff, Shoshana. (2019). *The Age of Surveillance Capitalism: The Fight for a Human Future at the New Frontier of Power*. Public Affairs.

Zürn, Michael. (2010). 'Global Governance as Multi-Level Governance', in Henrik Enderlein (ed.), *Handbook on Multi-Level Governance*. Edward Elgar Publishing..

Zwicky, Jan. (2015). 'Imagination and the Good Life', in *Alkibiades' Love: Essays in Philosophy*, 262–82. McGill-Queen's University Press. https://doi.org/10.1515/9780773596993-011.

Zwicky, Jan. (2019). *The Experience of Meaning*. McGill-Queen's University Press. https://doi.org/10.1515/9780773558502.

Zwitter, Andrej. (2013). 'From Needs to Rights - A Socio-Legal Account of Bridging Moral and Legal Universalism via Ethical Pluralism', *Politics and Governance* 1 (1). https://doi.org/10.17645/PAG.V1I1.86.

Zwitter, Andrej. (2023a). 'Introduction', in Andrej Zwitter and Takuo Dome (eds), *Meta-Science: Towards a Science of Meaning and Complex Solutions*. University of Groningen Press. https://doi.org/10.21827/648c59a2087f2.

Zwitter, Andrej. (2023b). 'Meta-Science: From a Science of Things to a Science of Meaning', in Andrej Zwitter and Takuo Dome (eds), *Meta-Science: Towards a Science of Meaning and Complex Solutions*, 17–41. University of Groningen Press. https://doi.org/10.21827/648c59a2087f2.

Zwitter, Andrej. (2025). 'Meaning as Interbeing: A Treatment of the We-Turn and Meta-Science.' *Open Philosophy*, vol. 8, no. 1, pp. 20250086. https://doi.org/10.1515/opphil-2025-0086 (Special Issue: We-Turn).

Zwitter, Andrej, Carole Bloch, George Ellis, Richard Hecht, Ariel Hernandez, Wakanyi Hoffman, Dean Rickles, Victoria Sukhomlinova and Karma Ura. (2025). 'Human Flourishing: An Integrated Systems Approach to Development Post 2030', *Earth System Governance* 23 (January): 100236. https://doi.org/10.1016/j.esg.2025.100236.

Zwitter, Andrej and Takuo Dome, (eds). (2023). *Meta-Science: Towards a Science of Meaning and Complex Solutions*. University of Groningen Press. https://doi.org/10.21827/648c59a2087f2.

Index

1 Corinthians 13 70
4IR *see* fourth industrial revolution
'2030 Agenda for Sustainable Development' ('Transforming Our World: The 2030 Agenda for Sustainable Development | Department of Economic and Social Affairs') 28–30

Abhidharma-kosa 176
Aboriginal Australians 215
academic and professional definitions of care 213–14
Acceptance and Commitment Therapy (ACT) 50–1
access 225
accomplishment 45, 46–7
Acquaviva, Graziella 124
acts of kindness 52
adaptation 201
adequate knowledge 80
Adler, Alfred 103–4
Adorno, Theodor W. 156–7
aesthetic stimulus 176–8
aestheticization of politics 107
African Ebola 205
African Indigenous Knowledge Systems 119–20
    ubuntu 119, 126, 128, 132–5, 215, 218–19
African Philosophy: Myth and Reality (Hountondji) 126
African-feminist knowledges 151–72
Agamben, Giorgio 23
agape 70, 72–3, 102, 213–14
agency 12–14, 17–18, 22, 226
    dynamics of transformation 241, 243–4
    GNH 185–6
    Homo curans (caring human) 226
    International law 245–60
    intersectionality 154–5
    justice 249–51

moral agency 241, 243–4
    prisoners, imprisonment and slavery 103–5, 114–15, 165–6
    ubuntu 119, 126, 128, 132–5, 215, 218–19
    Virtue Ethics 63–5
agential realism 196
Akan proverb 'Onipa nua ne onipa' 132
Akan proverb 'Woforo dua pa a, na yepia wo' 134
Alkire, Sabina 173, 268
Allah 108
    see also God
Allegory of California (Rivera) 110
Alvarez, Alejandro 248
American Psychological Association 43–4
amor dei intellectualis 81, 93
Anansi 126
Anaximander 89
ancient Greek and Roman religions 108–9, 230–1
Andean communities and knowledges 123
andreia 69
Andreotti, V. 122
anger 185–6
Angle, Stephen 146
animals 27–8, 176–8, 196, 205
    spiritual categories of mind 24
    see also evolution; flora and fauna
animate and inanimate nature 205
Anishinaabe communities and knowledges 122, 124, 126, 128, 131–2
Anselm 89
Anthropocene 197
anthropocentrism 197, 202, 210
anti-anthropocentrism 202
anti-capitalism 107
anti-essentialism 171
Antonovsky, A. 77
Anzaldúa, Gloria 155, 156
'appearances' 239

application contexts, positive psychology 51, 53–4
Aquinas, Thomas 67–9, 72–3
Arcades Project (Benjamin) 107
Archibald, Jo-Ann 121
Arendt, Hannah 233, 238–40, 243–4
Arican Philosophy Through Ubuntu (Ramose) 126
Aristotle 1–2, 16, 56
  ergon 62, 65–9
  see also *eudaimonia*
Armstrong, Paul B. 12–13
Article 2 (7) of the UN Charter 254
Article 49 (UN Charter) 248, 254
Article 50 (UN Charter) 248–9, 254–5
Article 70 (Code of the Civil Procedure) 252–3
Atmanspacher, Harald 84
atmospheric physics 13
Augustine 66–72
Ausbildung 90–2
Auschwitz Concentration Camp 103–4
authoritarianism 217, 238–9
  see also fascism
autonomy *see* agency
aviation 204
'Ayni' concepts in *Quechua* 125
Aztecs 111, 113

Bacon, Francis 91
Badiou, Alain *see* Neolithic social structure
Bank of Bhutan 190
Bantu communities and knowledges 126
Barad, Karen 196
Barth, Karl 107
Basic Facts about China (Zhao Yangfeng) 146
Bauer, Bruno 107
Bay Bridge, San Francisco 112
beatitudo perfecta/imperfecta 67–9, 75
beauty 238–40
Beck, Ulrich 230
beehaz'aannii (One Law) 122
being, flourishing as 97–192
  arc of hope 99–117
  Confucian language of flourishing 139–50
  feminism/radical democratic theory 151–72

folktales, storytelling, and traditions 119–38
GNH 173–92
Beloved (Morrison) 215
Benjamin, Walter 107
Bergen-Belsen Concentration Camp 103–4
Berkes, Fikret 127
A Berlin Childhood around 1900 (Benjamin) 107
Berlin Wall 203
Bhutan 54, 173, 186, 189–90, 266–7
  see also Gross National Happiness
Bhutan Power Corporation 190
Bhutan Telecom 190
bias 58
Bieri, Peter 90
The big picture: on the origins of life, meaning, and the universe itself (Carroll) 14–15
big state corporations 190
  see also organizations
Bildung 90–2
Bildungsroman 91
Bilge, Stirma 155
biographical sociology 168–9
biomass 203–4
Black and intersectional feminism 151–2
  see also intersectionality
Bloch, Ernst 101, 106–10
Blumenberg, Hans 34
Blumer, Herbert 169
Bohlmeijer, Ernst 49–50
Bolivar, Simón 112
Book of Documents 147–8
Book of Odes 139
Book of Rites 143
bosons 21, 22
boundary conditions 16–17
boundary objects 198–9
'Boy and the Eagle' 125
The Boy Who Lived with the Seals 123
Boyle, Robert 30
'Braiding Sweetgrass: Indigenous Wisdom, Scientific Knowledge, and the Teachings of Plants' (Kimmerer) 127
breeder-reactors 115
Broaden-and-Build Theory 45–6
Brown, John 112

Brümmer, Martin 46
Brutschin, Elina 262
Buddha, Buddhism 108, 176–8, 181–4, 185, 187–9, 213–14
Buen Vivir 128
Burtynsky, Edward 205

Cagui, Wang 146–7
Camus, Albert 186–7
cancer 29–30
canonic core 141–2
Capability Approach 268
capitalism 188–9
    see also economy, economic growth
Capps, Walter 106–7
carbon dioxide (CO_2) emissions 204
    see also global warming
cardinal sins 77–8
cardinal virtues 71
care 216–21, 283
    academic and professional definitions 213–14
    care-resistive power configurations 217
    in conservation, natural resources management and resilience 215–16
    disconnects in care mobilization 217
    Indigenous wisdom and holistic perspectives on care 214–15
    knowledge expansion and conventional scientific criteria 216–17
    manaakitanga (caring and hospitality) 130
    marginalized wisdom/inclusive solutions 217–18
    see also Homo curans
care of the self 218–19
carenon-Western knowledge systems 217
    manipulation under authoritarian regimes 217
Carter, Dudley 113–14
Cartesian dualism 79
Categorical Imperative 187
'categories', categorization 239
Catholicism 88, 101–3, 106–7, 115
causality 240
causation 16
CBT see Cognitive Behavioural Therapy

Chakrabarty, Dipesh 19-7
Chaldean Oracles 70
Chan, Wing-Tsit 141, 142
chance-configuration model 84–5
Chaplin, Charlie 112
character 63–5
    see also agency
charity 73, 74
Cheap Nature 180
cheng yi (sincerity of the will) 142, 145
Cherp, Aleh 262
children and the paranoid-schizoid position 242–3
China 204
    see also Confucian language of flourishing
CHM see Common Heritage of Mankind
Christianity 61–76, 108, 230–1, 235–6
    agape 70, 72–3, 102, 213–14
    see also Catholicism; God; Protestantism
Chroococcidiopsis bacterium 200–1
'circumastances' 45
citizenship and civic participation 227
City College of San Francisco 110–13
civil examinations 147
Civil War(s) 110–11, 114–15
Claparède 82–3
class discrimination 154–5
    see also poverty
climate adaptivity 271
climate change 204, 206
    IPCC 209, 214
    World Climate Conferences 257
    see also pollution
climate policy 220
    see also individual climate initiatives...
Coatlicue 113
Coccia, Emanuele 199
Cognitive Behavioural Therapy (CBT) 50
Cold War 106–7, 203
collective human flourishing
    planetary thinking 198–9
    self-care 218–19
    and storytelling 121–2
    see also interconnectedness
collective shame 31–2
collectivism 54–5
collectivity 154

Collins, Patricia Hill 154, 155
colonialism 247–8, 250–1
   see also decolonialism; slavery
colours 178
Combee River Collective 155
Common Heritage of Mankind (CHM) 255–7
Communism 110–11, 230–1, 232, 235–6
   see also Cold War
community
   and commons, GNH 180–2
   see also Indigenous knowledges
Community Cohesion Index 267
Community Inclusion Currencies 130
community with the universe 35
compassion and generosity 133, 181–2, 185, 186, 187–8, 213–14
   see also love
complementarity 48–9
complexity research 200
concentration 187–8
concept of reality 34–6
conditionalism 22
Confessions of a Nazi Spy (1939) 112
confidence and happiness 46–7
Confucian language of flourishing 139–50
   Daxue 140–2
      author's personal account 146–8
      history 143–6
      text 141–2
Confucius 108
Connes, Alain 87
conservation, natural resources management 215–16
   see also environmentalism
context
   application contexts, positive psychology 51, 53–4
   of discovery 83
   flourishing as being 97–192
      arc of hope 99–117
      Confucian language of flourishing 139–50
      feminism/radical democratic theory 151–72
      folktales, storytelling, and traditions 119–38
      GNH 173–92
   of justification 83–4

contextual relevance 273
Convention to Combat Desertification 257
COP *see* World Climate Conferences
Copernican involution 34–7
Copernicus 35
coral reefs 206–8
Cornell, Drucilla 171
Cornstassel, Jeff 129
cosmic participation 37–8
   see also universe
cosmology 21
cosmos *see* universe
cosmovivialism 203
courage 47–8, 71
   see also andreia
Covid-19 pandemic 27, 203–6
creative insight 81–7
creative insight in science 81–7
Cree story of 'Wîsahkecâhk and the Geese' 125
Crenshaw, Kimberle 154–5
Crick, Francis 15
Crilly, Mary 159
crimes against humanity 238–9
   see also human rights
critical history recovery 154
critical praxis 155, 160, 162–9
Crlenkovich, Helen 112
Crochet Coral Reef (Wertheim & Wertheim) 208
Csikszentmihalyi, Mihaly 43–4, 46
cultural and epistemological diversity 271
cultural expressions 56
Cultural Validity Index 267
culture(s), cultural aspects 265–6
   context, flourishing as being 97–192
      arc of hope 99–117
      Confucian language of flourishing 139–50
      feminism/radical democratic theory 151–72
      folktales, storytelling, and traditions 119–38
      GNH 173–92
   GNH 183–4
   harmony and the cardinal sins 77–8
   new American culture 110
   planetary thinking 198

positive psychology 54–6
  cultural bias 58
  universal virtues and cultural expressions 56
  rubrics 24
  see also diversity
customs 182
  see also traditions

Dahlbeck, Johan 90
Daoism 146
Darwin, Charles 34
Das Prinzip Hoffnung (*The Principles of Hope*) by Bloch 101, 106–10
Davis, Angela 154, 168
Davis, Nira Yuval 154
Dawkins, Richard 15, 33
Daxue 140–2
  author's personal account 146–8
  history 143–6
  text 141–2
De Revolutionibus (Copernicus) 35
Death of God Theology 106–7
deceit 63–4
decentralization 273
decision-making 269
  see also policy-making
Declaration of Independence 61–2, 112
decolonialism 171
Deep Institutional Change Project 171
Deep Institutional Innovation for Sustainability and Human Development (DIIS) 210, 231–3, 280–1
demos 170
Denmark 204
deontologists 63–4
depersonalization 105
'depressive' states 243
Descartes, Renée 134
Designs for the Pluriverse (Escobar) 130–1
Devenish, Stuart 265
Dickens, Charles 91
diffusion equations 14
dikaiosyne 69
Diné (Navajo) story of the 'Hero Twins' 124
Dirac, P. A. M. 84, 89–90
disconnects in care mobilization 217

discourses and entry points in sustainability transformation 212–13
discrimination 154–5
  see also marginalization
discursive social justice 164
disenchantment 30–4
Disordered Minds (Hughes) 236–7
diverse conceptions of happiness 55
diversity 271, 283–4
  see also inclusivity
divinely sanctioned violence 235–6
Dodaro, Robert 67
dogma 30
domestic violence *see* violent dominance hierarchies
domination, dominance 219
  see also authoritarianism; hegemony
Dowson, Martin 265
Druk Green Power Corporation 190
dual-aspect monism 78, 81, 84–5, 92–3
Dunbar, Robin 181
dynamics of transformation 232, 233, 241–4
  meshworks 241–2
  moral agency 243–4
  states of mind 242–3

Ebola epidemics 205
ECJ *see* European Court of Justice
ecological integrity 271
economic paradigms 223
economy, economic growth 29–30
  and care 221–3
  fascism and anti-capitalism 107
  GNH 179–80
  Grassroots Economics 130–1
  see also Gross Domestic Product; materialism, materiality
education 53, 90–2, 220, 225
Effective causation 16
effort 187–8
ego 102
Ehrenfeld, John 212, 218
Eichmann, Adolf 238–9
Einstein, Albert 81–3, 88
Eisler, Riane 219
electromagnetic field 14, 195
electrons 21

Elements (Euclid) 79
Elman, Benjamin A. 145
embodied anthropology 197
embodiment 30
Emejulu, Akwugo 154, 155
emergence 201
emotions 44–6, 242–3
    see also happiness
empathy 100
empirical self 240
empiricist materialism 5–6
    see also materialism, materiality
empowerment and participation 273
End Violence Against Women Coalition 159
energy 187–8, 224
engagement 45, 46
Enlightenment 25, 30, 229–30, 245
'Enneads' (Plotinus) 66
Environmental Stewardship Indicator 267
environmentalism 206, 215–16
environment-human axis 206
envy 77–8
epiphenomema 14–15
episteme 69
epistemic object 21
epistemic pluralism 278
epistemological diversity 271
epistemology 200
equations for the electromagnetic field 14
equity and social justice 272
Erel, Umut 153, 155, 156
erga omnes 253
ergon 62, 65–9, 70–1
Escobar, Arturo 130–1, 242, 243
Ethica Ordine Geometrico Demonstrata (*Ethics*) by Spinoza 78, 79, 84, 87–91
ethical directionality 232
ethics
    of care 282
    see also virtues, Virtue Ethics
ethno-mimesis 153, 162
EU see European Union
Euclid 79
eudaimonia 1–2, 44–6, 61–2, 64, 69–71, 74–5, 262, 265, 266–7
    and the *ergon* 65–9
Eudemian Ethics 65–6

European Court of Justice (ECJ) 251
European Union (EU) 247, 251, 253–4
Evangelicism 107
evolution 15, 35, 196–7, 202–3
    see also individual evolutionary scientists and species...
ex falso quodlibet 80
Existential physics: a scientist's guide to life's biggest questions (Hossenfelder) 14–15
existentialism, existentialists 101–3, 115
exnovation 196
Exodus 108
The Experience of Meaning (Zwicky) 92–3
extemporal nature of higher virtues 72–4

faith 62, 70, 71
    see also hope
Faraci 185–6
fascism 107, 110–11, 112, 235–6
    see also Communism; Nazi Party
Faust, Drew Gilpin 114–15
Fava, Giovanni A. 51
fear 185–6
Fehr 82–3
feminism 151–72, 232–3
feminist critical theory 152, 162, 163–4, 169
    see also critical praxis
'Feminist Epistemologies' seminar 153–4
Feminist Walk of Cork 153–4, 156–7, 160, 162–4, 168, 169–70, 171
Fernández-Llamazares, Álvaro 131
fields of sense 20–1
Final causation 16
finance and banking 190, 223
    see also economy, economic growth
First and Second Laws of Thermodynamics 17
Five Classics 144, 146
five-year plan (FYP) 189–90
'Fixed It' (Gilmore) 166
flora and fauna 127, 176–8
    see also coral reefs
Flournoy 82–3
folktales 119–38
Fonda, Jane 101–2
Foot 63
forgiveness 185

Formal causation 16
Foster 268
Foucault, Michel 218–19
fourth industrial revolution (4IR) 37
Fox, Claire F. 110, 112–13
framework(s) of reference 140
framing 31–2
Frankfurt School 107
Frankl, Viktor 101, 103–6, 115
Fredrickson, Barbara L. 49
free will 17
    see also agency
freedom 183
    see also human rights, human dignity
Frege, Gottlob 86, 87
Freud, Sigmund 103–4
Fry, Douglas P. 219
Funke, Cornelia 205
'Future Earth' network 199–200
future of work 222
FYP see five-year plan

Gable, Shelly L. 77
Gabriel 86
galaxies 21
Galilean style 34
Garden of Eden 108
Garfield, Jay L. 188
ge wu (investigation of things) 142, 144
Geist der Utopie (*Spirit of Utopia*) by Bloch 106
gender-based violence see feminism
General Assembly (UN) 248–50, 253
general intelligence 86
genes 15
    see also evolution
The Genesis of the Copernican World (Blumenberg) 34
genetic set point 45
Geneva Conventions of 12 Aug 1949 252
geotropic astronautics 197
Giaquinto, Bea 153
Gilmore, Jane 166
Global Flourishing Goals 265
global human-made mass 203–4
Global North 203, 206
Global Solidarity see sovereignty and solidarity
Global South 152, 202, 249
global warming see climate change

Glückseligkeit 75
gluttony 77–8
GNH see Gross National Happiness
God 88
    and agency 13–14
    Baruch de Spinoza 79, 89–90
    Ernst Bloch 108
    escaping the Neolithic 230–1
    eudaimonia and the *ergon* 65–9
    folktales, stories and traditions 123–4
    Gabriel Marcel 101–3
    higher and lower vitues 70
    knowledges of the soul 32–3
    worldviews and their ontologies 30
    see also religion
Goethe, Wilhelm 91
Golden Gate Bridge 110–11, 112
good health and well-being see well-being
good life, good life concepts 1–2, 69–70, 93, 128, 187
    see also *eudaimonia*; happiness; ubuntu
governance 182–3
    multi-level governance 264, 270–3
    see also *individual types of government...*
Grassroots Economics 130–1
gratitude interventions 52
Great Acceleration 199–200
The Great Dictator (1940) 112
Great Expectations (Dickens) 91
Great Work see universe
greed 77–8
Green, Rhys 216
Gross Domestic Product (GDP) 179–80, 186, 188–9, 219–20, 262–3
Gross National Happiness (GNH) 54, 173–92, 266–7
    community and commons 180–2
    culture 183–4
    emotions, psychological well-being and memory 185–7
    Homo curans 212
    living standards, market and ethics 179–80
    natural environment 176–8
    official application of GNH indices 189–91
    state and governance 182–3
    time 173–5
groundedness 29

'Growing Human Commitment to Religious Mystery, to Astral Myth, Exodus, Kingdom; Atheism and the Utopia of the Kingdom' (Bloch) 108
growth 223
   see also economy, economic growth
guo (nation) 146, 147–8
   see also zhi guo (ordering the nation)
Gyekye, Kwame 134

Haaken, Jan 153
Habermas, Jürgen 106
habitability and hospitality 202
Hadamard, Jacques 81, 82–3, 84, 86
Haida story 'Raven Steals the Light' 126
Haidt, Jonathan 45, 77
'hapiness formulae' 45
happiness 45, 69–70, 267–8
   Declaration of Independence 61–2
   diverse conceptions of 55
   GNH 173–92
   Homo curans 212
   life, liberty and estate 61–2
   'perfect good' 67–9, 75
   positive psychology 43–95
   see also eudaimonia; good life, good life concepts
Haraway, Donna 141
Harding, Sandra 155
'The Hare and the Lion' 125–6, 133
Harris, Sam 33
Harvard initiative on Global Flourishing Goals 265
Hasedzic, Fahira 153
Haudenosaunee (Iroquois) story of the 'Sky Woman' 125
health, health programs 225
   care settings 53–4
   planetary 195–208
   see also well-being
hedonic well-being 44–5
Hegel, Georg Wilhelm Friedrich 107
hegemony
   authoritarianism/fascism 107, 110–11, 112, 217, 235–6, 238–9
   decolonialism 171
   slavery 114–15, 165–6
   violent dominance hierarchies 151–72
   see also governance; violence systems

Helmholtz, Hermann von 83–4, 89–90
Heraclitus 89
Herget, Lauren 214
Heritage of Our Times (Bloch) 106–7
heritages 255–7, 265–6
   see also Indigenous knowledges
hermeneutics 200
'Hero Twins' 124
Hidalgo y Castilla, Miguel 112
hierarchy of needs 43
   see also Maslow, Abraham
higher virtues 69–71
Hinduism 108
Hitler, Adolf 110–11, 112, 232, 234, 236–7, 242–3
   see also Nazi Party
Hobbes, Thomas 185–6
Höffe, Otfried 197–8
Holocaust 101, 103–6, 238–9
Holocene 197
Homo curans (caring human) 209–28
   narratives and dimensions of transformation 221–7
   polycrisis, Sustainable Development Pathways 209–11
   reclaiming care 211–18
   and reimagining social institutions 218–21
Homo economicus 212
Homo economicus (subjectivity of modernity) 232–3
homo faber 115
homo ludens 115
Homo Sacer (Agamben) 23
Homo sapiens 19–7, 202–3
Homo sociologicus 212
Homo sustinens 212
Homo Viator: Prolégomènes à une métaphysique de l'espérance (Marcel) 102
honesty 63–4
hope 62, 70, 71, 240, 282
   arc of hope 99–117
   resistance and recognition in VAWG 167–9
   'wishful images' of hope 106–10
   see also meaning, meaning making
Hopi story of 'Boy and the Eagle' 125
hospitality 130, 202–3
Hountondji, Pauline 126

Howarth, Richard B. 213
Hudson, Barbara 163, 164, 168, 170
human development as flourishing 271
human motivation/potential 77
human rights, human dignity
    NGOs 226, 270, 273, 292
    state and governance, GNH 183
    traveller rights 157–8
    see also United Nations
human sacrifice 13
human-beingness 120
humanity 47–8
humanity under law 252–4
human-nature relationships 223–4
    see also nature
'Humans First' approaches 202
Humboldt, Alexander von 200
humility 56
Hursthouse, Rosalind 63
Husserl, Edmund 34
hybridity 198–9
hydro-power 190
hypermasculinity 237–8
hyperobjects 198–9

I think therefore I am 134
ICCAs see Indigenous Peoples' and Community Conserved Territories and Areas
ICJ see International Criminal Court
identity 226
    see also being, flourishing as
IDG see Inner Development Goals
IFMs see Integrated Flourishing Measures
Igbo proverb 'Igwe bụ ike' 132
Igbo proverb 'Onye kpọ ọbara, ọbara enyeghị ya aka' 134
'Igwe bụ ike' ('There is strength in community') 132
'Iktomi and the Ducks' 126
illumination 83, 84
    see also insight
imaginaries
    violent dominance hierarchies 169–72
    see also new imaginaries
imagination 79–80
'Imagination and the Good Life' (Zwicky) 93
imitation 196
immaterial well-being 221, 225–6
immigration 153
impersonal science 14–15
inadequate knowledge 79–80
inadequate sleep 175
Incas 111
inclusivity 217–18, 225
incubation 83
Independence, US 61–2, 112
indices, GNH 189–91, 267
Indigenous knowledges 23–4, 218–19, 232–3
    collective human flourishing 121–2
    common themes 124–7
    decolonialism 171
    folktales, storytelling, and traditions 119–38
    and holistic perspectives on care 214–15
    integration of 273
    interconnectedness 122–4
    radical democratic imaginary/violent dominance hierarchies 172
    science and SDGs 127–32
    ubuntu 132–5
    see also individual Indigenous knowledges...
Indigenous Peoples' and Community Conserved Territories and Areas (ICCAs) 129
indispensability thesis 19–26
individualism 54–5, 219–20
    see also agency
Industrial Revolution 37, 229–30
inequality 270
    see also marginalization
Inner Development Goals (IDG) 212, 265
innovation 196, 212, 225
    DIIS frameworks 210, 231–3, 280–1
    see also education; technology
Inquisition 88
insight 83, 84, 89–90
    metaphysics of 84–7
    phenomenology of 81–7
institutions 232
    and care 218–21
    institutional crimes 189
    see also organizations
Integrated Flourishing Measures (IFMs) 262–3, 266–70
integration of Indigenous knowledge 273

integrative approaches 16, 49–51
    treatment, positive psychology 49–51
intelligible care 103
'intentional activities' 45
inter and transdisciplinary research 152
'Interaction First' approaches 202
interconnectedness 125–6, 156, 272
    as inter-Indigenous belief system 122–4
    whanaungatanga (kinship and connectedness) 130
interdependencies, planetary thinking 199–201
Intergovernmental Panel on Climate Change (IPCC) 209, 214
International Criminal Court (ICJ) 248, 249–50
International Seabed Authority (ISA) 256–7
International sphere and law, sovereignty and solidarity 245–60
interobjectivity 198–9
intersectionality and feminism 151–72
    critical praxis and violent dominance hierarchies 162, 165, 169
    definitions 154–5
    neutralization and appropriation 156
interventions, positive psychology 51–3
intuition 69, 80–1, 282
    see also logic
IPCC *see* Intergovernmental Panel on Climate Change
Irish referenda 291–2
Irish travellers (*an lucht siúil* – the walking people) 157–8, 159–60, 164
Iroquois communities and knowledges 124
'Is life worth living?' (James) 43
ISA *see* International Seabed Authority
Islam 108, 235–6
Islamic zakat (charity) 213–14
isntrumentalism 210
Israel–Palestine conflict 235–6, 251
'It from Bit' 36–7

Jacques, Paul 159
James, William 43
Jaspers, Karl 89
Jefferson, Thomas 61–2, 112
Jesus Christ 73, 108, 230–1

Jewell, Jessica 262
Jewish emancipation 107
    see also Holocaust
Jigme Singye Wangchuck 173–92
jingxue (study of the classics) 140
joy 100
    see also happiness
Joyce, James 91
Judaism 108
Jung, C. J. 33–4, 37
justice 47–8, 64–5, 71, 73, 74, 99–100, 240, 248–51, 263, 264
    see also dikaiosyne; social justice
justification 83–4, 186–7

Kahlo, Frida 110
kami (spirits/deities) 215
Kant, Emmanuel 62, 103, 187, 202–3, 238–40
Kaptani, Erene 153
Kardashev, Nikolai 196
karma 176
Karma Ura 177
Karthikeya, R. 265
karuna (compassion) 213–14
    see also compassion and generosity
Kaufering III Concentration Camp 103–4
Keating, Ana Louise 154, 156, 163, 170
Kern, Margaret L. 58
Kernberg, Otto 236–7
Kikuyu people 123–4
    nicknaming 124
Kilifi, Kenya grassroots movements 130–1
Kimmerer, Robin Wall 121, 127
kindness, acts of 52
King, Martin Luther 99–100, 115
King Solomon 64–5
Ki-Wu (Lau 1967) 142
Klein, Melanie 242–3, 248–9
'knowledge is power' (*scientia potestas est*) 91
knowledge sharing 154
knowledges 47–8, 187–8
    adequate/inadequate knowledge 79–80
    Baruch de Spinoza 77–95
    care systems 216–18
    Immanuel Kant 239–40
    intersectionality 155
    phronesis (practical wisdom) 64–5, 69, 71

planetary thinking 198–9
of the soul 32–3
see also episteme; Indigenous knowledges; insight; science
Kobsev, Artem 146
Kokoska, Robert 200–1
Kovach, Margaret 121
Kuwait War (1990) 254–5

'La codification du droit international' 248
Laclau, Ernesto 171
Lakota communities and knowledges 123
Lakota story of 'Iktomi and the Ducks' 126
The Lancet 204–5
land and food 224
Language Vitality Index 267
Lao Tzu 108
Laozi, *Laozi* 89, 146
Lau, D. C. 142
laughing/communal laughter 181
law of gravitation 14
law, legislation 128–9, 163–4
of the International sphere 245–60
see also justice
Law of the Sea Convention (1982) 247, 255–7
law of war 252
laws of motion 14
Layton, Lynne 163, 171
Lebenswelten (life-worlds) 3, 4–5
Lederman, Harvey 146
Lee, Anthony W. 114
Legge, James 147–8
LGBTQ+ communities 151, 152, 289, 291
liang zhi (moral knowledge/intuitive comprehension) 145–6
Liaozhai Zhiyi 145–6
life, liberty and estate 61–2
The Life of Plants (Coccia) 199
Lifton, Robert Jay 156, 157
Liji (Book of Rites) 143
Lincoln, Abraham 61–2, 112
'little Classics', Chinese classics 141
living standards 179–80
see also happiness
Locke, John 61–2, 185–6
logic 62, 83–4, 238–40
logos 32
logotherapy 105–6, 115
Lomas, Tim 58
Lopez, Gloria 154

Lorde, Audre 218
love 62, 64–5, 71, 100
see also agape
loving kindness 181–2
lower and higher virtues 69–71
Ludwigshafen, Germany 106
Lunyu (The Analects) 140, 143, 146
lust 77–8
lying 63–4

McGregor, Deborah 131–2
McGroarty 66
Machiavelli, Niccolò 185–6
MacIntyre 63
Macrobius 67
MAGA 13–14
see also Trump, Donald
Magaña, Mardonio 112
Magokoro (true heart) 215
The Making of a Fresco Showing the Building of a City (Rivera) 110
malignant normality 156–61
Malignant Normality and the Dilemma of Resistance (Nicholson) 156–61
manaakitanga (caring and hospitality) 130
Mani 108
Man's Search for Meaning (Frankl) 101, 103–6, 115
Mao Zedong 232
Māori communities and knowledges 123–4, 130
Marcel, Gabriel 101–3, 115
marginalization 164, 171, 214, 217–18, 232, 244, 270
poverty 4, 130–1, 154–5, 268
racism/sexism 154–5, 157–8, 164
slavery 114–15, 165–6
stereotypes 24, 154–5
Margulis, L. 195
Marinus 67
maritime law 247, 255–7
market *see* economy, economic growth
Marsh, Gladis 111, 113
Marx, Karl 107
Marxism 106–10
Marxist Romanticism 106
masculinity studies 151, 152
Maslow, Abraham 43, 77
see also positive psychology

Material causation 16
materialism, materiality 5–6, 14–15, 32–3, 62, 221, 225–6
  and IFMs 269
  planetary thinking 195–6
  see also economy, economic growth; non-material conditions of flourishing
'Mathematical Creation' (Poincaré) 81–3
mathematical symbolism 14
Matthew 5:39 73
Matthew 22:39 73
maxims 134–5
Maxwell 14
Mayas 111
Mbiti, John 120, 132
Mead, Margaret 169
meaning, meaning making 21–3, 45, 46, 100, 168–9, 281–2, 283
  agency 18
  role of 55–6
  Viktor Frankl 101, 103–6, 115
Means, Alexander J. 170
meditation 181–2, 184
  see also Buddha, Buddhism
memory 185–7
Mengzi (The Mencius) 140
mental health 45–6, 265–6
  see also emotions
mercy 73, 74, 263, 264
mereological relationships 22–3
meshworks 241–2, 243
mestizaje 114
metabolic planet-human relations 195–6
Metamorphoses (Coccia) 199
metaphor 109
  see also symbolism
metaphysics of insight 84–7
meta-science 5, 277–8
  indispensability thesis and nature 19–26
  ontological security and worldviews 27–40
  understanding flourishing 9–40
    spectra, conceptual basis 11–18
Metz, Johannes B. 106–7
Mexican artists and craftsmen 111–12
Mexican independence 112
Mexico City 111
migrant women 153

Mijikenda communities and knowledges 122–3, 130–1
mind and meaning 21
  in nature 21–3
  spiritual categorization 23–5
  see also meaning, meaning making
mindfulness meditation 184
Mindfulness-Based Interventions 50
Miner, Maureen 265
Ming and Qing dynasties (1368–1912) 144
minglun tang (The Hall of Illustrious Talks) 144–5
Minima Moralia (Adorno) 156–7
mink, mass killings 204
Mino-Bimaadiziwin 128
  see also good life, good life concepts
Mitakuye Oyasin 123
Mixtec people 111
mobility 225
'Modern Man in Search of a Soul' (Jung) 33–4
modern-day slavery 165–6
Modernity 229–30, 232–3
  International law 245–60
  see also Enlightenment; technology
modi 79
Momaday, N. Scott 128
Moody, Bob 92
Moraga, Cherríe 155, 156
moral agency 241, 243–4
moral injury 156–61
morality 187–8
  see also virtues, Virtue Ethics
more geometrico 79
Morocco 250–1
Morrison, Toni 154, 215
Moses 108
Mottainai (regret over waste) 215
Mouffe, Chantal 171
Mountbatten-O'Malley, Eri 265
Mudarabah/Musharakah models 223
Multi-Dimensional Poverty Index 268
multi-level governance 264, 270–3
Mumford, Lewis 34–5
Mussolini, Benito 112
My Art, My Life (Rivera) 112
'Myth of Redemptive Violence' (Wink) 237–8
mythos 32

Naess, Arne 87–8
Nagarjuna 89
Nagel, Thomas 189
'Nanabozho and the Maple Trees' 126
Napoleon Bonaparte 236–7
narratives of agency 12–14
narratives and care 219–21
    dimensions of transformation 221–7
narratives of transformation 221–7, 232–3
National Socialism 235–6
    see also Nazi Party
Native American knowledges 122
    see also individual Native American knowledges…
natura 22
natural environment (nature) 185–7, 215–16, 239–40
    and care 221, 223–4
    expanded notion, indispensability thesis and 19–26
    and GNH 176–8
    planetary thinking (planetocentrism) 195–208
    see also animals; flora and fauna
natural resources management 215–16
natural virtues 71
natureculture 198
Navajo communities and knowledges 122, 124
Nazi Party 107, 110–11, 112, 232, 234, 236–7, 242–3
    see also Holocaust; World War II
negative emotion 45
Neiman, Susan 240
Neo-Confucianism 145
neoliberal academia 155
neo-liberal capitalism 188–9
Neolithic social structure 229–44
neologisms 198
Neo-Platonism 61–76
neurons 15
new American culture 110
New Enlightenment 25
new imaginaries 193–275
    Homo curans 209–28
    International sphere, sovereignty and solidarity 245–60
    planetary thinking 195–208
    a world fit for humanity 229–44
new indicators and measurement 264

New Institute 168, 209, 263
Newton, Isaac 14, 239–40
Ng, Weiting 55
Ngai (God) 123–4
Nicholson, Shierry 156–61
Nichomachean Ethics 69
Nietzsche, Friedrich 186–7
Nirvana 108
non-equilibrium 201
non-governmental organizations (NGOs) 270, 273, 292
non-material conditions of flourishing 41–95, 264–6, 281–3
    Baruch de Spinoza 77–95
    Christianity/Neo-Platonism 61–76
    positive psychology 43–60
non-reductive (middle) view 16–17
non-Western knowledge systems 217
    see also individual non-Western knowledge systems…
normativities, planetary thinking 201–3
Norren, Dorren E. van 214
not-yet-conscious (*Nocht-Nicht-Bewussten*) 108–10
noumena 239–40
nous 69
Novum Organon (Bacon) 91
Nussbaum, Martha 161, 185

Oades, Lindsay G. 58
objective idealism 24
objective time 175
occupants/objects 20, 21–2
OECD *see* Organisation for Economic Co-operation and Development
official development assistance (ODA) 188–9
oikopoiesis 197–8
oikos 197–8
Oliver, Mary 215
Olmecs 111
Olmstead, Frederick Law 114–15
'One Health' concepts 204–5
One Way Street (Benjamin) 107
Ong, Kuei Rong 55
'Onipa nua ne onipa' ('A human being's brother is another human being') 132
ontological pluralism 278

ontological security studies/popularizing disenchantment 30–4
ontological security and worldviews 27–40
ontologically committing to mindedness 22
'Onye kpọ ọbara, ọbara enyeghị ya aka' ('He who invites trouble will not be helped by others') 134
opinions 80
opium-priests 108
oral traditions *see* Indigenous knowledges
Organisation for Economic Co-operation and Development (OECD) 212, 266
organizations 54
  see also non-governmental organizations
Others, otherness *see* marginalization
Overy, Richard 234, 235–6, 241
ownership structures 224

Pacha concepts 123
Pact of 1939 between Hitler and Satlin 110–11
Pan-American Continental Radio Corporation 110
Pan-American Unity (Rivera) 101, 110–14
pandemics 203–4
  see also Covid-19 pandemic
paramita (excellences) 187–8
  see also virtues, Virtue Ethics
paranoid-schizoid position 242–3
Park, Stephen 114
Parmenides 89
pathway thinking 121–2, 209–11
patriarchy 168
Paul the Apostle 70
  see also Saint Paul
Pauli, Wolfgang 85–6
peacekeeping 248
Pedagogical College Bern 90
'perfect good', happiness 67–9, 75
perfected/imperfect society (*societas perfecta/imperfecta*) 73
PERMA framework 45–7
Perpetual Peace (Kant) 202–3
Perry, Rick 13
personhood *see* agency
Peterson, Christopher 47, 56, 265
Pflueger, Timothy 110–11

phenomena 239–40
  see also nature
phenomenology of insight 81–4
phenomenology and symbolic interactionism 169
philosophical materialism 62
philosophical revelation 89
philosophy 24
  see also individual philosophers & philosophies...
phronesis 64–5, 69, 75
  see also practical wisdom
physics 14–15
  middle view 16–17
  time 173–5, 240
  see also individual physicians...
Pindar 92
ping tianxia (bringing peace to the world) 142
planetary thinking (planetocentrism) 195–208
  interdependencies 199–201
  normativities 201–3
  planetary health 203–6
  praxis, coral reefs case study 206–8
'Planet-First' anti-anthropocentrism 202
Plato 32–3, 85–6
  see also Neo-Platonism
Pleistocene 197
Plotinus 61, 66, 67, 68, 69–70, 72, 89
pluralism 278
poiesis 197–8
Poincaré, Henri 81–3, 238
police society 185–6
policy-making 25, 190, 226, 263, 265–6, 269
  see also governance
Polisario 250–1
political ideologies 235–6
  see also individual ideologies...
politics and care-based narratives 220
pollution 179–80, 204, 206
polycrisis 209–11, 227–8, 229–34, 241, 280–1
Porphyry 61, 67, 70
A Portrait of the Artist as a Young Man (Joyce) 91
Positive CBT 50
positive education 53
positive emotion 44–6

positive psychology 43–60, 77
    critical reflections 56–8
        cultural bias 58
        lack of novelty 57
        methodological concerns 57
        neglect of systemic factors 58
        overemphasis on positivity 58
        theoretical/conceptual concerns 56
    cultural aspects 54–6
    features of 44–8
    interventions and applications 51–3
    and psychopathology 48–51
post-2030 development agenda 261–75
post-Neolithic 242
post-traumatic growth and resilience 49
post-traumatic stress disorder (PTSD) 105
potentia 91
potestas 91
poverty 4, 130–1, 154–5, 268, 270
    spiritual poverty 103–6
practical skill (*techne*) 69
practical wisdom 71
Prajnaparamita 185, 188
Pratt, Mary Louise 214
praxis 153, 160, 162–9, 206–8
Pre-Colombian Mesoamerica 111
preparation 83
pride 77–8
Prigogine, Ilya 35
prisoners, imprisonment 103–5
    see also slavery
Proclus 67
Prophet Muhammed 108
'Protection of Persons in the Event of Disasters' (UN GA) 253
protective factors 48–9
Protestantism 107
psychological strengths 47–8
psychological well-being 185–7
    see also well-being
A Psychologist Experiences the Concentration Camp see Man's Search for Meaning
psychopathology 48–51
Psycho-social Well-being and Happiness Indexes 267
public policy 54
    see also policy-making
purpose 55–6, 100
    see also meaning, meaning making

purpose-driven science 5
pursuit of happiness 61–2
pursuit of meaning 283
    see also meaning, meaning making
Putin, Vladimir 236–7
    see also Russia–Ukraine war
Putnam, Hilary 22

qi jia (regulation of the family) 142
qualitative approaches 200, 212
quality of contacts within a community 180–1
quantum theory 20
quasi-mathematical treatises 88
quasi-objects 198–9
queer theory 156
    see also LGBTQ+ communities
Quetzalcóatl 111, 112

racism 154–5, 157–8, 164
radical democratic theory 151, 169–72
radical relationalism via social justice 162–9
Ramose, Mogobe B. 120, 126, 133
rangatiratanga (self-determination and leadership) 130
Rata and the Tree 123–4
rational intuition (*nous*) 69
'Raven Steals the Light' 126
raw materials 224–5
raw openness 163
realist anthropology 197–8
'reality' 34–6
reason 62, 80
    see also logic
recentring planet-human relations 195, 197
reductionist physicists 14–15
referenda 250–1, 291–2
reflective social justice 166–7
refuge 35
regional and community specificities 269
Reich 107
    see also Nazi Party
Reichenbach, H. 83
Reimagining of Social Institutions 232
relational good 170–1
relational social justice 164–6
relationalism 200
relationality 283

The Relations of Ruling: a Feminist Inquiry (Smith) 155
relationships 46, 221, 223–4, 226–7
   mereological relationships 22–3
   see also collective human flourishing
relativity 20, 21
religion 55, 164, 226, 235–6
   divinely sanctioned violence 235–6
   see also individual religions…
renovation 196
This Republic of Suffering: Death and the American Civil War (Faust) 114–15
res communes 255
resilience 27, 215–16
   post-traumatic growth and 49
resistance, and recognition in VAWG 167–9
Resolution 59/193 248–9
Resolution 2703 250
resource provision 221, 224–5
   see also natural resources management
'Retractiones' (Saint Augustine) 66–7
Reynolds, Tracey 153
Richter, David 46
Rickles, Dean 84, 92–3, 100
Rivera, Diego 101, 110–14
Robinson, Edward G. 112
Rogers, Carl 170–1
Rohrer, Julia M. 46
root bridges of Meghalaya in Northeast India 199
root self-care 218–19
Rošker, Jana 140
Rothschild Hospital 103–4
Ruggeri, Kai 213
rumination and distraction 184
Russia–Ukraine war (2022) 251, 252
ruxue tradition 140

'Sacred Ecology' (Berkes) 127
sadness 185–6
Sagan, D. 195
Saint Augustine 66–72
Saint Paul 230–1
Saint Thomas Aquinas 67–9, 72–3
Salgado, Sebastião 205
salutogenesis 77
San Francisco 110–13
sanctions 248–9, 254–5
   see also divinely sanctioned violence

Sanskrit 187–8
SARS-CoV-2 *see* Covid-19 pandemic
savouring positive experiences 52–3
Schmukle, Stefan C. 46
scholia 79
Scholz, Sally J. 245–6
School of Hope 106–7
Schütze, Fritz 169
sciance, *see also* natural environment
science
   creative insight 81–7
   'Future Earth' network 199–200
   knowledge expansion and conventional scientific criteria 216–17
   power of impersonal science 14–15
   purpose-driven 5
   sense data and knowledge 239
   STEM 213–14
   see also meta-science
Science as a Vocation (1918 lecture) 33–4
scientia intuitiva 80–1
Scientific Revolution 229–30
scorn as dogma 30
SDGs *see* Sustainable Development Goals
sea law *see* maritime law
Security Council (UN) 248–9, 250, 254
self-care/self-cultivation 139, 218–19
   see also governance; ubuntu
Self-Determination Theory 47
self-determination/self-realization 154–5, 226, 248–51
   rangatiratanga (self-determination and leadership) 130
   see also agency
self-interpreting animals 27–8
selfishness 185–6
self-organization 201
self-perfection 61, 62, 70–1, 74–5
self-understanding 23–4
Seligman, Martin 43–8, 56, 265
Sen, Amartya 268
sense data and knowledge 239
sensitising concepts 169
sentencing disparities and racism 164
separation and techno-optimism 219–20
Sephardic Jews 88
serendipity 85
Sermon on the Mount 73
sex workers 153

sexism 154–5
Sexual Violence Centre Cork (SVCC) 158–60, 165–6, 167, 170
SFBT *see* Solution-Focused Brief Therapy
sharing, caring and sparing 216
Shay, Jonathan 160–1
Shintoism 215
signposts (*Wegweiser*) of hope 107, 108
Simmel, Georg 200
Simon, Andrea 159
Simonton, Dean Keith 84–6
sine qua non 1
situated knowledge 199
'Sky Woman' 125
slaughter 249
slavery 114–15, 165–6
sleep 175
Slote 63
sloth 77–8
Smith, Adam 72
Smith, Anne Marie 171, 172
Smith, Barbara 154
Smith, Dorothy 155
social actuality 156–7, 159–60
social cohesion 227
social innovation 212
social institutions 218–21
　see also institutions
social justice 162–9, 272
　as discursive 164
　as reflective 166–7
　as relational 164–6
socialization 200
societal development 212
　see also knowledges; Modernity; technology
society and relationship infrastructures 221, 226–7
　see also relationships
Socrates 80
Solution-Focused Brief Therapy (SFBT) 51
Song dynasty (960–1279) 143
　see also Tang and Song dynasties
Songling, Pu 145–6
sophia 65–6, 69
sophrosyne 69
South Africa and violence against women 161
sovereignty and solidarity 245–60, 272
　see also agency

Soviet invasion of Finland 110–11
space 240
Spanish Civil War 110–11
Spanish colonialism 250–1
spectra 11–18
Spinoza, Baruch de 77–95
　creative insight in science 81–7
　education 90–2
　Ethics 78, 79, 84, 87–91
　kinds of knowledge 79–81
spiritual animals 24
spiritual categorization, mindedness 23–5
spiritual poverty 103–6
spirituality 71, 226, 265–6, 282
　Shintoism 215
　see also religion; transcendence
Spivak, Gayatri 139–40
Stalin, Josef 110–11, 112, 232
state and governance 182–3
states of mind 242–3
　see also emotions
Steane, Andrew 15
Steinhoff Psychiatric Hospital 103–4
stereotypes 24, 154–5
Stewart, Maria 154
Still Life and Blossoming Almond Trees (Rivera) 110
storytelling 119–38
　biographical sociology, recognition and meaning making 168–9
　for collective human flourishing 121–2
　Liaozhai Zhiyi 145–6
Strange Stories from a Chinese Studio (*Liaozhai Zhiyi*) 145–6
Strauss, Anselm 169
strengths and virtues 47–8
　see also virtues, Virtue Ethics
strengths-based interventions 53
stress 175, 185–6
structural violence 237–8
sub specie aeternitatis 80, 91
subjective time 175
subjectivity of modernity 232–3
subject-object dichotomy 155
Sumak Kawsay 128
'Summa Theologica' (Saint Thomas Aquinas) 67
supernatural beliefs 71, 235–6
　see also spirituality
sustainability transformation 211–18

Sustainable Development Goals (SDGs) 2,
    3–6, 18, 62, 120, 259, 280–1
  braiding Indigenous science into
    127–32
  Christianity/Neo-Platonism 62
  Homo curans 209–28
  sovereignty and solidarity,
    International law 245–60
  see also United Nations
sustainable growth 29–30
SVCC see Sexual Violence Centre Cork
Swahili people 123–4
symbolism 11, 12, 14, 111–13, 207
  trickster figures 126
systemic positive psychology 58
systems theory 200

Tacana people of the Bolivian Amazon
    129
Tang and Song dynasties (1371–1905) 144
Tarascans 111
Tauli-Corpuz, Victoria 128
Taylor, Charles 27–8
techne 69
technology
  and care 221, 222–3
  Industrial Revolution 37, 229–30
  power of, impersonal science and
    14–15
  spiritual categories of mind 24
  'Technology First' posthumanism 202
  see also science
techno-optimism 219–20
Tedeschi, Richard G. 49
telescope metaphors 109
telos 67, 113
temperance 47–8, 71
  see also sophrosyne
Tenochtitlan temples 111
Teotihuacan 111
theology 67, 101–3, 237–8
  Death of God Theology 106–7
  see also Christianity
theoretical ecology 200
Theresienstadt Concentration Camp
    103–4
This Bridge Called My Back (Moraga &
    Anzaldúa) 155, 156
Thomist philosophy, Thomistic
    approaches 63, 67

Tillich, Paul 107
time 173–5, 240
  waking time 175
Tlingit stories 123
toleration 187–8
Toltec people 111, 112
tonglen meditation 181–2
totalitarianism see authoritarianism
totalizing thinking 238–9
toxic positivity 58
Traces (Bloch) 106
traces (*Spuren*) of hope 107
traditions 119–38, 283–4
  Indigenous wisdom and holistic
    perspectives on care 214–15
  ruxue tradition 140
  see also Indigenous knowledges
transcendence 47–8, 282
  see also Buddha, Buddhism
transcendental philosophy 239–40
transcendental well-being 221, 225–6
transformation 3–4, 283–4
  see also sustainability transformation
transformation wheel (*Homo curans*)
    221–7
transversal planet-human relations 195,
    198
Traveller Visibility Group (TVG) 158, 170
treatment approaches, positive psychology
    49–51
trickster figures 125–6
Trotsky, Leon 112
'true self' concepts 44
Trump, Donald 13–14, 236–7, 250
trust 170–1
Trusteeship Council (UN) 247–8
Truth, Sojourner 154
Tugade, Michele 49
Türkheim Concentration Camp 103–4
Tutu, Desmond 133–4
TVG see Traveller Visibility Group

ubuntu 119, 126, 128, 132–5, 215,
    218–19
  and human flourishing 133–5
'Umuntu ngumuntu ngabantu' ('A person
    is a person through other
    persons') 135
UN Covenant on Civil and Political Rights
    250

un-care systems 219–20
'unconditioned' 239–40
Unión de la Expresión Artística del Norte y de Sur de este Continente (The Marriage of Artistic Expression of the North and the South on this Continent) see Pan-American Unity
United Nations Convention on the Law of the Sea (UNCLOS) 247, 255–7
United Nations Department of Political and Peacebuilding Affairs 250, 254–5
United Nations (UN) 128, 206, 214, 245–60
　'2030 Agenda for Sustainable Development' 28–30
　General Assembly 248–50, 253
　ontological security studies/ popularizing disenchantment 32–3
　post-2030 development agenda 261–75
　Security Council 248–9, 250, 254
　sovereignty and solidarity, International law 245–60
　UN Charter 247–9, 254–5
universal law of the Categorical Imperative 187
universal virtues and cultural expressions 56
universe 14–15
　arc of hope 99–117
　Copernican involution 34–7
　GNH 176
　indispensability thesis and nature 19–26
　planetary thinking 200
　see also God
University College Cork 153–4
　see also Feminist Walk of Cork
University of Tübingen 106
Until the end of time: Mind, matter, and our search for meaning in an evolving universe (Greene) 14–15
utilitarianism 63–4, 213–14
　see also maxims; utopianism
utopianism 106–10, 114

Values in Action (VIA) Classification of Strengths 47–8
Vattel, Emer de 248
vegetarianism and veganism 205
verification 83
　see also insight
violence against women and girls (VAWG) 159
violence systems 233–8
　divinely sanctioned violence 235–6
　structural, Neolithic social structure 237–8
　war systems 234–7
violent dominance hierarchies 151–72
　dismantling of 162–9
　radical democratic imaginary 169–72
virtues, Virtue Ethics 47–8, 265
　agency and character 63–5
　Confucian language of flourishing 139–50
　dynamics of transformation 241–4
　extemporal nature of higher virtues 72–4
　GNH 179–80, 183, 185
　and being 187–9
　institutions and care 219
　lower and higher virtues 69–71
　reason, beauty and 238–40
　relationalism 200
　universal virtues and cultural expressions 56
　see also care; law, legislation
vital materialism 195–6
vulnerability factors 48–9

Wagner, Gert G. 46
waking time 175
Walker, Alice 154, 175
Wang, Huaiyu 142
Wang, Stephen 67–8
war systems 234–7
　dynamics of transformation 241
　see also individual wars...
Washington, George 61–2, 112
Washington's National Cathedral 99–100, 115
water ecosystems 224
Waters, Lea 58

wave equation 14
ways of seeing 218–19, 232–3
WBT *see* Well-Being Therapy
'We Are Free to Change the World' (Arendt) 238–9
Weber, Max 32
   Science as a Vocation (1918 lecture) 33–4
Weinberg, Steven 30–1, 34
well-being 44–5, 128, 206
   emotions, psychological well-being and memory 185–7
   Homo curans 209–28
   see also eudaimonia; good life, good life concepts; happiness
Well-Being Therapy (WBT) 51
Wenli Shuyuan 147–8
Wertheim, Margaret & Christine 208
Westerhof, Gerben 49–50
Western Sahara 249–51
Western-centric views 3
Weyl, Hermann 32–3
Whānau Ora of New Zealand 130
whanaungatanga (kinship and connectedness) 130
Wheeler, John Archibald 35–7
WHO *see* World Health Organization
Wild Geese (Oliver) 215
Wilhelm Meister's Apprenticeship (Goethe) 91
Williams, Paige 58
Wink, Walter 237–8
Wiredu, Kwasi 133
'Wîsahkecâhk and the Geese' 125
wisdom and knowledge 47–8, 187–8
   phronesis (practical wisdom) 64–5, 69, 71
   see also knowledges; *sophia*
'wishful images' of hope 106–10
Wittgenstein, Ludwig 19
'Woforo dua pa a, na yepia wo' ('When you climb a good tree, you are given a push') 134
World Climate Conferences (COP) 257

World Health Organization (WHO) 204–5, 206
world peace 142
World War I 106
World War II 242–3
   and the Holocaust 101, 103–6, 238–9
   see also Hitler, Adolf; Nazi Party
worldviews and their ontologies 28–30
worry 185–6
   see also stress
wrath 77–8
Wright, Frank Lloyd 91
Wuhan, Central China 204

Xi, Zhu 140–1, 143, 145
xiu shen (cultivation of the personal life) 142

Yanfeng, Zhao 146, 147
Yangming, Wang 145, 146
Yankee imperialism 110–11
Yaqui deer dancers of Northern Mexico 111
Yoruba tradition 133
Yuan, Ming and Qing dynasties (1279–1911) 143
   see also Ming and Qing dynasties

Zeitlin, Judith 146
zheng xin (rectification of the mind) 142
zhi (Chinese character) 141–2
zhi guo (ordering the nation) 142, 144
zhi xing he yi (the unity of knowledge and action) 145, 146
zhi zhi (extension of knowledge) 142, 144, 145
Zhongyong 146–7
Zhongyong (The Doctrine of the Mean) 140
Zhuli village 147
zoonosis 205
Zoroaster 108
Zulu folktales 125–6, 133
Zwicky, Jan 78, 86, 92–3